Samuel Edwin Solly

A Handbook of Medical Climatology

Embodying its principles and therapeutic application with scientific data of the

chief health resorts of the world

Samuel Edwin Solly

A Handbook of Medical Climatology
Embodying its principles and therapeutic application with scientific data of the chief health resorts of the world

ISBN/EAN: 9783337419585

Printed in Europe, USA, Canada, Australia, Japan

Cover: Foto ©berggeist007 / pixelio.de

More available books at **www.hansebooks.com**

A HANDBOOK

OF

MEDICAL CLIMATOLOGY.

EMBODYING ITS PRINCIPLES AND THERAPEUTIC APPLICATION

WITH SCIENTIFIC DATA OF THE CHIEF HEALTH

RESORTS OF THE WORLD.

BY

S. EDWIN SOLLY, M.D., M.R.C.S.,

LATE PRESIDENT OF THE AMERICAN CLIMATOLOGICAL ASSOCIATION.

ILLUSTRATED IN BLACK AND COLORS.

LEA BROTHERS & CO.,
PHILADELPHIA AND NEW YORK.
1897.

PREFACE.

THE non-existence of a systematic treatise on medical clima-tology, and the fact that this subject has engaged the author's at-tention for thirty years, may be advanced as the reasons for the preparation of this volume. The time, moreover, seems ripe for such an endeavor, for there is a growing appreciation both on the part of the profession and the public at large as to the value of climate in the prevention and treatment of disease, and fortunately to answer this awakening we have now accumulated accurate and extended meteorological observations upon which to found natural laws affording trustworthy conclusions.

In the endeavor to accomplish the purpose above outlined the author has collated and compared climatic observations from all parts of the world up to the present time, and has sought to sys-tematize the work of others with his own experience. It is hardly too much to say that it is possible to prescribe a climate with as much precision as a drug, and with far greater effect in appropriate cases. If the present volume contributes to the establishment of climatology in its proper place as one of the most definite and useful of the medical sciences, its purpose will be fulfilled.

European nations have long since recognized the therapeutic and financial value of their health-resorts, but Americans are only now coming to appreciate the equal resources of their own continent. The United States Government observations have been conducted with wisdom, and in connection with the special climatic studies of physicians have now covered practically the entire country. Its position and extent are such that climates of almost every variety and excellence can be found within its borders, and the rapid growth

of population and railroad facilities has, by creating improved accommodations, brought into practical use numerous resorts suited
even to the most fastidious health-seeker.

This work is, however, as before stated, not limited to the climatology of any country, but essays to report the typical climates of
the world. Comprehensive and comparative climatic studies have
been introduced, with tables affording opportunities of comparing
European and American climates. A twofold benefit is derivable
therefrom, for in the first place American Government investigations throw additional light on the principles of climatology, so that
physicians have a broader basis on which to found climatic prescriptions; and secondly, in comparing the better known European resorts
with those of America the physician can form conclusions from
data beyond the reach of local prejudice and interest.

Three classes of persons are interested in the subject of which
this work treats, but they have not hitherto been supplied with
adequate, modern and systematic information concerning it. The
present generation of physicians who prescribe travel, the travelers
themselves, and the rising generation of medical students should be
taught to recognize the principles of medical climatology and its
proper relation to general therapeutics. To meet the needs of these
three classes the author has sought far and wide for facts, has
endeavored to form theories on this foundation, and finally to
deduce reasonable conclusions from them. So far as is consistent
with clearness and accuracy, technicalities have been avoided.

The author takes this opportunity to express his gratitude to
his brethren in the profession, and especially to his fellow-mem-
bers of the American Climatological Association, who so cour-
teously responded to his request for information, and to Professor
W. M. Davis, of Harvard University, for his kind and valuable
criticisms upon certain of the meteorological portions of this work.

The section of the book which includes the meteorological tables
was under the charge of Mr. E. N. Peirce, at whose disposal were
placed all data concerning the various climates, together with criti-

cisms and details from the author's personal experience, and to him
belongs the chief credit of analyzing, digesting and setting forth
the information which was essential in regard to many of the
health-resorts and weather-tables.   Without such able co-opera-
tion, it would have been impossible amidst the claims of active
practice to carry out adequately this laborious portion of the book.

S. EDWIN SOLLY.

COLORADO SPRINGS, COL., May, 1897.

CONTENTS.

———

CHAPTER IX.

SECTION III.

CHAPTER X.

CHAPTER XI.

CHAPTER XII.

CHAPTER XIII.

CHAPTER XIV.

CHAPTER XV.

CHAPTER XVI.

CHAPTER XVII.

CHAPTER XVIII.

CHAPTER XIX.

CHAPTER XX.

MEDICAL CLIMATOLOGY.

IF we consider how great a sacrifice of time, money, inclination, and affection is involved when an invalid, under direction of a physician, leaves his home and journeys into another and perhaps a far country, we marvel at the small amount of thought and study that is bestowed by the majority of physicians upon the science of medical climatology; for without a fair knowledge and appreciation of this no rational selection of climate can be made.

The deficiency begins with the medical schools, which should teach at least the broad principles of climatology and the outlines of climatic therapeutics. What would be thought to-day of the physician who diagnosed and prescribed for a disease of some organ of whose structure and physiology he was ignorant, or of the surgeon who proceeded to operate upon parts the anatomy of which he had not studied? Why, then, should a physician presume, as so many do, to prescribe a climate without having acquainted himself with the meteorological facts and climatic data, and with their meaning and significance?

On turning to the mass of literature available upon climatological subjects we find it largely composed of empirical and biased accounts of various health-resorts, and that these reports differ little in their statements of the advantages to be derived. Commonly, each claims for its own resort the ability to cure all diseases, and the only invalids warned against coming are those in whom disease is far advanced. The facts given are few, and logical deductions from them are rare. In despair of making a choice from such sources, the physician is apt to take the casual opinion of patients or of other laymen who have visited certain resorts, and to select the climate accordingly. He cannot, however, form a correct judgment from such information unless he has previously grounded himself in the fundamental principles of climatology and studied the recorded facts.

2

In this desert of rubbish there are, nevertheless, bright oases of truth and reason, such as are to be found in the writings of Weber, Hirsch, Jourdanet, Lombard, Vivenot, Rohden, Copland, Davidson, Denison, Yeo, the Williamses (father and son), and others.

In choosing a health-resort for a given case it is not sufficient to select a locality the merits of which have been most loudly and persistently exploited by its advocates. That which the investigator discovers to be at once his chief necessity and his chief difficulty is the institution of a close comparison of weather from actual facts; therefore, accurate meteorological tables of comparison have been, as far as the limited data allowed, prepared for this treatise.

Unfortunately, meteorological data concerning many resorts are non-existent, incomplete, or scanty; and without a sufficient knowledge of these facts no *real* estimate of the climate can be made. As an illustration, the Adirondack region, so justly prized as a breathing-space and playground for the sick and weary of the Atlantic seaboard, is a melancholy example of neglect in meteorological accounting by the stewards of that noble heritage. Official reports, such as those of the United States Weather Bureau, are used when obtainable regarding any of the localities mentioned, and are supplemented by reliable reports of local observers, and much correspondence has been entered into to gain fresh knowledge.

Admirable as is the work of the United States Weather Bureau, unfortunately, through the misguided frugality of Congress, much valuable material has had to be omitted from the reports of the health-stations of the United States of America. The humidity-observations have not been taken at all at some resorts, and in no case have they been frequent enough during the twenty-four hours to allow of ascertaining accurately the comparative condition of the air with regard to moisture throughout both the day and night. This defect has been remedied as far as possible in these pages. For instance, by working the problem out, even with the imperfect material at hand, something is done to reconcile the conflicting assertions frequently made about the coast climate of Southern California, one being that the climate is dry, and the other that it is damp. Careful study of the meteorological material available shows that the term " moderately dry " can usually be applied to the midday hours, but from about sundown to about 10 A.M. the adjective " damp " is correctly applicable.

The night-temperatures have been given when possible, also the hours of sunlight in the notable mountain-resorts, all of which is most important in understanding what the customary weather is during the invalid's day, and what he may expect at night while under cover.

An effort has been made to present as many new facts as possible about useful but comparatively unknown resorts, and even about districts which, although not as yet readily accessible, possess valuable qualifications for the invalid. On the other hand, resorts, particularly in Europe, of which the literature is already ample and trustworthy, are briefly described, except the more characteristic and important. It seemed wisest to occupy most of the space with new and useful information not contained in other books, and to state briefly the essentials of what can be found elsewhere at greater length.

In order that this book may be more conveniently studied with the least possible expenditure of time and labor I have divided the subject-matter into three sections. The first of these deals broadly with the principles of medical climatology, and describes the close connection of this science with physics, meteorology, ethnology, geographical pathology, etc. The second section treats of the therapeutics of climate in relation to disease. No attempt has been made under this head to prescribe special climates for special diseases, and it is obvious that as considerations so many and so different enter into the fitting of the individual case to the particular climate, this would be a futile effort. The third section is devoted to a description of special climates as typified in selected resorts, and includes comparative and other tables.

SECTION I.

CHAPTER I.

THE PRINCIPLES OF MEDICAL CLIMATOLOGY.

In order to appreciate the relations of medical climatology to general climatology it will be well to give some brief definitions to explain the sense in which certain terms are used throughout this treatise.

Weather consists of the individual atmospheric conditions experienced from day to day.

Climate consists of the average values of the current weather-conditions, with their ranges, in a given locality, taken in connection with its latitude, elevation and topography, soil and vegetation.

Thus the weather in a health-resort may be good, bad, or exceptional at any given time or season, while its climate may remain the same for the season or year as the constant elements modify the transitory peculiarities of the weather.

Meteorology, a term often used interchangeably with climatology, is the study of both weather and climate; but it may be understood to deal more particularly with the detailed methods of recording and predicting weather and the formation of deductions from the summarizing and comparing of the records of the various climates.

Climatology is the study of the climates of the various parts of the earth's surface and their relations to each other; it is either pure or applied. It may be considered to embrace not only the meteorology of the various climates, but the meteorological data in connection with the elevation, latitude, topography, soil, and vegetation of a region.

Pure climatology is the study of climates irrespective of their effects upon life, either vegetable or animal.

Applied climatology is the science of climatology with respect to the influence of climate upon some particular form or order of life.   When applied to the growth of crops, edible herbs, or fruits, it may be termed agricultural climatology; when to ornamental shrubs or plants, horticultural; and so on through the various departments of vegetable life.

With respect to the applications of climatology to the animal world, they may be divided into various groups, those with which we are concerned being the physiological, pathological, and therapeutical.   The combination of these three, limited in their application to man, constitutes *medical climatology.*

Medical climatology, in the arrangement of its various departments, may be compared to a pyramid, the base of which is formed of the study of climatic physics; this includes the *essence* of meteorology and certain divisions of geography, geology, botany, and zoölogy.

The
Indi-
vidual Case
and its Appro-
priate Climate.

Study of Special
Climates and Regions.

Individual Climatotberapy.

General Climatotherapy.

Classification of Climates.

Geographical Pathology, or the Distribution of Disease.

Ethnology, or the Distinctions of Race.

Physiology : the General and Particular Influences of Climate
upon the Human Organism.

Climatic Physics : including portions of Meteorology, Geography, Geology, Botany,
and Zoölogy.

Resting upon this foundation is the tier formed by the physiologic effects of climate; this is the study of the influence of the various physical elements of climate, first separately and then in combination, upon the normal human being; the primary considerations being the influence which it exerts upon special organs and functions, and the secondary that brought to bear·upon the human organism as a whole.

The next department, or tier, in the pyramid is ethnology, which deals with the distinctions of race; how their location has been determined through their selection of a climate on account of their requirements or by reason of their necessities, and how their characteristics become modified by climatic influences, first as to physiologic peculiarities and next as to pathologic. Thus we have placed upon the layer of physiology one of ethnology, and upon this again a tier of geographical pathology, by which we are taught the special tendencies and dangers of each climate.

Next, the classification of climates should claim our attention.

We have now arrived at a sufficient elevation in our pyramid of knowledge to continue the building up of our structure with climatic therapeutics, beginning with a tier composed of general climatotherapy; that is, a consideration of what diseases and what stages and forms of them are likely to be benefited by climatic change, and the particular climate suited for special diseases. Then follows the study of the influence of climate upon the various temperaments and diatheses; next the study of special climates and regions ; and, finally, the apex is reached of the individual case and its appropriate climatic treatment.

PHYSICS.

Six elements—namely, earth, air, water, sunlight, temperature, and electricity—by their various proportions and modifying influences upon each other, go to make a climate. In order to appreciate the fundamental causes of the varieties and results of climate it is necessary to disentangle arbitrarily what are always closely related in nature, and so make certain subdivisions of these six elements.

The first climatic factor of the elements enumerated is earth. This admits of two chief divisions, viz., composition and configuration, its composition giving rise to two subdivisions, viz., soil and vegetation, or inorganic and organic elements, and these being again divided as shown on page 24:

EARTH
- Composition
  - Inorganic
    - Rock.
    - Garden-soil.
    - Clay.
    - Sand and gravel.
  - Organic
    - Trees.
    - Grass.
    - Plants.
- Configuration . . . . .
  - Plains.
  - Mountains.
  - Valleys.

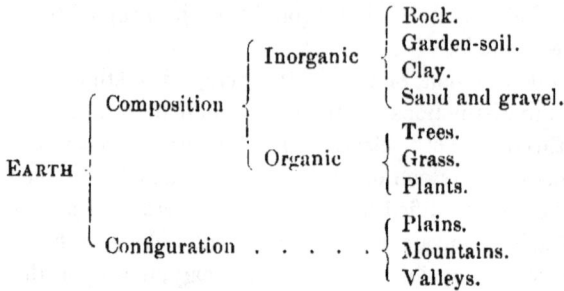

Air is a mixture of nitrogen, argon, carbonic acid, and oxygen, and contains also water-vapor, ozone, and varying amounts of other gases, germs, dust, fogs, cloud-particles, etc.

Position. Latitude, that is, distance from the equator (temperature decreasing and barometric pressure increasing from the equator to latitude 30° and then generally decreasing toward the poles).

Altitude. Height above sea-level (barometric pressure decreasing whilst rising from the sea-level).

Distance from the ocean, giving rise to divisions into:

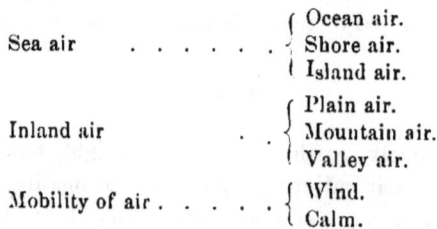

Sea air . . . .
- Ocean air.
- Shore air.
- Island air.

Inland air . .
- Plain air.
- Mountain air.
- Valley air.

Mobility of air . . . . .
- Wind.
- Calm.

Water. Composition: hydrogen and oxygen chemically combined, accompanied by more or less salts and frequently vegetable and animal impurities.

Localized Water. As in seas, lakes, streams, and springs.

Mobilized Water. As in clouds, mists, fogs, rain, snow, hail, and dew.

Vaporized Water. That is, the water-vapor of the atmosphere.

Absorbed Water. As present in solid bodies, soils, rocks, and vegetation.

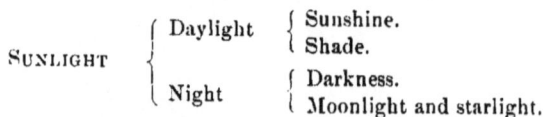

SUNLIGHT
- Daylight
  - Sunshine.
  - Shade.
- Night
  - Darkness.
  - Moonlight and starlight.

TEMPERATURE . { Heat.
                { Cold.

ELECTRICITY { Constant . { Negative on the earth's surface.
            {   { Positive in atmosphere.
            { Inconstant { Thunderstorms and other electri-
            {   { cal disturbances.

EARTH.

The **ground** exerts a most important influence upon the temperature and humidity of the air immediately overlying it, which is nearly of the same moisture and warmth as the ground itself. The various soils receive and retain heat and air very differently, and also absorb and hold moisture in very different proportions, and a common defect in climatic reports is the absence of accurate statements about the soil, which often varies greatly even in different parts of the same resort, both in its quality and state of cultivation. Ground-moisture, as is well known, tends to promote the development of phthisis and rheumatism, and, when accompanied by heat and decaying vegetation, produces malarial fevers.

Vegetation influences climate, increasing, when luxuriant, the humidity of both soil and air. The degree to which this influence is exerted depends, however, upon the character of the vegetation, evergreens fostering much less dampness than deciduous trees, and meadows less than ploughed lands.

Plains are most open to the sweep of storms; when dry they heat and cool rapidly, the difference in temperature between day and night being very marked.

Mountains. Clouds form quickly in the cooled air around mountains, and, remaining for the most part stationary around the peaks, diminish radiation and so lessen extremes; damp winds, however, bring clouds which, instead of rising and hanging on the peaks, tend to blow across the slopes and valleys and so produce rain and snow and lowering of temperature upon the windward side of the mountain, while the other, or lee side, is warmer and drier. Valleys vary much in climate as they happen to be sheltered or exposed to winds, storms, and sunlight, and even opposite sides of a valley sometimes show marked contrasts. The question of the number of hours of sunlight is always an important one to consider in choosing a valley as a place of residence, and what it gains in shelter it may sometimes lose in ventilation. Humidity is increased by the

proximity of glaciers and snow-fields, and this is especially noticeable in the spring when they are melting.

AIR.

Air brings to us most of the other elements and is the chief sup-
porter of life, and yet, while we make minute inquiries into our
water-supply, we are apt to neglect the question of air, which we
use in far greater quantities. This is all the more to be wondered
at because water, before and during its passage to the stomach, is
subjected to the qualifying influences of cooking, association with
food, etc., but air goes through far less modification in its passage to
the lungs.

Oxygen and Carbonic Acid. A very slight deficiency of oxygen or a very small increase of carbonic acid in the air we breathe
is of much importance.

Ozone, which is allotropic oxygen, is never present in the atmosphere in a greater proportion than 1 in 700,000. It is a powerful
disinfectant and is most abundant in open places, such as the
ocean or desert-plains, and on mountains, and is also plentiful
in pine-woods; it is increased by strong sunlight and thunderstorms.

Dust, Germs, and other Foreign Bodies. These, in a greater
or less amount, are always found floating in the air of cities, offices,
and workshops, while germs are only scantily present in the atmosphere of the ocean, or of arid plains and mountain-slopes; this
absence of poisonous and irritating substances is doubtless one of
the chief merits of open spaces as resorts for invalids.

Latitude—that is, the relative distance of a given resort from the
equator—largely influences a climate, both because the temperature
declines with distance from the equator (more rapidly in the
northern than in the southern hemisphere), and because of the
variation in the density of the atmosphere—that is, the barometric
pressure. In consequence of the heating of the air at the equator it
expands, and thus, becoming lighter, it is raised into the upper
regions and flows toward the poles, where, being cooled, it again
contracts, and becomes heavier, sinks, and moves from the poles
toward the equator, to be once more warmed and to mount upward.
Thus there are always two constant, steady currents of cool air
flowing over the earth's surface from the poles toward the equator,

one from the north pole and one from the south. Both are modified by certain causes, which will be spoken of later.

Altitude. As the air-pressure steadily diminishes with the height of the land, the column of air above an elevated place weighs less than at sea-level. For instance, at Potosi, Bolivia, 13,300 feet elevation, it is only one-sixth as dense as at sea-level. But this difference in barometric pressure is modified by local causes which produce periodical and accidental variations.

Distance from the Ocean. As we leave the ocean and approach the centre of a continent the land rises, and, therefore, the weight of the atmosphere diminishes, as does also the humidity by reason of our being further removed from the influence of the evaporation of the sea-water.

These general conditions are, of course, often modified by local causes: as where the land, in small areas, is depressed, or where the humidity of any section of country is increased by the presence of great lakes or inland seas.

Currents and Winds. It was stated in the definition of latitude that the movement of the air which surrounds us is southerly in the northern hemisphere, and northerly in the southern hemisphere, while the higher currents move in the opposite direction. This general direction is, however, greatly modified by the interposition of seas and mountains, and, moreover, the usual daily course of the wind is often controlled by local influences which temporarily overcome these general causes.

Besides these main directions of the winds, which are caused by permanent differences in the barometric pressure and the temperature at the equator and at the poles, the rotatory movement of the planet causes more or less obliquity in the direction of the wind. The seasons, by reason of the varying position of the earth and the sun, also modify the direction, while the difference in conduction and convection[1] over land and water brings about further modifications.

[1] *"Convection* is a process by which unlike temperatures are partially equalized in liquids or gases. This is of great importance in the atmosphere. It may be first illustrated by a simple example in the case of water:

"When a vessel of water is heated at the bottom the warmed layer is expanded and thus made lighter than an equal volume of cooler water above it. In consequence of this unsteady arrangement the heavier overlying water is drawn downward by gravity, displacing the bottom layer, which then rises to the surface. It is our common habit to say that the warm lighter layer ascends; but it must not be forgotten that its rise is a passive process, and that the really active process is the descent of the overlying water, which is drawn down by gravity. By coloring the bottom layer its ascent through the overlying layer may be easily perceived. If

However, it will be best for our purpose to pass to a brief consideration of the local diurnal variations of winds.

"Over the oceans the velocity of the wind shows no distinct diurnal period; but over the lands, and particularly in clear, warm weather, the winds are distinctly stronger about noon than in the night. There is a slight diurnal variation in the mean direction of the wind. The wind tends to veer a little to the right in the northern hemisphere as the day passes and to turn back again as night comes on."[1]

"Although winds from some westerly point prevail at most places in our country (United States of America), yet their direction is variable. They generally blow more from the land to the sea in winter, and more from the sea to the land in summer. The winds change their direction and strength also with every passing storm-eddy. Smoke from forest fires commonly spreads eastward. Clouds, especially the higher ones, generally drift from some western point. Thunderstorms usually travel from west to east. Even the great eddying storms move eastward. The variable winds which we feel are chiefly in the lower layers of the air.

"The winds of the eddying storms are seldom destructive on land. They are of great service in bringing most of our rainfall. On the sea and the lakes, however, they are stronger and cause many shipwrecks. The form and movement of these storms are well shown on the daily weather-maps issued by the National Weather

the temperature be at first uniform throughout, it will be noticed that the warmed water from the bottom is raised to the very top of the liquid, maintaining its higher temperature all the way, except for a slight loss by conduction and mixture during ascent; while the rest of the water settles a little distance toward the bottom. Then the new bottom layer repeats the process; and so a circulatory motion is established. This is called a convectional circulation, and by its means the entire volume of water will be warmed to almost as high a temperature as is maintained at the bottom. It depends essentially upon the disturbance of a condition of rest by the introduction of a change in the temperature and a consequent change in the density of the water, which is, therefore, followed by motion under the action of gravity. After this deliberate explanation of the convectional process, its further statement may be made more brief by speaking only of the ascent of the warm under-layer, with which we are generally most concerned.

"*Conduction* in the atmosphere may be illustrated by the cooling of the lower air at night, when it loses heat chiefly to the colder surface of the ground beneath. This change of temperature is not followed by convection, for it leaves the heaviest layer of air at the bottom, and does not give gravity any opportunity to cause motion. In the daytime, however, conduction is followed by convection, which then becomes an active process. Let us consider the case of the air over a dry plain, beneath an unclouded, torrid sun. The ground warms rapidly in the morning, and soon becomes hotter than the air which rests upon it. Conduction, aided, as at night, by radiation, increases the temperature of the surface-stratum of air. This stratum then expands, and lifts up the overlying air by a small amount, thus reversing the process of the night before. A peculiar optic effect may then be produced, called a mirage."
—DAVIS, *Elementary Meteorology.*

[1] Davis: Elementary Meteorology, p. 132.

Bureau. The changes in weather which the storms produce are generally correctly predicted.

"As the winds become cool, cloudy, and wet when they ascend mountain-slopes, so they become warm, clear, and dry when they descend the leeward slopes. For this reason a mild, dry wind, called the *chinook*, is often felt in winter on the plains along the eastern base of our northern Rocky Mountains. The chinook is of great service in *drying away* the snow so that the cattle can find grass."[1]

Sea- and Land-breezes. At the seashore there is usually in fair spring or summer weather a day-breeze blowing toward the shore and a night-breeze blowing toward the ocean. The causes of these are that during the day the land is heated more quickly than the water, and, therefore, the air lying over it, being warm, is pushed upward by the colder air which flows in from the ocean to take its place. This condition is reversed after sundown. Where great plains meet high mountain-ranges, as in the eastern foot-hills of Colorado, the same phenomena ensue, owing doubtless to the greater radiation of heat from the mountain-sides than from the level plains. Perhaps, also, the composition of the mountain-sides, being rocky, may increase the radiation; this would be furthered wherever the slopes are exposed to the morning sun, because the sun as it rises shines less obliquely, and, therefore, with greater power on sloping ground than on level. Thus the air on these slopes is warmed first and rises, being displaced by a current of air from the plains. There is a reasonable theory that the valleys in the mountain-slope, by concentrating the current of air, intensify the draught, as does a flue; and therefore these breezes are stronger on a seashore or wide plain which is bordered by a mountain-range.

Mountain- and Valley-winds. In valleys, during the hours of sunshine, the lower layers of air are warmed first, and so rise along the mountain-sides, while toward night the cooler air descends as the evening-wind. For this and other reasons a hotel is often more pleasantly situated a little way up the mountain than at the bottom of a valley.

The **effects of winds** on climate are very marked, for they are the great ventilators and purifiers of the atmosphere; winds carry

[1] Frye's Complete Geography.

with them from a distance the temperature, humidity, and impurities of another place, and so partially transport one climate to another. Therefore it is necessary in selecting a resort to know the character and causes of these foreign winds as well as those of the local breezes, and for these reasons the shelter and aspect of the resort should be examined. Wind, in proportion to its lack of humidity, withdraws moisture from the surface of the soil and heat from the body, and, though it is often beneficially stimulating, it requires a certain power of reaction in the individual in order to become so.

WATER.

All water, apart from possible impurities and more or less animal life, contains some salts, the quantity varying from the very large proportion contained by the dense water of the salt lakes to that held in rain-water, which is water naturally distilled and possessing a minimum amount of salts and impurities.

Seas, lakes, and rivers modify climate, increasing its humidity and generally its rainfall, and when they are extensive they provide the neighboring land with an abundance of pure air which blows from off their surfaces. As has been shown, the difference in the radiation of heat between the land and the water causes the day-breezes, which are naturally stronger on the surface of the water than on the shore.

Rainfall may be great, while the humidity is small, or *vice versa.* When no cold breeze is blowing vapor remains in the air, and the air is damp; but if the air becomes chilled by cold winds or other causes, the vapor is condensed and is precipitated in the form of rain or snow or fog, and the air is drier immediately afterward. Therefore a resort with a light rainfall may yet have a damp air. An apt illustration is found in the coast climate of Southern California, where, while rain is infrequent, the air is usually moist. A heavy annual rainfall does not always mean that there is necessarily a large number of rainy days, as it may rain heavily for only a very short time, and the rest of the day be clear and dry, or there may be a short rainy season followed by a long dry one. Both the amount of rain and the degree of humidity diminish, as a rule, with the distance from the ocean. The general law is that the number of rainy days and the rainfall increase in proportion to the proximity to the equator. With the rise in the height of the land the precipitation usually increases on the European continent, while on the American continent the reverse is generally true. On the Atlantic coast of North America the rainfall increases from north to south, and on the Pacific coast from south to north. Thus the normal precipitation is one-half more on the coast of Florida than on that of Maine, while the Southern California coast has less than one-sixth the rainfall of the coast of Washington State.

Rain purifies the air and increases the ozone, and, for the time at least, decreases the humidity. Thus the effects of a moderate rainfall, not so frequent or continuous as to limit outdoor life, are such as to revive and benefit the invalid.

The **evaporative power of the air**, which varies greatly according to the temperature, relative humidity, density, and rate of motion, modifies considerably the effects of rain in a given locality.

Clouds consist of small particles of water, either liquid or ice-crystals, floating in the air, usually at considerable elevations. In a general way a cloud may be said to be formed through the cooling of a saturated stratum of air, so that the vapor, being condensed into minute drops or solid crystals, is rendered visible. While clouds are often of service to the invalid in lessening the heat and glare of the sun in summer, their prolonged presence is depressing, especially in winter, at which season they usually increase the sense of chilliness during the day. On the other hand, by limiting the escape of the heat radiated from the ground and reflecting it back, they often render warmer what would otherwise be a cool night, and also lessen the range of temperature.

Fogs and mists are of the same composition as clouds, but rest on the earth's surface.

Humidity is the condition of the atmosphere with respect to the water-vapor which is always present in varying degrees. When this degree is great the humidity is said to be high; when small, low. Water-vapor is formed by the evaporation from the seas and lakes, which together constitute three-fourths of the earth's surface; the great change in temperature and the breaking up of this water-surface by land cause marked variations in the humidity of the air in different localities and at different times of the day. This vapor is derived from a wet surface and slowly mixes with the air. Its distribution through the atmosphere is chiefly influenced by the movement of the air, both local and general. Thus wind,

when it is not itself laden with moisture, tends to make a resort less
damp than it otherwise would be.

"It is commonly the case that the air over the land does not
possess as much vapor as it might, but on cool, damp nights the air
near the ground is often saturated. This is not because more vapor
is present then than in the daytime, but because of the fall of temperature from day to night, by which the capacity of the air is
reduced so far that the amount of vapor already present saturates
the air. On the ocean, particularly in the calms of the doldrums,
the air is nearly saturated all the year round. This is because the
inflowing trades, blowing over the warm surface of the ocean and
warming slowly as they advance from either side, are continually
supplied with vapor, so as to maintain them in an almost saturated
condition. While loitering in the doldrums their vapor is even
more increased. Even the slight cooling of the air over the equatorial
ocean at night is, therefore, sufficient to make it excessively damp.

"On the other hand, in desert regions the supply of moisture is
so small that the quantity of vapor present is far from satisfying
the capacity of the air. There is very little moisture present compared to that which might exist; the air is then relatively dry. The
same condition occurs frequently in our cold northwest winds of
winter. These come from a northern interior region, where their
temperature is very low, and where but little moisture is present.
They then advance rapidly into milder latitudes, warming as they
come, but proceeding so quickly that their increasing capacity for
vapor is not satisfied. They frequently do not contain half as much
vapor as they might, and sometimes this fraction falls as low as a
third. The upper regions of the atmosphere are also found to be
prevailingly dry. This seems to be the case even over the ocean,
and must be regarded as the result of the remoteness of the great
volume of the atmosphere from the ocean surface, and of the
obstruction that the air presents to the upward diffusion of vapor.

"The humidity of the atmosphere exercises a strong control over
our bodily sensation of the temperature of the air. The body does
not act like a thermometer, readily accepting the temperature of
the surrounding medium, but attempts to maintain an internal temperature of about 98°, known as ' blood-heat,' at all seasons. We
prevent an uncomfortable reduction of temperature in cold air by
sheltering the body from loss of heat by a covering of clothing ; if
the air is windy, more protection is needed than when it is calm;

if it is damp as well as cold and windy, it abstracts all the more heat from us, probably by means of the better conductivity given both to the air and to the clothing by the moisture; hence the difference between the bracing though severe cold of our dry northwest winter-winds and the penetrating, searching chill of our damp winter northeasters. The difference between the so-called ' dry cold ' of the interior and the ' damp cold ' of the New England coast is thus explained. On the other hand, when the air is warm our bodily temperature would rise too high if it were not for the cooling of the skin by the continual evaporation from its surface. In very hot and very dry air the evaporation is so much hastened that the skin is parched and burned; in hot and very damp air evaporation is checked and the air feels sultry and oppressive. Moderately dry, hot air is less uncomfortable than at either of the extremes of dryness or dampness. The oppressiveness of our ' dog-day ' weather in July and August depends as much upon its humidity as on its heat.''[1]

Absolute humidity is the actual number of grains of vapor in a given quantity of air, usually estimated by the weight of vapor in each saturated cubic foot of air. As air expands with a rise of temperature its capacity for holding vapor increases rapidly. Thus when the temperature is at zero each cubic foot of air can hold only 0.54 grain, this being its point of complete saturation. At 32° F., however, it can hold 2.13 grains; at 50° F., 4.09 grains; at 80° F., 10.95 grains; and at 100° F., 19.79 grains.

Relative humidity is the percentage of vapor actually contained in a given quantity of air as compared with its capacity; its value as a test of dampness is, of course, entirely dependent upon the temperature, and cannot be considered without it. An amount of vapor sufficient to cause a high relative humidity at a low temperature would cause only a low relative humidity at a high one. For example, if the relative humidity of the air stood at 60 per cent. while the temperature was 40° F., it would, in the same air, drop down to only 10 per cent. if the temperature was raised to 80° F.

Dry and Moist Climates. In classifying climates as dry and moist it is important to notice whether absolute or relative humidity is intended, and this must often be inferred from the connection in which the words are employed. Relative humidity is important in respect to the liability of the formation of fogs and dew, while phys-

[1] Davis : Elementary Meteorology, p. 144.

iologically the absolute humidity has often a greater significance. When statistics of relative humidity alone are given it is necessary to take some standard temperature to make a fair comparison. Wendt computes the mean annual temperature of the United States as 55° F., which is probably nearly correct. Assuming, then, a standard temperature of 55° F., we can say that less than 50 per cent. of relative humidity is *dry;* from 65 to 75 per cent. is *medium;* from 75 to 85 per cent. is *moist;* and above 85 per cent. *very moist.*

The importance of the temperature should never be forgotten in estimating the value of the relative humidity. For instance, on an ordinary sunny day in winter with the shade-temperature at noon at 40°, we should consider under 70 per cent. of relative humidity as *very dry;* from 70 to 80 per cent., *dry;* from 80 to 90 per cent., *moist;* and above 90 per cent., *very moist.* On the other hand, on a day in summer when the shade-temperature at noon is 70° F., we find less than 30 per cent. of relative humidity at that temperature is *very dry;* from 30 to 40 per cent. is *dry;* from 40 to 50 per cent. is *medium;* from 50 to 70 per cent. is *moist;* and over 70 per cent. is *very moist.*

Dew-point is the temperature at which the atmosphere deposits dew. As has been stated in the paragraph upon absolute humidity, the capacity of the air for holding water-vapor is increased by a high temperature and diminished by a low one. For this reason dew is formed usually in the night, when the air is cooled, and it is absent during the hours of sunshine, when the air is warm. As dew can only form by the cooling of air which is charged with vapor to the point of saturation, the damper the air the sooner the dew-point is reached, so that in a damp climate the dew-point is high and in a dry one low.

The dryness of a resort may be judged not only by the relative and absolute humidity, but also by the dew-point when taken in connection with the mean temperature, because when the mean dew-point and the mean temperature are near together dews, fogs, and mists are sure to be frequent, and when they are far apart such manifestations of humidity must be of rare occurrence.

In speaking of dew as being formed from the water-vapor of the atmosphere, it must also be remembered that vapor will arise from the soil, drawn up by capillary attraction from the subsoil; if the dew-point of the air is high, the air soon becomes saturated and the moisture drawn from the soil forms an additional deposit of dew.

Vegetation also gives off into the cooling air the water that it draws from the soil, or has previously received from the atmosphere, and so further contributes to the dew-formation. Thus dryness of the soil and scarcity of vegetation lessen the frequency of dew. So we see that a low dew-point and a low relative humidity are an evidence of the rarity of dew, mist, and fog.

As evaporation is increased by heat, dryness, and wind, it is greater in summer than in winter, in sunshine than in shade, in wind than in calm, at noon than at night. This readiness of the air to absorb moisture keeps it continually supplied with the vapor which is necessary to all living matter. The water-vapor thus drawn up and diffused in the air exerts a most powerful influence over climate: first, because by its absorption of heat it lessens the radiation of heat from the earth's surface; and, secondly, because when condensed, as in clouds, it prevents the free passage of the sun's rays, and thus regulates light and heat. The activity of electricity and the amount of ozone are controlled by it, while rain, dew, fog, and snow are formed from it.

Consequently, in dry climates and during dry seasons, there being little vapor to hold the heat in the air and few clouds to prevent its radiation into space, the sun's rays are but slightly obstructed; the sunshine is hotter, the shade cooler, and the extremes between day and night more marked than in damp climates and during damp seasons. On the other hand, in damp climates the power and brilliancy of the sun's rays are lessened, and the moist air, while soothing, is apt to be enervating, especially when warm; excessive rain is still more so, and is more objectionable because it interferes with outdoor life.

SUN.

Light. The direct effects of light, apart from heat, are difficult to demonstrate, but it has been proved that it checks the growth of bacteria and low organisms, while it increases oxidation. Weber and others have observed that a want of sunlight will develop symptoms of general depression and a condition analogous to intermittent fever. Sunlight is more intense in dry than in damp climates, and lasts longer in summer in high latitudes, while during winter the reverse holds good. This question of the number of hours of sunshine is very important in estimating the value of health-resorts situated in mountain-valleys.

Heat and its General Causes. The cause of the warmth of the atmosphere and the earth's surface is the sun, which raises the temperature:

1. By its direct heat or radiation.
2. By return-radiation from the earth.
3. By conduction from the earth.
4. By atmospheric currents.

1. Direct radiation gives but little warmth to the air, since pure, dry air is diathermanous to the sun's rays, and it is only by means of the aqueous vapor-clouds and dust that the rays part with any heat on the way, so that the damper the air and the more dust-particles it contains the more easily is it heated by the sun.

2. Indirect return-radiation, which is chiefly reflection from the earth's surface, is influenced by the nature of the solid and liquid surfaces and also by their color; the nearer white it is, the more the sun-heat is reflected back; most of the heat is reflected back at once, the rest is retained for a longer or shorter time on the ground; then all, or at least a greater part, is radiated back into the atmosphere. For this reason the character of the soil in a health-resort should be taken into consideration, sand and gravel being best.

3. The earth gives up a portion of the warmth which it has taken up, to the layer of air which is in immediate contact with it; this is pushed up and replaced by a heavier and cooler layer, which in its turn rises and gives place to another, and so on, and thus a large amount of air is warmed.

4. Winds are the most important form of movement in the aërial ocean; by them atmospheric changes originating in one place are carried to distant localities. The heating of a layer of air by conduction and radiation causes it to become lighter, and the surrounding cold air, pushing it upward, takes its place, giving rise to a current which, when marked, is called a wind. By such currents heated air is carried from the equator toward the poles.

Cold. As an opposing influence to the heating process first named there are constant cooling-influences at work:

1. Continual radiation of heat into space.
2. A fluctuating amount of heat rendered latent by evaporation from the ground, waters, and plants.
3. When the earth, as it always does, has radiated heat more quickly than the air, it becomes cooler than the atmosphere, which then gives up heat by radiation as well as by direct conduction.

Local causes of heat often raise the temperature of a locality above that which is normal to its latitude (*i. e.*, distance from the equator). These causes are many, and we may instance the following : the neighborhood of warm-water currents, as the Gulf-stream; warm-air currents blowing from off warm seas or deserts; or the shelter afforded by mountains from cold winds.

Local Causes of Cold. On the other hand, elevation above sea-level; the intervention of wide seas between the locality and the equator, or their absence toward the poles; the existence of cold currents; mountain-ranges which divert warm winds; the frequency of clouds in summer and their absence in winter, and like causes, frequently make a climate colder than is normal to its degree of latitude.

The **mean annual temperature** is often the same in widely differing climates, and is, therefore, of little importance therapeutically; while the temperature of the various seasons, and of day and night, also the frequency of changes of temperature from day to day, are of the greatest consequence as affecting the amount of time and frequency of opportunity for outdoor life.

ELECTRICITY.

The electricity of the earth is negative and that of the atmosphere positive. Electric disturbances are most common between 3500 and 6500 feet above sea-level. The positive electricity of the atmosphere is increased at altitudes and when the sky is clear, which may partly explain the exhilaration experienced by visitors to medium elevations. During fogs the atmospheric electricity is highly positive, according to Schübler, while negative electricity, on the rare occasions when it has been found at all, has been present in the air during heavy showers of rain.

" The electricity of the atmosphere is stronger in winter than in summer, and is strongest in January, decreasing from this month to June, and then increasing again to January. Like moisture and warmth, with which it is closely connected, it is subject to diurnal periodical variations—that is, to a double maximum and minimum (Saussure and Schübler). It rises from daybreak to the first maximum between 6 and 8 A.M. in summer, and at about 10 A.M. in winter; it then falls to the first minimum between 4 and 6 P.M. in summer, and at about 3 P.M. in winter; it rises to a second maxi-

mum one and one-half to two hours after sunset, and then decreases
to a second minimum, which is reached about daybreak.   The
sources of electricity are said to be evaporation, vegetation, oxidation and other chemical processes, and friction.   According to Pel-
tier and Lamont, however, the negative electricity of the earth's
surface seems to be the chief agent in the production of the electricity of the atmosphere when it contains aqueous vapor."[1]

While it is probable that the effects of atmospheric electricity
upon both healthy and diseased persons are very decided and important, yet our knowledge is, at present, too limited to enable us to
form definite conclusions regarding them.

[1] Ziemssen's System of Therapeutics, vol. iv. p. 56.

CHAPTER II.

PHYSIOLOGY.

Heat promotes growth, though a smaller amount of food is consumed or required, but it lessens muscular power; when humidity is present these effects are intensified.

Cold. In a cold atmosphere carbonic acid is more freely given off and there is greater waste of substance. Cold limits growth, but tends to preserve what may have been gained and to increase muscular power. If, however, a cold climate be also humid, these effects are not so marked. Cold is more healthful than heat, as shown by the fact that the rate of mortality lessens with distance from the equator.

Humidity. *Absolute humidity*—that is, the actual amount of vapor in a given quantity of air—is of importance in breathing, as the amount of water-vapor exhaled with each breath is much greater when the air is dry than when it is damp. Further, as there is always more vapor in warm than in cold air, not only is more water evaporated from the lungs in dry than in damp air, but also in cold than in warm air; thus both absolute and relative humidity modify the air we breathe and influence the processes of respiration. As a result, cold, dry air lessens secretion from ulcerated lung or bronchial tissue. It also reduces the body-temperature by the increased evaporation.

Perspiration. In consequence of the same laws it must follow that there is an increased secretion from the skin in dry air; but the evidence of perspiration is less noticeable because it is immediately converted into vapor by the dry air, instead of condensing upon the surface of the skin in drops of sweat. Thus, in a *warm, dry air* much of the perspiration is insensible; while in a *warm, damp air* it is readily perceived. Thus through the increased evaporation, and its cooling effect upon the body, the fact is accounted for that a higher temperature can be borne with less discomfort in a dry than in a moist air, as is shown in the Turkish or hot dry-air bath and the Russian or hot-vapor bath.

Cold as well as heat is more endurable when the air is dry.

Evaporation from the skin occurs also in *cold, dry air*, but to a limited extent only, and the consequent loss of heat can be largely obviated by clothing, provided a high wind be not blowing. There is also very little loss of heat by conduction in cold, dry air and a very great loss in *cold, damp air;* when the wind is also blowing this loss of heat is far greater than in *cold, dry air.*

Increased barometric pressure up to an addition of a half to two atmospheres has been shown by experiment to lessen the pulse-rate, although it increases its strength and volume. The number of respirations is diminished, while their depth is increased. A larger amount of oxygen is absorbed, and consequently more carbonic acid is exhaled. The desire for food is increased and also the ability to assimilate it. When the pressure of two or three additional atmospheres is suddenly removed, as has been the case with workmen returning from labor in caissons and deep mines, the following effects have been observed: vertigo, nausea, pains in the joints and ears, even paresis, which in most cases was temporary, but in some fatal. In one of these an autopsy showed peculiar lesions of the spinal cord, which appeared as if grooved and torn.

When atmospheric air is compressed the oxygen-tension is increased. In experiments upon animals placed in chambers containing compressed air it was found that there was a slight lessening of the rate of respiration, and, if the pressure was very high, convulsions arose and death ensued from asphyxia. This was due to retention in the blood of carbon dioxide, which could not escape owing to the great air-pressure, although there was a superabundance of oxygen to be absorbed. As might be expected, the effects of compressed air are exactly contrary to the effect of reduced barometric pressure as exhibited under bell-glasses or in high altitudes. These experiments did not cause any of the peculiar nervous effects arising in some persons after leaving caissons in which they had been working; which symptoms are chiefly due to congestion and frequently to degeneration of the lower half of the spinal cord. Autopsies have revealed congested areas with spots of degeneration, even rents in the tissue, escaping bubbles of gas (asserted by Paul Bert to be nitrogen), black blood, etc. During life various paralyses were observed. It is probable, as stated by Andrew Smith and also by Gilman Thompson,[1] that the cause of these nervous

---

[1] Medical Record, February 3, 1894.

phenomena is the exhausting nature of the work in the caissons, followed by too sudden a change back into ordinary air.

Upon the effects of compressed air Dr. Theodore Williams writes as follows :

" Some years ago I placed two remarkably healthy house-physicians of the Brompton Hospital in the compressed air-bath, where they remained three hours and a quarter, and most of my conclusions as to the effect of compressed air on normal subjects are the results of observations carefully taken by these gentlemen on each other and afterward checked by myself. The effect on respiration of the compressed-air bath is that the individual finds that he breathes slower, deeper, and more easily. The respiration-rate, according to my observations, falls from 16 or 20 to 14 or 15 at least. Von Vivenot found it fall to 4 or even 3 a minute. Inspiration becomes very easy, but expiration is less easy; the ratio between them undergoes considerable modification, expiration being sometimes twice or three times as long as inspiration. The increased depth of the inspiration is shown by Lowne's spirometer, which invariably gives an increase in the amount of air expired. It would appear that breathing compressed air increases lung-capacity, probably by opening up more alveoli, which had previously not been brought into use, and we must suppose that the diminished number of respirations means that their amplitude makes up for their smaller number. The effect on the circulation is that the pulse is slower, smaller in volume, but of increased arterial tension; the capillaries smaller, and the veins less full of blood. The pulse-rate diminishes 4 to 20 beats a minute, but on returning to the outer air it returns at once to the normal. Sphygmographic tracings show a lowering in the height of the tidal and dicrotic waves.

"The effect of the pressure on the circulation was admirably shown by Von Vivenot's observations on a white rabbit in the bath. Under normal pressure, with the rabbit quiet and at liberty, the ears were full of blood, the conjunctival vessels injected, and the iris tinted deep red ; but in a compressed-air bath the vessels of the conjunctiva became finer and more pale, and in one experiment they alternately filled and emptied. When pressure was maintained at the maximum the iris and pupil became decolorized, and the ears, seen by transmitted light, showed empty vessels, and the larger vessels were scarcely visible.

" From these latter observations the conclusion is that compressed air exercises an intropulsive influence on the circulation, affecting those surfaces most exposed to it, such as the skin and lungs. The blood is thus drawn into the organs protected from air-pressure, namely, the brain, heart, liver, kidneys, and spleen. The pressure is exerted more on the capillaries and superficial veins than on the deeper veins and arteries, and its tendency would be to reduce pressure in the right side of the heart and to increase it on the left. Dr. Burdon Sanderson thus accounts for the slower pulse-rate: ' The effect of the diminished fulness of the venous system is to retard the filling of the ventricles during the period of relaxation, and consequently to lengthen the diastolic period and thus diminish the frequency of the pulse.'

" The introduction of a larger amount of oxygen causes greater absorption by the lungs and leads to further oxidation and tissue-change; this being proved by the bright color of the blood seen during bleedings in the bath, by the increase of carbonic acid exhaled from the lungs and of urea excreted by the kidneys. Muscular power is augmented, appetite generally improves, and weight is almost invariably gained. The temperature is not materially affected."[1]

Depressed climates, which are examples of naturally increased barometric pressure, are found in places on the earth's surface which are below sea-level. As no complete or satisfactory scientific observations have been made with regard to the peculiar physiological effects of increased pressure, they cannot be here discussed; but certain of the resorts are described and the clinical experience given in a later chapter.

Decreased barometric pressure is best discussed in its physiological effects under elevated climates. At high altitudes the special effects of decreased pressure are not directly produced by the scarcity of oxygen in the atmosphere, but by the diminished oxygen-pressure; for even at the greatest heights ever reached by man the amount of oxygen in each breath is always in excess of that needed to sustain animal life. It has been demonstrated by experiment that blood can absorb only a certain percentage of the total amount of oxygen present in the air to which it is exposed, and so, when the barometric pressure is reduced, the blood may be unable to extract sufficient oxygen from the air because the oxygen-pressure is reduced below the required point.

[1] Williams's Aërotherapeutics, p. 102.

Mountain-sickness is a malady caused by this oxygen-starvation. If this were all, it would follow that when the oxygen-pressure was sufficiently reduced animal life would be impossible from continual mountain-sickness; but there is developed a wonderful compensatory process whereby the blood's power of absorbing oxygen is increased, so that a given weight of blood in a living animal can absorb more oxygen in proportion to the reduction of the barometric pressure. This is brought about by a growth in the number of red corpuscles in the blood and of the hæmoglobin which is contained in them. The hæmoglobin is that portion of the blood contained in the red corpuscles through which oxygen is absorbed.

While these blood-changes, which need some three or four weeks for their completion, are progressing, the breathing becomes more rapid, so that while less oxygen is taken in at each breath, it is received into the blood more frequently; and with this more rapid respiration there is increased heart-action, the heart pumping more blood through the lungs in a given space of time.

This increased rapidity of heart-beat and respiration is, however, only temporary, and gradually disappears. The amount of air taken in at each breath becomes greater as the chest expands and the air-cells, many of which are, at lower altitudes, often unused, are dilated. The heart's cavities, having been stretched, are also dilated, so that more blood is propelled at each stroke. Thus the blood's capacity for absorbing oxygen, the lung's capacity for taking air, and the heart's capacity for pumping blood are increased; the rapidity of respiration and pulse diminishes, but they become normal again in rate as soon as this process of compensation has effected a balance.

These changes in the blood, lungs, and heart continue during a residence at high altitudes, but disappear again upon a return to low ground. However, they are occasionally so incompletely carried out in certain individuals, owing to age, feebleness of reaction, or disease, that attempted ascent into the upper air is exceedingly dangerous and continued residence on high ground impossible.[1]

Balloon-ascents. Rapid diminution of barometric pressure, as in balloon-ascents, is much more trying than the more gradual change consequent upon the ascent of a mountain. The phenomena of mountain-sickness may appear at lower elevations in those who

[1] On this subject see Twentieth Century Practice, vol. iii. p. 216.

climb than in those who are carried in balloons or chairs, on account of the fatigue attendant upon the exertion. If the aëronaut remains for a short time at a moderate elevation, the symptoms in his case may not be so severe; but the results of ascending rapidly to a great height are often alarming and may be fatal, as in the case referred to by Whymper, as follows : "On April 15, 1875, Messrs. Croce-Spinelli and Sivel left the earth at 11.35 A.M., and in two hours more hovered about 26,000 to 28,000 feet. At the end of this time both were found suffocated, with their mouths full of blood ; but neither the time nor the elevation at which they died is known exactly, as M. Tissandier, the sole survivor of the party, was rendered insensible and was thus unable to give a complete account of the affair."

The cause of this disaster was undoubtedly the same as that which gives rise to mountain-sickness in mountain-climbers, namely, the diminished barometric pressure; but in a gradual ascent this cause operates simply through diminished oxygen-pressure, while with the rapidly ascending aëronaut not only does diminished oxygen-pressure distress him, but also the sudden change of atmospheric pressure, which causes for a time an inequality in the mechanical pressure of the atmosphere upon the outside and the inside of his body. The mechanical effect of sudden change in the density of the air surrounding an individual is also exhibited in caisson-disease.

It may be said, then, that in mountain-ascents the cause of mountain-sickness, leaving out the effects of exertion, is simply diminished oxygen-pressure, while in balloon-ascents it is this *plus* rapidly diminished atmospheric pressure. The inhalation of oxygen has been resorted to by aëronauts when at great elevations, with marked relief to their unpleasant symptoms.

The greatest height known was reached by Coxwell and Glaisher in a balloon on September 5, 1862; they lost consciousness at an elevation of 27,000 feet, and, continuing rapidly to ascend, reached a much greater but unknown height, not recovering until they had again descended to 27,000 feet.

Mountain-climbing. Healthy persons, ascending to elevations, experience a more or less pleasant stimulation of body and mind ; but while the impulse to exercise is increased, the ability to do so is diminished with the altitude, though in the strong tolerance is soon established, and feats exhibiting great powers of endurance have been accomplished by mountain-climbers. In the feeble or diseased, or when considerable heights have been reached, irregularities of

breathing and heart-beat are experienced, and symptoms, due doubtless to anæmia of the brain, such as giddiness, nausea, sleeplessness,
loss of memory, and even hallucinations, may occur; while at a
height of over 25,000 feet unconsciousness, hemorrhages, cyanosis,
and symptoms of paraplegia have been noted.

The observations taken by Whymper during his ascents of the
Andes,[1] and those of Conway while climbing the Himalayas,[2] with
their several experiences, are interesting and instructive, especially
when studied together. They illustrate the physiological effects of
diminished barometric pressure upon mountain-climbers in normal
health. Both Whymper and Conway, and most of their com-
panions, had been accustomed to mountaineering at lower elevations
in the Alps. Both experienced mountain-sickness the first time,
within three weeks of their arrival in the country, Whymper at an
elevation of about 15,000 feet, ten days after landing. Neither
felt it appreciably again at as low a level. Whymper, a week
later, after climbing to a height of 16,664 feet, was more affected
than on the first occasion. However, after being four weeks in the
country, he reached the summit of Chimborazo (20,608 feet), the
highest mountain he scaled, with very little discomfort. Conway,
about three months after his arrival, reached 18,600 feet, also without any mountain-sickness. About two weeks later, however, on
reaching the top of Pioneer Peak, 22,600 feet, he experienced very
serious symptoms. This was the greatest height he climbed, and he
doubted if they could have gone any higher.

The fact that both Conway and Whymper were affected within
three weeks of their beginning to climb heights between 15,000 and
16,000 feet, and later were only slightly disturbed at various elevations up to 20,000 feet, points to an improved capacity for climbing
due to the increased power of oxygen-absorption of the blood, which
the diminished pressure usually effects in the first four weeks, as was
proved by Regnard's experiments, mentioned elsewhere. This explanation is confirmed by the fact that they had very little mountain-
sickness after the first four weeks, except when Conway reached a
much greater height. They both speak of headache, dizziness, gasp-
ing for breath, rapid pulse, an occasional slight rise of temperature,
and difficulty in exerting themselves. Freedom from nausea and
vomiting seemed to depend somewhat upon how and what they ate;
the best plan being to eat sparingly but frequently of light food, such

[1] Travels among the Great Andes of the Equator. Edward Whymper.
[2] Climbing in the Himalayas. W. M. Conway.

as kola-biscuits, chocolate, meat-peptones, and also to drink a little brandy (Conway). There was some difficulty in sleeping after the climbing had been especially hard. Whymper says they became somewhat accustomed to low pressures, and while they were never able to work as fast as at sea-level, they improved much in this respect. They had no hemorrhages, vomiting, or nausea, and the absence of the latter he attributed to careful and spare dieting.

Whymper, in order to see if he could make as good time at a height as at the sea-level, walked six miles on a flat road at an altitude of 10,000 feet, and compared it with a walk of the same distance previously taken at sea-level. He knew his usual pace, and kept, as far as he could, the same stride on both occasions, and the other conditions he considers were about equal. He found that it took him 54 seconds more, on the average, to walk a mile at 10,000 feet elevation than at sea-level. His pulse before starting in both places was nearly the same, 73; but at the finish at sea-level it was raised to 96, and on the high ground to 101.

With respect to the ability of athletes to perform their accustomed feats as well at a high altitude as at sea-level, I have made some inquiries about the racing of well-known professional runners on the track at Denver (altitude, 5280 feet). I find that when they attempt to equal their record within the first few weeks after arriving they usually fail to do so in long-distance races; but in short ones they sometimes exceed it, as did H. M. Johnson, on August 18, 1889, when he ran a hundred yards in $9\frac{3}{4}$ seconds, beating his previous record of $9\frac{4}{5}$ seconds (Cleveland, July 31, 1886), and making the best time on record.

With few exceptions, neither the best professional runners nor the wheelmen of the same class have raced on the Denver track; but the information that I have collected as to the records which have been made there seems to render the following results certain: in sprinting, both on foot and on the wheel, better time has been made at Denver than elsewhere, as the appended evidence shows.[1] For

[1] Bicycling records, from the New York Referee Calendar for 1896:
880 yards, paced, 51 seconds. B. B. Bind, against time, Denver, Col., Oct. 19, 1895. The best world record.
Unpaced, 59 seconds. Harry Clark, against time, Denver, Col., Nov. 20, 1895.
In competition, $58\frac{1}{5}$ seconds. C. M Murphy, Denver, Col., Oct. 19, 1895.
$\frac{2}{3}$ mile, unpaced, $1.21\frac{1}{5}$. H. Clark, Denver, Col., Oct. 17, 1895.
1 mile, unpaced, $2.00\frac{2}{5}$. W. W. Hamilton, against time, Denver, Col., Oct. 12, 1895.
1 mile, paced (Class A), $1.54\frac{3}{5}$. P. Becker, Denver, Col.. Oct. 19, 1895.
1 mile, $2.02\frac{2}{5}$. C. C. Collins, in competition, Denver, Col., Oct. 17, 1895.
2 miles, unpaced, $4.46\frac{1}{5}$. H. Clark, Denver, Col., Nov. 21, 1895.
3 miles, unpaced, 7.15. H. Clark, Denver, Col.. Nov. 21, 1895.

distances from half a mile to three miles there is a gradual falling off, and for distances over three miles the Denver records are behind those made at places nearer sea-level.

Conway found that when he was at a great height the discomfort of breathing was relieved by increasing the rate of respiration, sometimes to 38 per minute, and was made worse if a cramped position was taken while climbing. The symptoms were much slighter when he was at rest than while he was exerting himself. Both he and Whymper found smoking enjoyable except when they were suffering from mountain-sickness. Conway further says that when he was at 22,600 feet he had to walk very slowly, and found any exertion difficult. His breathing, however, was very little affected at that time, especially when he was in repose. He had then been about three months in the country. " But the sphygmograph showed their hearts were sorely tried and that they had reached the limit of their powers." He could not sleep that night, and " felt his heart racing like a screw out of water." These feelings lasted even after he got down to camp at 20,000 feet; but the next day they descended to 8640 feet, where they slept long and heavily, ate heartily, and felt much relieved. Conway found that the symptoms of mountain-sickness came on at lower levels than usual when the sun was hot or when they were travelling through a close ravine. After a few weeks of climbing he noticed that the rate of the pulse and of respiration as well as the body-temperature had again become normal.

Whymper refers to the difference between climbing a mountain gradually and rising rapidly in a balloon, which latter, he says, is much more dangerous.

In the chapter upon Elevated Climates the subject of barometric pressure is dwelt upon at greater length.

CHAPTER III.

IT is now generally admitted that the chief characteristics of race are mainly due to climatic causes. This view is supported by the fact that, with the exception of the Mongolian and Ethiopian races, the progeny of natives of one climate who become permanent residents of another exhibit the peculiarities of the people among whom they are living and lose those of their ancestors. This even applies to a stronger race which may settle as conquerors of a new country. As to the two races mentioned above, it is suggested that they having been for so long domiciled amidst unchanged climatic influences, with little or no admixture with other races, are merely slower to lose their original racial traits.

The normal being can accustom himself to and thrive in any climate, the inhabitants of temperate climates naturally excelling in this capacity. In spite or rather because of their rapid and unexpected changes, impossible to foresee or guard against, variable climates have a most favorable influence upon the robust development of the bodily and mental powers. The natives of northerly inland countries, while they suffer from the extremes of temperature, are less exposed to those endemic sources of disease which produce so much illness and death in low or level countries and in more southerly climates, and the very vicissitudes of climate, by training the system to endure severe physical conditions, must react favorably upon the mental attitude. As Copland has said, " The physical and moral history of the British Isles, Denmark, Sweden, and the more continental districts of western Europe demonstrates this fact."

The inhabitants of the countries of eastern Europe, Central Asia, and North America, on the contrary, have been led, by the greater regularity of seasonal changes, to the adoption of precautionary measures and a more luxurious mode of living, which depress the vitality, encourage disease, enfeeble the frame, and shorten the mean duration of human life. The practice of living in overheated houses unfits the frame to bear the rigor of the cold, dry, external atmosphere, and hence the alternation from one to the other produces diseases of the thoracic and abdominal viscera.

In natives of northern climates the functions of the lungs and kidneys are very extended, while those of the liver and skin are much more limited. In natives of temperate countries all the emunctories are stimulated to a great and more equal degree of activity, but in the warmer districts of such countries the respiratory changes are lessened, while those brought about by the intestinal mucous surfaces are increased, and this is especially true of districts the atmosphere of which is damp as well as warm.

It may be said, in a general way, that in intertropical countries the conditions given for temperate climates are reversed. The skin of the white man performs its functions to a very limited degree compared to that of the negro, for instance, which differs not only in color but also in texture. Through this latter peculiarity it is enabled to perform excretory functions in aid of the respiration and biliary secretion, and its action compensates, to some extent, for the diminished action of lungs, liver, and kidneys which has been observed in the natives of intertropical climates. It gives forth more of the aqueous fluid and carbonic acid from the blood and elaborates an oilier secretion than does the skin of the white, thus counteracting the extreme effects of the sun's rays and carrying off superabundant caloric.

"The most conspicuous difference in the external aspect of men and of races of men is in color; and here comparative pathology would lead one to look for some corresponding differences in susceptibility to disease, for the experience of horse-breeders and veterinarians is pretty clearly expressed on this point. Thus Youatt says that the dark chestnut, as a rule, yields to no other color in any quality; but that the light chestnut, which appears to be the analogue of the sanguine blond-man, is spirited, but irritable and delicate in constitution. Black horses, again, number among them some of the very finest of their species; but many of them are heavy and dull in temperament, and there is an idea afloat that they are particularly liable to malignant disease. Here we may be led to think of the choleric and the melancholic temperaments. Among breeds of sheep the black-faced have the reputation of being hardier than the white-faced. Certain black pigs, according to Darwin, can eat with impunity what would be poisonous to white ones on the same pasture; and like differences are seen in black and white rats.

4

On the whole, however, the deposition of pigment in the skin and hair of mammals would seem to be the result of processes which connote or accompany health and vigor rather than the opposite.

" The statistics of morbidity and mortality, which alone could yield a sound foundation for generalizations on this subject, are unfortunately imperfect or altogether wanting in the regions where the material would be most valuable—those regions, namely, where nations of different colors and constitutions of body live side by side under comparable conditions. In fact, we have hardly any trustworthy statistics except from the most civilized of the countries whose populations are compounded from more or less distinct divisions of the human race (vide Dr. Billings's article on ' Medical Statistics ')."[1]

If, then, it be admitted that the effects of climate upon the inhabitants of different countries are so definite and so obvious, it must be plain that they are closely associated with the nature of disease and with its treatment.

Let us now consider the relations between food and the soil and climate which produce it, together with their combined operation upon man, who must be acknowledged to be in a degree the creature of both. His moral and physical development is modified and limited by both, and therefore the natural history and the diseases of the inhabitants of a country cannot be comprehensively viewed without taking account of the productions of the soil by which they live.

It must be remembered, however, that man is as much the moulder of circumstances as he is their product; that he can accustom himself to bear alike an equatorial heat or the rigors of an arctic temperature. No one sort of diet is necessary to his well-being, nor is a mixed diet required. He must to some extent eat as his environment demands. " The Russians who winter in Nova Zembla, according to Dr. Aiken, imitate the Samoieds and eat raw flesh and drink the blood of reindeer in order to preserve their health in these arctic regions."[2]

Tropical Regions. In these regions the extreme and constant heat tends to stimulate the vascular and nervous systems, the effects of high temperature being, however, largely modified by the amount of humidity.

---

[1] Allbutt's System of Medicine, vol. vi. p. 21.
[2] Copland's Medical Dictionary, vol. i.

It has always been stated that the increased function of the skin of intertropical races tends to modify the extreme effects of the sun's rays and to give off superabundant animal heat. A heat-making diet of flesh would, then, be entirely unsuitable to the needs of these people, and even a mixed diet, such as is required by the inhabitants of temperate climates, would be harmful here. But tropical countries, while they offer the most lavish supply of productions of the vegetable kingdom, provide few of those gregarious animals which are used as food, and some of these are pronounced sacred by the dominant systems of religion. Therefore, the natives, except perhaps in cool, elevated districts, are driven to adopt, because it is easily procured, a vegetable diet, by which their health is maintained. In conjunction with this diet the inhabitants of intertropical regions are accustomed to consume a large quantity of the hot spices indigenous to these localities. These stimulate the system, largely counteract the effects of the lowered temperature and unhealthy emanations consequent upon the rainy seasons and monsoons, and give, to a certain extent, immunity from intestinal worms and other parasitic animals.

It is certainly a fact that the causes of disease are nowhere more energetic than in tropical countries, but comparatively few maladies arise from the nature or wrong use of food, the prevalent diseases being chiefly such as proceed directly from the soil and climate, and the same region which produces a disorder produces also its most powerful remedy or prophylactic.

Polar Regions. In high latitudes man's requirements are the direct opposite of those necessitated by warm climates. The lowered temperature and the lack of sunlight during the greater part of the year tend to depress the vitality, lessen nervous and vascular energy, and lower the tone of the system.

In order to counteract the prejudicial effects of the climate the inhabitants have sought and found the best possible food for their needs, viz., an animal diet especially rich in fat and oil, which are the great heat-producers. By these means their nerve-force, richness of the blood, and bodily heat are maintained at a very high level, and they are enabled to preserve their health in spite of the extreme rigors of the climate.

Temperate Climates. *The north temperate zone,* possessing the greatest diversity of climates, provides a plentiful supply of both animal and vegetable food, and man is thus enabled to adapt his diet to the seasons or to the climate of the locality in which he lives. If the population of any district should increase through the demands of commerce or manufacture, until its requirements exceed the means of support furnished by such locality, the necessary food should be imported from a similar climate, otherwise it is apt to prove unwholesome. The consumption of unwonted luxuries brought from distant countries often produces disease, and this remark applies also to the unusual and unsuitable preparation of food which may in itself be harmless.

The warmer the climate the more necessary is it to approximate to the conditions suitable to life in tropical countries, whereas cooler climates demand a freer adoption of those rules necessary to existence in high latitudes.

From these statements the following obvious conclusions may be drawn: first, it is generally true that man finds, indigenous to the country in which he may chance to be, the food most suitable to the climatic conditions; secondly, the nature of the food consumed by him unites with the climate to modify as well as to support his constitution.[1]

The Effects of Climate. If the foregoing pages have made clear the action of cold or warm climates upon the human frame and constitution, it will be easily understood that the effects of climatic change are important in direct ratio to the suddenness and degree of such change, and that they include some disturbance of the normal action of the system, temporary or otherwise, frequently, though not necessarily, giving rise to fever.

---

[1] Much of the material for this chapter may be found in the article upon climate in Copland's Dictionary of Practical Medicine.

CHAPTER IV.

GEOGRAPHICAL DISTRIBUTION OF DISEASE.

It would appear, speaking broadly, that the tendency to disease in an organ, particularly if of an inflammatory character, is in proportion to its functional activity. Though all the organs act more equally in temperate climates than in extreme ones, yet the chief labor devolves upon the kidneys, and so renal diseases are more common; while in *hot climates* the liver, bowels, and skin being more active, such diseases as hepatitis, dysentery, and leprosy abound. In cold climates the secretion from the kidneys, liver, and bowels is moderate, and that from the skin especially limited, so the work is chiefly performed by the lungs. Therefore, pneumonia, bronchitis, etc., are frequent. It has been shown elsewhere that humidity greatly modifies these tendencies. And, further, within certain limitations, it is often true that a climate which, by its stimulating qualities, tends to provoke acute disease, through the same causes mitigates or removes certain chronic affections which are the results of previous illnesses.

Germ-diseases. We find that contagious diseases are naturally most prevalent in crowded districts, and that contagion may prove stronger than climatic influences; the latter may modify its manifestations and progress, but cannot always prevent them. This accounts for the fact that such diseases as scarlatina, measles, diphtheria, etc., are found in all climates, and it is also true, in a somewhat less degree, of tuberculosis, as will be shown later.

However, as some germs grow more vigorously or solely in high temperatures, for instance, those of yellow fever, and some more vigorously in low temperatures, as those of smallpox, climate does modify many of the germ-diseases. The degree of humidity in the air has also an influence on germ-life, but not nearly so much as has the amount of moisture in the soil. This is notably the case with the germs of malaria, and in sending patients to damp climates the condition of the soil with respect to its moisture must be particularly inquired into. Speaking broadly, it may be said that germ-diseases are more active in hot than in cold climates, and in damp than in dry ones.

Some germ-diseases are contagious, some spread only by infection through the air, some only through water and food, and some in all these ways, as does malaria.

DEATH-RATES PER 1000 DEATHS FROM ALL CAUSES.[1]

Countries.	Periods.	Smallpox.	Measles.	Scarlet fever.	Diphtheria.	Diphtheria and croup.	Whooping-cough.	Enteric fever.	Pneumonia.	Consumption.	Diarrhœal diseases.	Cancer.	Childbirth.
England and Wales	1880–89	2.5	23.1	19.6	8.1	15.3	23.6	10.6	53.6	90.9	39.5	29.8	7.8
Ireland	1880–89	0.7	10.7	13.8	3.8	14.6	16.8	9.1	29.4	115.8	18.1	21.8	9.3
Scotland .	1880–89	0.2	18.9	16.9	10.9	21.9	31.1	12.8	54.0	104.7	25.6	29.6	9.1
Belgium .	1880–87	15.7	25.1	12.4	......	27.1	32.8	26.1	......	......	73.6		11.2
Sweden .	1880–89	1.0	11.0	30.3	31.1	12.3	10.7	15.2	......	......	25.4	......	6.8
Prussia	1880–89	0.6	16.7	18.3	......	65.7	20.8	15.4	56.5	122.6	41.8	14.4	8.4
Austria	1877–86	19.9	16.4	20.0	......	53.7	35.7	23.4	......	126.0		13.9	
Saxony	1881–90	......	9.4	13.8	......	55.0	9.9	7.7	......	85.7		26.3	......
Norway	1880–89	0.3	6.3	21.8	33.0	......	10.4	6.4	46.5	82.7		29.7	12.3
Rhode Island .	1880–89	0.1	5.3	26.6	29.9	45.1	7.9	27.6	75.7	133.7	86.2	29.0	5.7
Connecticut	1880–89	0.5	5.4	15.8	32.4	47.2	6.9	22.9	75.3	124.5	80.3	25.3	5.8
Massachusetts .	1880–90	0.4	5.7	12.2	34.2	48.3	6.5	23.7	82.1	151.3	74.7	28.4	4.9
New Jersey	1880–89	3.0	5.8	25.5	......	53.4	7.9	25.8	......	131.3	116.5	21.3	......

The almost universal distribution of certain diseases, even in places and under circumstances which make it impossible to believe that the virulence of the germs can have been derived from another case (there having been no previous cases of illness of any kind in the neighborhood), suggests the theory that the germs of these diseases may be almost universally present in an innocent state, and only become virulent when they are brought into contact with certain conditions inside or outside of the body.

Take, as an instance, typhoid fever, which has been proved not to spread through the air, but only through food or water in which the germs are present; while, without doubt, it usually arises from the ingestion of germs in water or food infected by other typhoid cases, yet it is probable that it does sometimes occur without being derived from another case. The body of the patient has perhaps been subjected to an unusual waste of tissue, such as comes from prolonged or violent muscular exertion, and there has been imperfect elimina-

---

[1] Allbutt's System of Medicine, vol. vi. p. 17.

tion due to a chill or other cause, and thus a soil is furnished within the body fitted to convert the harmless germ into a virulent one.

It has lately been suggested that the bacillus coli communis is identical with the typhoid bacillus, as the former is present in the intestines of healthy persons. Some also think that the comma-bacilli, which are believed to be the cholera-germs, may sometimes be innocent, because they are found in healthy persons. The two diphtheria-bacilli, identical in appearance, one innocent and one viru-lent, furnish another instance.

In accordance with this theory, one may have chosen a health-resort possessing good sanitary conditions and yet cases of typhoid may present themselves, and some of the problems afforded by the connection between germ-diseases and climate may thus be explained. As we are only at the beginning of the scientific study of both germ-life and climate, it is not surprising that we have to resort, in dis-cussing them, more to speculation than to proof.

Fortunately, most of the germ-diseases are not those maladies for which change of climate is sought, and therefore are not of material importance in a practical treatise such as this aims to be. The dis-eases which chiefly concern the invalid and his advisers will be discussed under their different heads in the part devoted to climato-therapy.

The Modification of Disease by Season and Weather over and above the General Influence of the Different Climates. Regular variations in the curves of mortality and sickness are noted as the seasons change, and these normal variations may be influenced by exceptionally bad or good seasons. The changes of weather from day to day also modify the normal seasonal curves. Dr. Weber has demonstrated this subject so clearly that I quote as follows :[1]

" A mere glance at the health and mortality statistics for different seasons, and under different conditions of weather, shows the influ-ence of different climatic factors on the state of health of the popula-tion, and the different seasons represent to a certain extent different climates. Thus, when the weather is moderately cool, being at the same time moist and liable to sudden changes, we observe the prevalence of rheumatic affections and of catarrhal and inflammatory states of the respiratory organs. At a still lower temperature of the air, without

[1] Von Ziemssen's Handbook of General Therapeutics, vol. iv. p. 4.

moisture, but more so with it, we notice that in temporarily or permanently weak subjects most of the functions of the body suffer, and that in aged people apoplexies become more frequent. In hot weather there appears a tendency to diarrhœa and to other diseases of the abdominal organs. On the other hand, we find that during the mild or moderately warm and not too damp weather of the second half of spring and beginning of summer, many chronic affections, particularly chronic catarrhs and chronic rheumatisms, improve considerably, and that weakly people gain in appetite and digestive power, blood-formation, and muscular force. We notice also that persons with chronic catarrh and emphysema suffer less during the prevalence of warm or moist weather, and are more able to undergo fatigue, while many subjects with chronic dyspepsia and tendency to depression of spirits and hypochondriasis always feel better and seem like quite different persons when the temperature is moderately cold and there is a clear sky with sunshine. To remove persons to climatic conditions in which the influences of certain seasons hurtful to them are as far as possible absent, but where the favorable influences of other seasons prevail, is the chief object of climatic treatment."

The geographical distribution and prevalence of phthisis are the most important matters to be considered in connection with change of climate for invalids, and they are discussed in the chapter on Phthisis in Section II.

CHAPTER V.

CLASSIFICATION OF CLIMATES.

THIS is a task which it is impossible to carry out with scientific accuracy because of the intimate relation which the various climatic factors sustain to each other, the local modification of general laws, and the seasonal variations. On this account it can only be treated in a broad and general way.

It is admissible to classify climates according to their physiological effects (as stimulating or sedative, etc.) or therapeutically; but such division should be preceded by a classification based upon their physical peculiarities. The tabulation should be made, first, in reference to their position, by separating them into sea and land climates, with their subdivisions (as in Table A); and, secondly, according to temperature, these groups being again divided by humidity (as in Table B). All of these may be regarded as subdivisions of Table A.

TABLE A.—POSITION.

Sea climates . . . {
  Ocean climates.
  Island climates.
  Coast climates.
}

Land climates . . {
  Low climates (up to 2500 feet).
  Medium climates (elevation up to 4500 feet).
  High climates (elevation from 4500 feet upward)·
}

TABLE B.—TEMPERATURE AND HUMIDITY.

Cold climates {
  Moderate {
    Dry { Moderate climates. / Extreme climates. }
    Moist { Moderate climates. / Extreme climates. }
  }
  Extreme
}

Hot climates {
  Moderate {
    Dry { Moderate climates. / Extreme climates. }
    Moist { Moderate climates. / Extreme climates. }
  }
  Extreme
}

Ocean. The air over the ocean is always damp, owing to the constant evaporation; the rainfall is usually large, and fogs are not infrequent. In southern latitudes, however, and at certain seasons there is often very little rain and fog, which may be partly explained by the fact that the constant evaporation, with the presence of an excess of water-vapor in the atmosphere, modifies the power of the sun's rays and makes the change between night and day and between winter and summer less marked, equability of temperature being one of the most notable features of an ocean climate. The air is impregnated with salt and iodine and bromine, and is strongly ozonized. There is much movement of air, and winds from a distance, not being diverted by shelter, often bring with them the climate of distant lands, and so modify natural equability.

The ocean-currents, such as the Gulf-stream, the Kurosiwo (the great Japan current), or the cold waters from the arctic regions, also modify the climate in different parts of the ocean, the Gulf-stream in some regions increasing the atmospheric humidity.

Physiological Effects. The radiation of heat from the body is therefore greater, and warmer clothing is needed. Metabolism is more active, and the bodily weight increases, as do usually the appetite and the inclination to sleep. The nervous system is soothed. In short, on those with whom it agrees, the air of the ocean acts as a *sedative tonic.*

Sea-bathing. The uses and effects of this usually agreeable and valuable therapeutic agency are admirably put by Dr. W. M. Ord, of London, who says: "The great benefits derived from the inhalation of fresh sea-air and from sea-bathing cannot be too highly appreciated, but, as in the case of all other remedial agents, their use has its bounds and its qualifications. People accustomed to a non-invigorating inland atmosphere cannot with impunity expose themselves to the often keen air of the seaside. As a rule, they require warmer clothing than at home; and when want of strength reduces the power of taking exercise the sense of drinking in health with the air does not justify sitting for long in exposed positions and without shelter. In respect of bathing we may speak more strongly. Even for robust persons of good swimming-power a prolonged immersion is productive of exhaustion. Doubtless strong people and perhaps even weakly ones can stay in the stimulating salt-water longer than in fresh without feeling the bad effects of the lowering of the temperature of the body ; and it must be admitted—nay, urged—that every individual body has its own rule. In use, even for healthy persons coming from the enervating air of large cities, the first baths should certainly be of short duration. They should include, if possible, a plunge into water sufficient to cover the shoulders, and, if possible, a short swim. The water should be quitted in a few minutes, before depression has followed stimulation. The condition of the bather after the resumption of his clothes will soon afford a test of the exposure which he may undergo with advantage. This will consist, on the one hand, in a sense of warmth, refreshment, and readiness for muscular activity ; on the other hand, subsequent feeling of nausea, of chilliness, of headache, or of palpitation will show that the just measure has been exceeded."[1]

Island Climates. Moderate-sized islands away from the coast enjoy the benefits of ocean-air without the drawbacks of ship-life ; but their climates are, of course, modified by currents, and by coasts and their configuration.

Coast climates, while resembling those of the ocean and of islands, are less equable, owing to the more rapid diurnal radiation and stronger irregular changes, the latter being truer of eastern temperate coasts than of western, which are sometimes more humid and their rainfall greater; but where large deserts lie adjacent and where the prevailing wind is seaward these conditions are often reversed.

Inland climates of low elevation (under 2500 feet) have less ozone and less purity of atmosphere than the ocean ; but, like it, they have high barometric pressure, which, however, diminishes as the land ascends. They have also medium humidity, which decreases with distance from the ocean, though this is modified by the intervention of mountain-ranges on which the humidity is partially precipitated, as on the Coast Range of California; or it may be increased by inland seas or lakes or by the character of the soil or vegetation. The mean temperature usually diminishes with distance from the equator, and the changes between day and night and the seasons are more marked as the humidity lessens with the distance from the ocean.

Extreme hot and cold climates (whether moist or dry) are, for obvious reasons, unsuited for therapeutic purposes.

Moderately moist, warm climates are often of service because of their sedative effects upon the nervous system and mucous membranes; in the proper season the even temperature and moderate precipitation permit of a pleasant outdoor life. Such climates are found on the Riviera, in Italy, and in southeastern Georgia, Florida, and Southern California. They are useful to those who wish to

---

[1] The Climates and Baths of Great Britain. Royal Medico-Chirurgical Society, London, 1895.

avoid the rigors of their home-winter and also to enjoy more open-air life, or for invalids to whom climatic extremes are unsuited.

Moderately moist, cool climates have no positive effects or characteristics, but certain of the resorts in such climates are valuable because they afford change of air combined with pleasantness of surroundings and accessibility. In these climates mineral springs are common, as at Baden-Baden and Saratoga. Many mountain-resorts of from 1500 to 3000 feet elevation, such as are found in the Adirondacks, Allegheny, and Cumberland Mountains, come under this head, and these are of more service to a certain class of invalids than resorts of less elevation because they are sparsely settled and on account of the pine-woods and opportunities for camp-life and sport.

Moderately dry, warm climates are rare in Europe. Egypt and Algeria are the most resorted to, being especially noted for these qualities; while parts of the lower lands of Arizona, New Mexico, and the inland regions of southeastern California furnish examples on this continent. They have a positive *tonic* and *stimulating* character, and are of much value where the more extreme effects of higher ground are not desirable. They would be more frequented were it not that their chief qualification, that of dryness, has prevented their being settled, and so they are apt to be deficient in accommodations and amusements. For permanent residence they have the objection of being too hot in summer.

Moderately dry, cold climates, such as the winter climates of parts of Canada, Dakota, Nebraska, and Minnesota, are healthful and stimulating to the robust, but are too severe for all but the strongest class of invalids.

Depressed Climates. These are the climates of the comparatively few spots in the world which are below sea-level, such as the so-called sinks near the Caspian Sea, the valley of the Dead Sea, and southeastern California. These are all in desert countries, and are, therefore, hot and dry; they will be more fully described later.

Elevated Climates (from 4500 feet up) all come under the head of dry climates, while they are both cool and warm—that is, the air itself is cool, while the differences between sunshine and shade and day and night are very marked; seasonal changes are, however, usually not extreme, and the contrasts referred to are specially noticeable in winter.

With this dryness of air there is often considerable precipitation, much rain falling at a time, so that the number of rainy or snowy days is generally small while the number of clear days is very large.

The direct rays of the sun have exceptional power, both for light and heat. The air is usually in motion, and winds are frequent but rarely extremely high; germs are few, and the air is aseptic and highly charged with ozone and positive electricity. The soil is commonly dry and well drained; but the chief peculiarity, and the one upon which probably much of the therapeutic merit depends, is the lessened barometric pressure. The temperature decreases on an average of 1° F. for every 300 or 400 feet of elevation.

General Physiological Effects. The most important of these effects is that caused by the rarefied air, which, on account of its deficiency in oxygen, causes an increase of red corpuscles and hæmoglobin in the blood, as is elsewhere explained. This deficiency in oxygen, or rather in oxygen-pressure, at first compels quicker respiration, but later causes a deeper breathing, which permanently enlarges the chest. As a consequence of this rapid breathing the heart's action is increased ; and when, later, the respiration becomes normal in rate, though increased in depth, the heart's action returns to its accustomed rate, but each beat is more forcible and the cardiac muscle is permanently strengthened.

"At altitudes varying from 5000 to 8000 feet the inconvenience of the unacclimated seems to have its origin in the increased respiratory activity, and this, in turn, leads to increased work of the heart, which, by its overwork, causes at first an active hyperæmia and an irritability of the nerve-centres. Later the nerve-centres suffer from imperfect nutrition induced by a poor blood-supply, which results from passive hyperæmia. We have, then, the ' irritable weakness " of the old pathologists. After the nerve-centres have become irritable, neither the respiratory act nor the heart's action is performed as regularly and methodically as in the normal condition ; and in consequence various unpleasant symptoms are experienced by the unacclimated. Persons possessed of considerable vigor and capable of adapting themselves to greater changes in their environments go to high altitudes and live almost as they had done at sea-level, and find no appreciable inconvenience in doing so ; but for those advanced in years and for those of feeble health the consequences are far different."[1]

[1] Nervo-vascular Disturbances in Unacclimated Persons in Colorado. J. T. Eskridge. Transactions of the American Climatological Association, 1891.

The foregoing, in conjunction with what appears under the head of decreased barometric pressure, gives the chief points concerning elevated climates. Nothing has been said concerning climates in which the elevation was between 2500 and 4500 feet. Though there are many good resorts within these limits which exhibit more or less the characteristics of elevated climates, still such characteristics do not begin to be markedly manifested at less than 4500 feet above sea-level.

Summer and Winter Climates.

Summer. In the summer of 1894 Professor M. W. Harrington read before the annual meeting of the American Climatological Association an original and instructive paper entitled "Sensible Temperatures," in which he called attention to the importance of considering the agency of evaporation in lowering the actual temperature, in order to judge fairly of the value of a summer climate.

The published temperatures for the different weather-stations are the readings of the ordinary dry-bulb thermometer. The influence of evaporation is shown by what is called the wet-bulb thermometer, the bulb having a covering of cotton or muslin which is kept moistened. The consequent evaporation from the surface of this wetted bulb is similar to that of the human body from which the perspiration is evaporated, thus causing coolness. The temperature shown by the wet-bulb thermometer is called the "sensible" temperature, and is supposed to be the temperature felt at the surface of the skin. As a matter of fact, it is probably lower, because the cloth covering the wet bulb is continuously saturated with water, while the surface of the skin is usually but slightly moistened, and is not subjected to such rapid evaporation. The wind is an important factor in sensible temperature, because if the air is in motion that portion which is in contact with the human body is continuously replaced by dry air, while if the air is stationary it becomes slightly warmed and more humid from heat and moisture of the body, and the amount of evaporation from the surface of the skin is necessarily less. The amount of the reduction or cooling of temperature is in direct ratio to the dryness of the air; it will be greatest where the air is driest, least where the air is cool, moist. The greater the depression of the dew-point below that of diary or shade temperature, the less the relative humidity; the drier the air, the more rapid the evaporation and the greater the consequent reduction of temperature

In Washington, Philadelphia, or San Francisco, where the moisture is usually abundant, the dew-point is generally not far below the shade-temperature, evaporation is relatively small, and hot weather *feels* hot. On the other hand, at Denver, Santa Fé, or Prescott, where the moisture is usually scanty, the dew-point is much lower than the shade-temperature, especially in hot weather, and the reduction is great—the greater the higher the temperature.[1] This is true of all arid regions, where the difference between the dry and wet bulbs during the warmest and driest portion of the day will range from 20° to 40° F. or more.

During the summer the extremes of heat sometimes felt in Eastern cities are almost unbearable. In New York, for instance, there may be several days when the ordinary shade-temperature reaches over 90° F., and the relative humidity may be 80 to 85 per cent. or over, which means that the difference between the wet and dry bulbs is only from 4° to 9°; that is to say, the sensible temperature is 81° to 86°, which is oppressive, as there is so little evaporation. With an ordinary shade-temperature of 90° in arid regions, and a difference of 30° between the dry and wet bulbs, the relative humidity will be 32 per cent. The air will not *feel* hot because the sensible temperature, which is the temperature the body endures, will be but 60°, and the dryness in the air insures rapid evaporation.

The advantage of this increased evaporation in the arid country of the United States is not merely theoretical. It is very real, and makes impressive the fact that 88° at Charleston and 88° at Santa Fé are two different things.

To convey an idea of the distribution of such temperatures, Professor Harrington, at the meeting of the American Climatological Society in 1894, exhibited charts, one of which gave the normal " sensible " or wet-bulb temperatures for the month of July. He says: " It is an instructive chart, and clearly shows t   the temperatures which one actually feels in the dry West and Sou hwest are decidedly lower than the corresponding temperatures in   Eastern States."

A " sensible " temperature-line (wet b {   ) of 65° F. on the Massachusetts coast, starting inland r m Bof , goes by Albany, up to St. Paul, then sweeps in a wi .e cur   ittle below El Paso and passes by Tucson, Phœnix, Los Angch is d San Diego.

[1] Harrington : Transactions of the American Climatological Association, 1894.

The July sensible temperature-line of 60° F. starts from the cool Maine coast, and goes almost due west across the Great Lakes and near Duluth ; then, after first sweeping up to Canada, it curves almost southward across the Dakotas, western Nebraska, and eastern Colorado, until it turns in a northwesterly direction near Eddy (N. M.), and passes by Silver City and Prescott, over the Sierra Nevada to Lake Tahoe, where curving first to the southwest it then returns through the San Joaquin Valley and by Santa Barbara to the sea.

The sensible temperature-line of 55° F. starts from Medicine Hat, in Canada, and runs a little east of south over the mountain-ranges west of Denver, turning near Santa Fé to go west by Albuquerque, up through southern and western Nevada into Northern California, and then into the Pacific a little north of San Francisco.

This chart explains the comfortable summers which are usual throughout a great portion of the West. It shows that apparent high temperatures do not cause the prostration and discomfort felt in Eastern cities.

During the month of July El Paso is actually more comfortable than Iowa, and the vicinity of Denver has the coolness of Canada near the Great Lakes, of the St. Lawrence, or of the Maine coast near Mt. Desert.

Professor Harrington observes : " To obtain the beneficial effects of the reduction of temperature by evaporation the shade must be sought and the direct sun's rays avoided. The effects may be heightened by a natural or artificial breeze or wind, and for parts of the body covered by clothing they may be obtained by adapting the clothing to the free passage of air and moisture. For hot weather, and in the shade, the color of the clothing is of less con- sequence than its texture, together with sufficient looseness to permit of the free access of air."[1]

In a valuable and practical paper in the Denver *Republican* of January 1, 1896, on the subject of sensible temperature-readings and their importance as a basis in judging climate, Captain W. A. Glassford, U. S. A., spoke very forcibly as to the value of using the readings of the wet-bulb thermometer. The ingeniously arranged chart which accompanied the article showed, among other things, the surprising summer coolness of the country south of Denver, running into the middle of New Mexico and Arizona, which is no

---

[1] Trans. Amer. Climatol. Assoc., vol. x. p. 373.

hotter than on the Canada-line or along the shores of the Great
Lakes and the coast of Maine. Captain Glassford has kindly
granted permission to use the following extract from his table :

Tabular statement of the average metallic (or ordinary shade)
temperature-record and the sensible temperature-record with their
difference at selected signal-service stations in the United States
for the months of July and January. Compiled by Captain W.
A. Glassford, Signal Service, U. S. A., from observations taken at
7 A.M., 3 P.M., and 11 P.M.

Signal Service Stations.	July temperature.			January temperature.		
	Metallic record.	Sensible temp.	Difference.	Metallic record.	Sensible temp.	Difference.
Atlantic City, N. J.	72.3° F.	68.6° F.	3.7° F.	31.1° F.	29.3° F.	1.8° F.
Baltimore, Md.	76.5	68.6	7.9	32.4	30.9	1.5
Boston, Mass.	70.6	64.1	6.5	25.6	23.5	2.1
Chicago, Ill	70.6	64.3	6.3	20.7	19.1	1.6
Denver, Col.	72.1	57.0	14.2	27.9	22.9	5.0
Eastport, Maine	60.6	56 7	3.9	21.2	19.5	1.7
El Paso, Texas	82.6	64.8	17.8	42 3	35 5	6.8
Los Angeles, Cal	70.0	63.0	7.0	53.0	47 0	6.0
Nantucket, Mass.	70.2	67.3	2.9	28.9	27.4	1.5
New York City, N. Y. . .	72.9	66.6	6.3	28.4	26.7	1.7
Philadelphia, Pa.	75.4	68.8	6.6	30.0	27.8	2.2
Prescott, Arizona	72.5	59.5	13.0	34.0	29.5	4.5
St. Louis, Mo.	77.6	71.2	6.4	26.4	24.4	2.0
St. Paul, Minn.	70.2	64.0	6.2	7.4	6.4	1.0
Salt Lake City, Utah . . .	75 6	60.0	14.7	27.1	23.3	3.8
San Diego, Cal.	67.5	62.9	4.6	53.7	48.9	4.8
San Francisco, Cal. . . .	59.4	55.8	3.6	49.4	46.4	3.0
Santa Fé, N. M.	68.8	55.9	12.9	26.3	22.6	3.7
Washington, D. C.	76.0	69.0	7.0	31.0	29.0	2.0
Yuma, Arizona	91.1	75.0	16.1	52.8	45.3	7.5

Winter. When the frosty nights of autumn begin to merge into
the sharper cold of winter the invalid, whose thin blood and low
vitality cause him to dread the period of icy blasts and snow, begins
to think of the warmth and sun of the South.

Besides the records of temperature, other important essentials in
judging of a climate are pure air, dry soil, pure soft water, abundant sunshine, moderate rains, and satisfactory accommodations
and food.

Temperature should not be the only consideration.

There is a difference in the value of similar temperatures in
winter and summer.

Where very equable temperature with its accompanying moist air
is not desired, a valuable index of the amount of actual humidity
in the air is shown by the "absolute humidity," which is computed

5

by the number of grains of vapor to the cubic foot at the temperature and relative humidity or dew-point given.

The amount of absolute humidity will usually indicate the *feeling* of the air, which in a *very moist* climate may be chilly at 56°, while in a *very dry* climate at the same temperature no feeling of chilliness will be noticed.

Heat expands the air and raises the point of saturation. The drier the air, however, the more rapid is the evaporation and, therefore, the greater is the feeling of coolness.

On the other hand, cold contracts the air, the vapor is taken out of it by condensation, and it is capable of holding but a small amount of moisture at saturation, the capacity decreasing with the temperature.

Dry, cold air is stimulating, producing a feeling of vigor when the temperature is not too low. Of course, in the selection of a satisfactory winter climate only a moderate amount of cold is desirable.

Another reason why dry, cold air does not feel so "chilly" is that the conductibility of the air depends on its moisture. Cold air is necessarily somewhat dry, and the body does not (without conduction) readily part with its heat.

A bar of iron *feels* colder than a stick of dry wood, because it is a better conductor, just as *moist, cold* air is a better conductor than *dry, cold* air, and the feeling of chilliness produced by moist air, even when it is only moderately cold, is well known.

The temperature of the air for either summer or winter should not be considered independently of its humidity, as cold, *dry* air is as essential to the comfort of some delicate persons in winter as is *warm, dry* air in summer. The cause of this is not difficult of explanation. It has already been stated on a previous page that, with a low humidity, many invalids can endure comfortably a much lower temperature than when the humidity is high.

General W. H. Greely, in an article in *Scribner's Magazine* (November, 1888), entitled "Where Shall We Spend Our Winter?" said: "Next in importance to the temperature is the humidity of the air, a subject to which the public does not generally pay due attention, partly through inadvertence and partly through lack of accessible data. . . . While at all times sensations of dryness or moisture (and in summer rapid evaporation,[1] which lowers tem-

---

[1] As indicated by the wet-bulb thermometer.

perature and promotes comfort) depend largely on the *relative* humidity, yet, during the winter season, the absolute humidity becomes a most important and potent factor in determining the fitness of any particular climate as a sanatorium." Accompanying the article is a chart prepared by General Greely, showing the condition of absolute humidity for the United States for the month of January from ten years' observations. By the chart the least amount of absolute humidity is seen to exist in the extreme cold of Canada, whence it goes down the elevated plains to southern New Mexico, east and west of which the humidity increases to the coast. On the Atlantic coast the absolute humidity south of Charleston increases steadily from 3 grains to the cubic foot of air to 4 grains a little south of Jacksonville; 5 grains across the Florida peninsula from Tampa to Indian River; 6 grains at Miami, and 6.6 grains at Key West. Pensacola and New Orleans have about $3\frac{1}{2}$ grains; Galveston, $5\frac{1}{2}$ grains; San Diego, $3\frac{1}{4}$ grains.[1]

A *cold, dry atmosphere* will be found near the Rocky Mountains above latitude 35°. A *warm, dry atmosphere* is found in western Texas, southern New Mexico and Arizona, and southeastern California.

The greatest softness and mildness of climate in the United States are found along the California coast, where General Greely says the *daily variation* of temperature during the winter months scarcely exceeds 2° F. Arizona and the interior valleys of California average 3°, as do the Gulf coast and Florida south of St. Augustine. The rest of the country varies from 4° to 6°, increasing toward the Canadian border; from the Gulf coast northward the variability increases one or two degrees during the months January to March.

The resources of the United States in the matter of climate are very varied, from the mountains, lakes, rivers, and elevated plains along the Canadian border to the extreme southern limits of Florida and Texas.

In spite of the increasingly good work of the Weather Bureau, detailed meteorological reports have been generally meagre, and it must be many years before all of the attractive though comparatively unknown districts of the country can be brought into general notice.

The best climatic advantages of the Alps can be found in an im-

---

[1] For details of absolute humidity, see Table VI.

proved form—because available all the year round—in the ranges, valleys, and plains of the Rocky Mountains; similar climatic conditions to those of Egypt prevail in the deserts between El Paso (Texas) and Palm Springs of Mohave (California); while the equable climate of the Riviera, with the same advantage of the protection afforded by the mountains above the coast, with a smaller amount of annual rainfall and less discomfort from harsh, cold winds, is to be found in Southern California. Florida, too, offers an asylum for winter visitors, especially on the Atlantic coast at St. Augustine, and southward to Lake Worth; while Key West is an excellent example of a warm, equable, and very moist climate. Unfortunately, there are no complete meteorological records relating to St. Augustine. There is, however, a weather-station at Jupiter, just south of latitude 27°, where there is a lighthouse. Jupiter is eight miles north of the upper end of Lake Worth, on which is the new winter-resort of Palm Beach.

Among interior health-resorts in the Southern States, Aiken, Thomasville, and Asheville have many advantages, which are referred to in the descriptions of those places.

In that portion of the country known as the Great Southwest, the winter climate of Colorado, northern New Mexico, and Arizona may be characterized in a phrase as *cold, dry;* that of New Mexico and Arizona south of latitude 35° as *warm, dry;* and that of the coast of Southern California (west and south of the mountain-ranges) as *warm, moist.*

In Colorado during the winter months (December, January, February) the day-temperature on the plains between 12 M. and 2 P.M. averages 40° or 45° F., and the relative humidity at the same time will probably range between 30 and 35 per cent.[1]

As the reported daily means include the lowest temperature during the twenty-four hours, the extreme cold of 2 to 4 A.M. greatly reduces what is termed the mean monthly temperature, which, for Denver, Colorado Springs, Pueblo, and Santa Fé (classifying the latter with the Colorado climate), is about 30° F., with a mean relative humidity of 54 per cent. and a winter rainfall of from one to two inches. Cañon City has about the same mean winter-temperature, 34°, but is more sheltered from the wind.

A resort can hardly be said to have warm winters if the monthly

---

[1] See comparison of Colorado Springs with Guadalajara ; also description of the climate of Colorado.

mean temperature for that season is less than 45° F. It need not be considered a disadvantage—especially on the seacoast—if it is over 60°.

We see, then, that to get the benefit of greater warmth it is necessary to go south of latitude 35°.

Arizona affords the best desert climate of the United States, and, as winter-residences for health-seekers, it possesses the three towns of Tucson, Phœnix, and Yuma. As the entire territory of Arizona is west of the great Continental Divide, and slopes steadily toward the Pacific, it is subject to a certain amount of ocean-influence—not to the extent to which the winter-rains in California bear witness, but sufficiently to be subject, between December and March, to occasional rains or snows on the high plateaus and on the southwest slopes of the mountains. There are few cloudy days in Arizona, and the spring weather is usually dry.

The greater part of New Mexico—about four-fifths—is situated east of the Continental Divide, and enjoys almost rainless winters. It has also a dry spring. Colorado also has little rainfall during the winter, but it is one of the disadvantages of the climate that this cannot be said of the spring. The normal rainfall at Denver and Colorado Springs for the two months of April and May is 4.8 inches and 4.6 inches respectively; while for both El Paso and Phœnix it is less than one-half an inch for the same period.

The entire country south from Denver and then west in a wide curve to Los Angeles has a great amount of sunshine, and on the elevated inland plains east of the 115th degree of longitude the air is cool, balmy, and very dry.

In southern Texas a *warm, moist* climate is found at San Antonio, the location of which is about as far south as that of the city of Chihuahua in Mexico. Chihuahua has, however, a much greater altitude and a climate more resembling that of El Paso.

(For details of winter climates, see Tables VI. and X.)

Mexico should be included in this brief review of winter climates, as that country has practically no winter rainfall, while there are abundant sunshine, moderate dryness, and a temperature about the same as that of the resorts in Southern California and Florida during the same period. In the cities on the great Mexican tableland the nights are usually cool on account of the elevation.

Going south from Zacatécas the country is more thoroughly

cultivated, the cities are more beautiful, and the air is warmer and more humid. The *plazas* and gardens have a greater wealth of verdure. Even the patios of the hotels exhibit more plants and flowers.

Among the unpleasantnesses associated with travelling through Mexico the prevalence of fine dust is an annoyance that cannot be avoided. The doubtful quality of much of the drinking-water is a matter of still greater importance, and in most of the Mexican hotels proper sanitary arrangements are badly neglected.

The seasonal winter-temperature for a few Mexican cities is as follows: Mexico, 54° F.; Guadalajara, 60°; Monterey, 55°; Durango, 50°; Zacatécas, 55°; Aguas Calientes, 57°; San Luis Potosí, 57°; Querétaro, 59°.

SECTION II.

INTRODUCTION.

THIS section treats broadly of the ailments to which climatic treatment is applicable or for which it is commonly used, and of the way in which the values of climates and meteorological factors influence them, each group of diseases being considered in turn. Change of air is rarely indicated in acute affections, but very frequently in convalescence and in chronic disorders. *Phthisis* being by far the most important of such diseases, is first considered, and is treated at greater length and more elaborately than other chronic affections. Much concerning the effects of climate upon this disease is, however, applicable more or less to certain of the other maladies, and can be studied with advantage in such connection, as it has not seemed best to extend the length of the treatise by repeating these points so fully under the head of each separate disease.

As has been stated in the general introduction, particular climates are not recommended for particular diseases—in short, no readymade prescriptions are given ; but if the different sections of the book are studied together, a climate can be chosen for an invalid upon rational grounds ; and although the results may not be entirely successful, owing to individual peculiarities in the patient, to unexpected or untoward circumstances in his private affairs, to the prevalence of unusually bad weather in the chosen resort, or to some change for the worse in the accommodations provided for invalids, yet the physician will have the satisfaction of feeling that he has at least, so far as the data allowed, used the scientific in place of the customary empirical method in selecting a climate, and has, consequently, done something toward putting medical climatology abreast of the other branches of scientific medicine, behind which it has lingered for so long.

It is very much to be regretted that so few physicians have reported the results of climatic treatment upon their patients, particularly in phthisis, though a careful search through the literature enabled me to find reports which together reached a total of over eight thousand cases, and these have been tabulated and considered in the chapter on Phthisis.

CHAPTER VI.

Definition and Nature of Phthisis. What is phthisis? This term is simply a Greek word for wasting, and it is literally the same as consumption. It is applied to chronic lung-disease which is accompanied by progressive emaciation. Since Koch's discovery of the tubercle-bacillus it has been found that, with very few exceptions, chronic diseases of the lungs in which there is a tendency to a consuming of the body-strength and substance are tuberculous in origin, although, after the tuberculous process is arrested, the fibrosis—that is, the change of the lung-substance into fibrous tissue generally provoked by the resistance of the lung to the irritation of the tubercle—may go on and cause death by destroying the elasticity and interfering with the blood-supply of the lung.

Cases of phthisis without tuberculosis, or those in which the tubercle-bacillus cannot be discovered, are doubtless in most instances caused by some other micro-organism. These non-tuberculous cases are comparatively few in number, and, when once established, tend to run along lines similar to the tuberculous cases. During their course they all exhibit fibrosis; but this, instead of being provoked by the bacillus, is caused by the inhaling of some irritating substance, such as the dust arising in the manufacture of grindstones, emery-powder, etc., by previous repeated or extensive inflammations of the lung-tissue or bronchial tubes, or by special poisons, such as syphilis or cancer.

As such cases are very similar to the true tuberculous cases, it will suffice for the purposes of this inquiry to limit the meaning of the term phthisis to that which is accompanied by pulmonary tuberculosis.

Phthisis may then be defined as a disease of the lungs, occasionally acute throughout its entire course, but usually chronic, or, at least, tending to become so soon after its inception. It is dependent, as its primary source of irritation, upon the presence of the tubercle-bacillus, which, being deposited either in the lung-tissue or in the air-cells, or in the bronchial tubes leading thereto, produces a low form of inflammation, *provided* that the tissue in which it has secured lodgement is susceptible to its influence—*i.e.*, that it contains the food upon which the bacillus is nourished. This inflammation is called *tuberculosis*, because of the tubercles—minute, rounded bodies, composed of bacilli, which have become surrounded by a mass of degenerated lymph. These tubercles may be either grouped or scattered, and sooner or later they almost invariably give rise to an irritation and consequent inflammation of the structures in which they are imbedded. The irritation thus set up usually results in consolidation of the affected tissue, which is thereby rendered impervious to air and therefore useless. This is the *first stage of phthisis*, and this condition may endure for a long time, or it may, on the contrary, be cleared up by reabsorption into the circulation. Again, the diseased tissue may become fibrosed and shrink up like a scar, or it begins, perhaps, to ulcerate, a condition which indicates that the *second stage of phthisis* has been reached. At this point it occasionally happens that the lung clears up entirely, the arrest of the disease being effected in one or both of two ways: either a partial discharge of the tubercles takes place through expectoration, or the ulcerations dry up, and the tubercles, after losing their moisture, contract into calcareous mases, the surrounding tissue becoming fibrosed. Otherwise the ulcerations break down *en masse*, and the *débris* is expelled by expectoration, leaving cavities surrounded by a wall of tissue, chiefly fibroid in character. This condition constitutes the *third stage of phthisis*. By this process the bulk of the affected portion of the lung may be eliminated and the further spread of the disease be limited by the cavity-wall. These cavities themselves may in time heal and contract, or, after remaining quiescent for a period, they may finally break through the wall and extend; or the secretion from them may, without the breaking down of the wall, be carried to the healthy portions even of the opposite lung, in any of which events new foci or centres of disease originate without direct contact between the diseased and healthy tissues. Again, the purulent matter may be absorbed into the blood and cause pyæmia—that is, a poisoning of the system by the carrying through the bloodvessels of blood impregnated with pus.

During these various morbid processes there is almost always more or less loss of flesh and strength, frequently accompanied by fever, night-sweats, and, in 50 per cent. of the cases, by slight or severe bleeding. In many of these stages, but especially in the third, the fibroid process is the method whereby the healthy lung is protected from the further encroachment of the disease, and what remains is encapsulated. Fibrosis, although primarily a benevolent process, sometimes becomes a malignant one, because, when its extent is considerable, it renders the lung so inelastic that breathing cannot be carried on efficiently; thus the pulmonary circulation is interfered with and body and strength waste from lack of sufficient oxygenation and healthy blood.

Tuberculosis. Tuberculosis is an infective disease, dependent upon the growth of tubercles in living tissue, a tubercle being a minute body composed of degenerated lymph surrounding a microbe called the tubercle-bacillus, as has been before stated.

It is the general, and probably the correct, belief that tuberculosis *cannot* exist without the bacillus;[1] but it is possible and indeed probable that the bacillus can and occasionally does exist in the body without tuberculosis—that is, without becoming the centre of a globule of peculiarly degenerated lymph, and so forming a tubercle and causing tuberculosis. Bacilli have been found in the air-passages and in other parts of the body, although there was no tuberculosis, and no irritation or disease had so far been induced by their presence. Sir Hugh Beevor has shown that the tubercle-bacillus, which is a vegetable growth, will develop outside of the body at ordinary temperatures in Europe, and he states that it occasionally appears as a saprophyte. It would therefore seem as if the tubercle-bacillus, and indeed the other micro-organisms which always accompany special diseases, such, for instance, as typhoid and diphtheria, must be brought into connection with degenerated lymph or some decomposed matter in order to become virulent and infective.[2] In

[1] Though cases of a supposed pseudo-tuberculosis have been described as occurring very occasionally, but what germ, if any (other than the bacillus) is the cause, has not yet been discovered.

[2] With reference to the influence of environment and food upon the virulence of germs, the following extract from a paper by Dr. G. Sims Woodhead, upon " The Relation of the Morbid Conditions Dependent on, or Associated with, the Presence of Streptococci," will be found pertinent:

" It is evident from the wide distribution of streptococci under normal conditions, and from the great variety of the disease-processes with which it is to be found associated :

"1. That it must itself undergo great modifications as regards its power of growing and of forming its special pus or inflammation-producing products.

"2. That it is so frequently found associated with other organisms in widely different conditions that the modifications mentioned under 1 may well be due in part at least to its symbiotic existence.

"3. That it produces even when present in pus, say, as a pure culture, such very different

the case of the tubercle-bacillus it is the degenerated mass *plus* the bacillus which constitutes a tubercle ; and a tubercle, and not the bacillus alone, is necessary to cause tuberculosis.

It is true, then, that the tubercle-bacillus may occasionally be present in living tissue without causing irritation, and when found in small numbers it can readily be destroyed by some element of the blood or tissue. Just what this element is is as yet uncertain. Moreover, these bacilli die or are consigned temporarily or permanently to a state of innocuous desuetude when they become surrounded by a protective envelope of hardened exudation, and in this condition they may lie dormant till called into destructive activity by an inflammation of the tissue in which they are imbedded. This fact probably accounts for many of the tubercular joint-affections following an injury ; for in these cases nests of bacilli, apparently of long standing, are often found. In regard to this question of latent tuberculosis the extract given below is interesting.[1] When ulcerative processes arise in connection with

degrees of reaction, when inoculated into different species of animals, and animals in various states of health, that the state of the tissues themselves must play a most important part in determining the life-history and functional activity of the parasitic organism.

"We now know that even outside the body a virulent streptococcus may rapidly lose its pathogenic power, and it is only by exercising the greatest care, by growing it in serum-bouillon, and then passing the organism through rabbits from time to time, that this pathogenic activity can be exalted or even maintained. The exaltation of virulence that has been obtained from comparatively harmless streptococci by using Marmorek's method is, however, so remarkable, that the discrepancies observed by earlier workers at once become explicable, on the theory that the same organism may at different times adapt itself to a saprophytic mode of life on the one hand, or, as the conditions under which it exists become gradually altered, to a parasitic mode.

"It is evident, of course, that during the saprophytic life the activity of the organism becomes diverted along the lines most favorable to the reproduction of its species in large numbers, as the agents inimical to its existence are comparatively few and of a different kind from those with which it has to contend during its parasitic existence. It leads, therefore, what may be called a vegetative life, devoting its energies to reproduction and to withstanding the action of light and similar destructive agents

"During its parasitic existence, on the other hand, the organism is waging warfare for its very existence ; in its new surroundings, with the different food now at its disposal (proteids of various kinds), and contending against living cells, much of its vegetative activity is diverted to the production of substances which will exert a paralyzing influence on the living tissues. The streptococcus is living, as it were, at a higher level, and is forming substances which in its merely vegetative existence it is incapable of producing.

"We must not assume from this that the micro-organism is still capable of going back suddenly to its saprophytic existence, and of at once reproducing its progeny as rapidly as before. This is not the case ; the streptococcus does not grow so easily outside the body as before, and it is only as the higher, or toxin-forming, function becomes modified that the power of vegetation returns in full force."—*London Medical Journal*, October 3, 1896.

[1] "It is a well-known fact, long observed, that tuberculous lesions are often found after death when their existence during life had not attracted especial attention or had even escaped observation. So well established is this fact that the Germans have adopted an axiom that every one ultimately becomes infected with tuberculosis. The multiplication of the tubercle-bacilli and the generation of toxins occasion only local effects until the intensity of the process

tuberculosis the local destruction or death that follows would appear to be due immediately at least to what is termed mixed infection—that is, to some of the other micro-organisms, such as the various forms of the streptococci which accompany the inflammation, and these micro-organisms flourish because of certain defects in the tissues or blood of the individual. Such defects may be of a temporary or a permanent character, and are generally found in conjunction with the presence of the tubercle-bacilli, but always with some form of inflammation or catarrh. On the other hand, it would appear that when the bacilli are absorbed or inoculated in vast numbers the phagocytes, or other germicidal elements of the blood or tissues, are overwhelmed, and bacillary tuberculosis is implanted over a considerable area near the point of entrance. It may be limited to this area, at least for a time, if the individual attacked has the ability to rally and to set up a wall of defensive lymph ; but if such individual have special susceptibility, enlarged lymph-spaces, or other peculiarity characteristic of the tuberculous or scrofulous diathesis, which is commonly inherited, and if he be also without efficient

has reached such a degree that the resulting products gain entrance into the circulation and thus give rise to constitutional manifestations.

"Maragliano, in an address recently delivered, discusses the question of latent and larval tuberculosis and offers a number of interesting and valuable considerations bearing upon that subject. When tuberculosis is present without subjective or objective symptoms, he goes on to say, the latency may pursue one of three courses: (a) it may persist indefinitely ; (b) it may be limited in duration ; or (c) it may be intermittent in occurrence. When the latency is persistent the infection is beyond the range of certain detection, the processes of autotherapy or autoserumtherapy sufficing to control the advance of the disease. Late in the history of the case there may be some impairment of resonance, in consequence of the presence of new-formed cicatricial connective tissue. When the latency is limited in duration the infection—for a variable period not manifest—suddenly makes its appearance. In this group belong cases in which, without previous symptoms, hæmoptysis occurs ; also those in which manifestations of tuberculosis make their appearance in connection with some acute infectious process. The duration of this limited latency is variable and uncertain. The translation from latent to manifest tuberculosis may be viewed as an evidence of increased intensity of infection or of diminished bodily resistance, or perhaps a combination of the two. The developed disease may (a) progress, (b) remain stationary, or (c) subside, perhaps permanently, perhaps to recur.

"Larval tuberculosis is that in which typical manifestations of infiltration are wanting, although other symptoms of the infection are present. This type of the disease may appear in one of two forms ; (1) dystrophic, (2) typhoid. The first is characterized by progressive disturbance of nutrition. The patient gradually fails, anæmia develops, the heart becomes enfeebled and the pulse rapid, the appetite is lost and the digestion impaired, debility ensues, and mental depression results. As a rule, there is an absence of fever, and physical signs may appear only late. The typhoid form of larval tuberculosis is from the beginning attended with fever, to which derangements of innervation are early added. The fever is at first intermittent, later becoming remittent or subcontinuous. The general strength may be maintained. Exacerbations closely resembling attacks of typhoid fever are repeated from time to time. In some cases both types of the disease may be present.

"The manifestations of larval tuberculosis are to be attributed to intoxication with the products of bacterial activity, and vary as one or other poison predominates. The symptoms of tuberculosis may be masked, whatever the localization of the lesion ; but this is most often the case when the lungs and the lymphatic glands are involved."—*Medical Record*, July 11, 1896.

power to build a wall, the bacilli will be carried all through the organ, and through the entire body unless the system too quickly succumbs. The experiments of Trudeau showed that when guinea-pigs were inoculated with moderate doses of bacillary matter and kept under good hygienic conditions the bacilli would disappear entirely or the consequent tuberculosis would be limited. Large doses, however, produced general tuberculosis, proving that even in healthy persons the absorption of great quantities of bacillary matter can cause general tuberculosis, though it doubtless occurs oftenest in those who have a special susceptibility.

Tuberculosis is usually followed by phthisis, but not necessarily so. If the tissue which is inoculated with the bacilli is not susceptible and is healthy, the tuberculosis does not spread, and in due course the bacilli are killed and eliminated ; but if, as has been said, a very large amount be introduced, the infection will spread even through healthy tissues, and will destroy life by affecting the organ or organs attacked so that they cannot carry on their functions. This process is known as acute infective tuberculosis.

Thus, in recapitulating, we may fairly assume from experimental and other good evidence that the following statements are true :

1. That phthisis and tuberculosis, while usually allied, are not the same thing.

2. That phthisis may exist without tuberculosis.

3. That tuberculosis may exist without phthisis ; but it is probable that all forms of phthisis are dependent for their essential characteristics upon the development of a neoplasm resembling true tuberculosis. By true tuberculosis is meant the common form of the disease —that which is always connected with the presence of the special micro-organism known as the tubercle-bacillus of Koch. In cases of real or apparent tuberculosis in which Koch's bacilli are not found it may be that the bacilli have become destroyed or eliminated, or they may not be in a condition to take the stain and so become visible under the microscope, or perhaps such cases are caused by some micro-organism other than Koch's bacillus.[1]

---

[1] This at present is mere speculation, but it would explain the cause of the difference of opinion between those bacteriologists who contend that there is no tuberculosis without the Koch bacillus, visible or invisible, and such eminent clinicians as the late Sir Andrew Clark, who believed in the occasional existence of phthisis independently of bacillary tuberculosis. The ulcerative processes of all cases of phthisis are always accompanied by the presence of destructive and irritating micro-organisms other than the tubercle-bacillus, such as the cocci of pus, etc. It is readily conceivable that, in a person of depraved health, a chronic inflammation of pulmonary tissue, arising from one of many causes, such as a simple inflammation

4. Phthisis is a destructive lung-process attended by a general wasting of the body; it is dependent for its progress upon the irritation set up by the intrusion of some foreign substance into the respiratory organs, and is modified in its character by the resistance of the individual. Such resistance may be too feeble to hinder the increase of the irritation, or it may be so energetic as to be locally destructive, the effect in either case being eventually to further the disease. Of the several causes, bacillary tuberculosis is so much the most common that it is rare to find a case of phthisis which has originated in any other way.

5. Tuberculosis is an infective disease dependent upon the deposit of tubercle in the lungs or other organs; it is usually but not always followed by phthisis.

6. A tubercle is a minute, rounded body composed of a specific germ, the tubercle-bacillus, surrounded by a mass of degenerated lymph; a tubercle cannot arise without the bacillus, though the bacillus may later disappear from tuberculous masses, or at least may not be discoverable by the usual tests.

7. The tubercle-bacillus is found occasionally without the surrounding tuberculous mass, and it is apparently unable to produce the phenomena of tuberculosis while in this condition. Further, it may be accepted as proved that, in the laboratory, progressive and fatal tuberculosis can be produced in healthy animals even while they are under good hygienic conditions, provided that they are inoculated with a sufficiently large amount of tuberculous matter.

Causes of Phthisis. In man it is doubtful if tuberculosis is ever established unless there be some lesions of the membranes to form an entering-point, and usually a lowering of the normal resisting-power also. For this reason it is rare to find tuberculosis running its course without the consequent existence of more or less phthisis. On the other hand, cases of phthisis which have not resulted from tuberculosis are few compared with those in which tuberculosis is the primary cause, and of this limited number not many are free from tuberculosis during their entire course. It is therefore best, in order to avoid confusion, to discuss phthisis in connection with climate as if it were always tuberculous.

or one of the dust-diseases, may form a soil for the lodgement of some destructive micro-organism other than the tubercle-bacillus, which surrounds itself with an envelope similar to that of true tuberculosis. Osler, in the second edition of his *Practice of Medicine* (page 215), in giving a lucid and succinct description of the formation of the tubercle, says that, in the initial stages, it does not differ from certain other inflammatory processes.

Prevalence of Phthisis. In considering the causes of the prevalence of phthisis the prominent points appear as follows:

1. It is a germ-disease, and therefore is most prevalent in densely populated cities and populous countries, where infection is naturally rife.

2. As the disease can be acquired only by inhalation of the bacilli from the dust of the dried sputum, by the consumption of food and drink which have been contaminated by tuberculous matter, by direct contact of bacillary material with abraded surfaces or by direct inheritance, and as the life of the bacillus is very brief when it is exposed to pure air and sunlight, it follows that *where the air is impure phthisis is most common.*

3. As tuberculosis may be directly transmitted from the parent to the child (although more commonly merely a susceptibility is inherited), phthisis is most frequent among the oldest-established races. Viewed, therefore, in its character as a germ-disease, it is not surprising that the researches of Hirsch, Lombard, Copland, Woeikof, Weber, and others have demonstrated that phthisis is least frequent among those who dwell on mountains or on open and comparatively uninhabited plains, and that it is relatively rare in the following order : on certain seacoasts, on certain islands enjoying an ocean climate as nearly pure as possible, in certain desert-places of considerable extent—such as the great American and African deserts and the steppes of Tartary—in the polar regions, and, finally, at elevations, the rarity of the disease increasing with the altitude.

Having considered phthisis apart from the direct influence of climate, and shown that as a germ-disease with a certain limited power of contagion it can establish itself under favorable conditions in any climate, we will pass to a consideration of the climatic factors which influence its development and progress.

The Various Climatic Factors which Affect the Prevalence of Phthisis. Pure Air. The experiments of Brown-Séquard, Stokes, Trudeau, and others show that when animals which have been successfully inoculated with bacillary matter are kept in confined spaces with imperfect ventilation they become tuberculous, while those which, after similar inoculations, are kept in open-air quarters escape the disease. This goes to prove that the presence of pure air not only makes infection from the bacilli less likely, but that even when they are received through inoculation or otherwise it aids the body's natural resistance and promotes recovery. Therefore, the first essential of a climate for a consumptive is pure air, because it assists to a cure and lessens the dangers from infection. This is why the risk of the spread of disease from the assembling of large numbers of consumptives in health-resorts, where the air is pure, is very much less than in the cities whence they came, provided cleanliness and ventilation of the dwellings are not neglected. The danger is always greater in the cars, streets, alleys, hotels, and houses of cities than in country towns and open places, even if similar unclean conditions prevail, for it is possible for the bacillus to maintain its virulence outside the body only in dirty, ill-ventilated holes and corners. .

Sunlight. Next in importance to pure air is sunlight. Where sunlight is abundant, pure air must be present. Its power to nullify or to mitigate the results of unsanitary conditions is shown in the healthiness of the air in such sunny climates as Tangier, where, although much foulness and filth lie rotting in the streets, an invalid who avoids impure water, admits the sun and air to his apartments, and lives a healthy life, often derives benefit in spite of the lack of sanitation. The power of sunlight to purify water, as shown later, accounts no doubt for the comparative lack of odor and harmlessness of the open sewers which extend through the centres of the streets in old European and Oriental cities. The power of sunlight to destroy the life of the tubercle-bacillus and other bacteria has been demonstrated by numerous experiments. In direct sunlight a few minutes suffice to kill them ; and if air which has been sunned is admitted to apartments infected by bacillary sputum, they may be safely occupied at the end of three to five weeks, especially if the bedding, carpets, and hangings have been taken out and exposed to sunlight.

"The credit of first bringing to notice the fact that direct sunlight kills or stops the growth and action of bacteria is, so far as known, due to Dr. A. Downes and Mr. Thomas P. Blunt, who, in 1877, communicated to the Royal Society of London their researches on the effects of light upon these micro-organisms."

The results of their numerous experiments are stated by them as follows :

"1. Light is inimical to the development of bacteria and the microscopic fungi associated with putrefaction and decay, its action on the latter organisms being apparently less rapid than on the former.

"2. Under favorable conditions it wholly prevents that development, but under less favorable it may only retard.

"3. The preservative quality of light, as might be expected, is most powerful in the direct solar rays, but it can be demonstrated to exist in ordinary diffused daylight.

"4. So far as our investigation has gone it would appear that it is chiefly but perhaps not entirely associated with the actinic rays of the spectrum.

"5. The fitness of the culture-liquid to act as a nidus is not impaired by insolation.

"6. The germs originally present in such liquid may be wholly destroyed and a putrescible fluid perfectly preserved by the unaided action of light."

Downes and Blunt noticed also that the germicidal action of light was held in abeyance when the experiments were performed *in vacuo*.

The general retarding and destructive influence of sunlight upon bacteria has since been verified by many other investigators.

Tyndall confirmed the fact that light restrained the development, but failed to find that it impaired the vitality of bacteria. About ten years ago Duclaux experimented with pure cultures of several different bacteria, and found that after direct exposure to sunlight their spores failed to germinate. Airlong and Momont have proved by their independent researches that the moist spores of anthrax-bacilli are incapable of development after an exposure of two hours to the direct solar rays, and Momont has also demonstrated that sunlight has no effect on these spores if air be excluded.

Koch, in an address delivered before the International Medical Congress at Berlin in 1890, stated that tubercle-bacilli were killed by an exposure to direct sunlight varying in length from a few minutes to several hours, the time depending upon the thickness of the layer exposed. He also stated that tubercle-bacilli were destroyed in from six to seven days by exposure to ordinary diffuse daylight, such as is found near windows in fairly lighted houses.

In 1892 Geisler observed that the development of the typhoid bacillus was retarded by insolation, and recent investigations, conducted by Billings and Peckham, have shown that the typhoid bacillus is destroyed by a direct insolation of from three to six hours.

Buchner and Minck, in studying the influence of light upon various bacteria suspended in water, the bacilli of typhus and of cholera being among the number, found that sunlight exercised a powerful germicidal action. Water containing 10,000 germs to the cubic centimetre showed no living germs at all after an hour's direct insolation, while a control-specimen kept in the dark at the same temperature showed a slight increase in the number of bacteria. These investigators deduce from their experiments that sunlight must be considered one of the important factors in the auto-disinfection of rivers and lakes, and for this reason they consider the exposure of these sources of public water-supply to sunlight of great sanitary importance.

Though the results obtained by numerous other investigators might be cited as cumulative evidence, yet enough has probably been adduced to establish the value of sunlight as "one of the most potent and one of the cheapest agents for the destruction of pathogenic bacteria, and that its use for this purpose is to be remembered in making practical hygienic recommendations,"[1] and also to show that "the popular idea that the exposure of articles of clothing and bedding to the sun is a very useful sanitary precaution is fully sustained by the experimental data relating to the action of heat, desiccation, and sunlight."

The same data have likewise emphasized two important limitations of the germicidal power of sunlight: first, that its destructive activity extends but little beyond the surface of opaque bodies, such as soil, clothing, and the like, and that its power gradually decreases with the increasing thickness or depth of transparent bodies, as water and glass; and, secondly, that the presence of oxygen is essential for its activity.

Apparently the germicidal action of light is chiefly due to a direct oxidation of the micro-organisms accomplished at low temperatures, the oxygen being stimulated to activity by the presence of light probably in much the same manner as occurs in photography. But there is also another effect of light that may play a not inconsiderable part in certain cases. Recent experiments of Roux and others show that the conclusion of Downes and Blunt, that light did not affect the fitness of culture-liquids as *nidi*, is not true in all instances, and that light *does* in some cases modify and change either the physical or chemic state of the culture-fluids so that the bacteria-growth is considerably restrained, and in some instances is incapable of taking place at all."[2]

[1] Sternberg's Bacteriology, p. 154.
[2] Climate and Health, No. 3, p. 62 (Weather Bureau, Washington).

It has been suggested that perhaps the Röntgen rays may be used to destroy micro-organisms, such as the bacillus tuberculosis, in the living body through the germicidal properties of the violet rays.

Apart from the direct destructive effects of fresh air upon the germs, its indirect effect, through supplying a pure food to the lungs, the chief seat of the disease, is most important, as without this no improvement can take place. Thus, by assisting the healthy working of the lungs it increases their power of resistance as well as that of the blood. Sunlight, as has been said in the chapter on Physiology, also exerts a powerful beneficial influence upon the body, promoting healthy growth and increased oxidation of tissue, and revivifying and gently stimulating the mental and nervous systems.

Sun-heat. An adjunct to the direct sanitary effects of sunlight is sun-heat, when not too intense or prolonged. Its direct physiological effects are to increase growth, to lessen the amount of food needed, and to diminish muscular force. An important indirect effect of sun-heat is to induce patients to live more out of doors, and to keep their doors and windows open.

Temperature. The indirect effects of the sun's heat in raising the temperature of the air must be considered independently. It does not follow that a climate must be hot because it has abundant sunshine; the heat may be great under the direct rays of the sun, and the air, if its humidity be low, may yet remain cool, for air only receives and holds heat through the watery vapor which it contains. In regard to the question as to whether consumption is most prevalent in hot or cold climates, it must be remembered that the important point is not the degree of heat or cold possessed by a climate, but the amount of its humidity.

Humidity of the Soil. The influence of a damp soil upon phthisis has been proved by Bowditch, Buchanan, Milroy, Pepper, and many others to be even worse than that of damp air, hot or cold.

Humidity of the Air. It is found that, other conditions being equal, consumption is most prevalent in climates in proportion to the temperature and humidity as follows: first come damp, cold climates; second, damp, hot climates; third, dry, hot climates; and, fourth, dry, cold climates; and, in promoting recovery from consumption, as a broad statement the reverse order holds good. Thus, in a general way, it may be said that dampness is harmful to the consumptive and dryness beneficial, while the relative effects of heat and cold depend upon the humidity. Therefore, in a damp climate heat, except when tropical, is less harmful than cold; while in a dry climate cold, on the contrary, is more beneficial than heat— that is, to the vital resistance, though not always to the disease.

The fact that damp, cold climates (such as the winter climates of New England and the Great Lake region, and of the British Isles) are more harmful to the consumptive than damp, warm ones (such as the winter climates of Florida and Madeira) is not so apparent as it should be, partly because reports of consumptive cases are not sufficiently abundant, and partly because, consumption being a germ-disease and undeniably much influenced by density of population and unhygienic conditions, more is laid to these causes than in any way belongs to them. Nevertheless, statistics and the reports of observers confirm this statement. At the same time, while the disease arises more frequently in cold, damp climates, it runs a more rapid and fatal course in hot, damp ones. Hirsch demonstrated that, while damp heat is not shown to have a positive influence on the production of consumption, it has a decidedly unfavorable one upon its progress when once established. He also proves, as do others, that damp cold is a positive factor in the production of phthisis and is highly injurious to those in whom the disease is developed.

Apparently humidity of the air, apart from other factors, does not in itself produce phthisis; the comparative immunity from consumption among the men of the British navy contrasted with those of the army, and the rarity of the disease in many islands, such as the Faroe, the Shetlands, the Hebrides, and Iceland, show this. The influence of these climates and of sea-voyages on the disease when developed and active has, however, not been shown by the evidence of others nor by my own observation to be advantageous, removal from the sea being generally of most benefit. Where advantage has been derived from a sea climate it would appear probable that it was owing to the great purity of the air or the elimination of unsanitary conditions and hurtful occupations, as when, for instance, an overworked citizen takes a sea-voyage, or a Bostonian is sent to such a climate as that of the Isles of Shoals, or a Philadelphian to Atlantic City. Therefore, it is fair to assume that when a humid climate, apart from other climatic factors, proves beneficial to phthisis it is purely negative in its action. This does not, however, prevent the possibility of its being of positive benefit in the exceptional cases of phthisis in which the accompanying conditions of the circulation, nervous system, or mucous membranes demand a damp climate for their relief.

Dryness of the air, on the contrary, is known to be of positive benefit to the consumptive. The excellent results obtained from desert-air (apart from great altitude, which will be considered later) are too well known to quote at length. The benefit derived on the steppes of Tartary is often mentioned, and I have myself noted while on the spot, and also in patients I have sent there, remarkable results produced by the climate of Upper Egypt and by that of Arizona and of the Mojave Desert, lying behind the littoral of Southern California. Some years ago, while practising in London, I also observed in several cases of consumption which were tending to deteriorate great improvement in the general and local signs of phthisis after they had been employed for several months as shampooers in the Turkish baths, where they lived for most of the twenty-four hours in a very dry and superheated air. That the dryness and not the heat was the cause of the improvement may be fairly inferred from the fact that hot, *moist* climates are not beneficial; and I may add that from inquiries I have made in Egypt and elsewhere where the so-called Russian and hot vapor-baths are in vogue, I find that consumptives are very properly warned against their use.

In crediting dryness of air *per se* with a beneficial influence upon phthisis we must not forget that some share of the credit at least belongs to the other necessary accompanying factors of a dry atmosphere, namely, more powerful sunlight and heat, less depressing cloudy weather, cooler nights and shade, and a higher electric tension of the body. What is also of prime importance is the greater opportunity for exercising and resting in the open air, and the free access of fresh air to the house by day and by night while an invalid is indoors.[1]

With respect to the rarity of phthisis in dry, low climates, Hirsch writes as follows: " There are few countries of the world so characterized by uniformity of temperature and comparative dryness of the air as the inland districts of Lower Egypt and the valley of the Nile in Central and Upper Egypt—regions in which phthisis, according to all observers, is very uncommon. On the other hand, as we

---

[1] Author's article upon Climate. Hare's System of Therapeutics.

have already seen, localities on the coast, such as Alexandria, Damiett, and Port Saïd, with a moist climate and a great range of temperature, are much more subject to the disease. The same relation between the sort of climate and the number of cases is found in the interior districts of Algiers, on the one part, and the coast-belt of that country, on the other. In India, says Hunter, the localities specially distinguished by dryness of climate (and uniformity of temperature), be they on the plains or among the hills, are least affected by phthisis; and the same relationship may be discovered in Java, in the Gulf States of the American Union, in Mexico, in Guiana, and in many of the West Indies. It is a probable conjecture that the way in which the climate affects the amount of disease is through the particular states of the weather (high degree of atmospheric moisture along with great variations of the thermometer), causing catarrhal affections of the mucous membrane and the after-effects of the same, that the climatic influence, accordingly, is in all probability a predisposing factor in the development of phthisis. It is from the same point of view that we may explain the beneficial effects of the climatic treatment of consumption—a treatment which consists in withdrawing the subjects of phthisis or those who are threatened with phthisis from these harmful climatic influences."[1]

Variability. Severe and sudden changes of temperature from day to day have no more determining influence *per se* than has the absolute height of the temperature. Hirsch refers to the evidence furnished in abundance from numerous places in the more elevated regions of North and South America, where the disease is rare, although these districts are subject to very extreme fluctuations in temperature. The evidence shows that *variability, with dryness of air*, has at least no prejudicial influence on phthisis, except, of course, in some special cases; and there are many theoretical reasons, and I believe good, practical ones also, as will be shown later, for believing these qualities to be beneficial.

Hirsch has shown that *variability, with dampness of air*, when conjoined with frequent oscillations of temperature, predisposes to phthisis; while, on the other hand, *damp heat* has been shown not to have a positive influence in its production, but a decidedly unfavorable one upon its progress when once established, doubtless because it favors the growth of the germ.

[1] Hirsch : Handbook of Geographical and Historical Pathology, vol. ii.

Density of Population. In studying the causes of the geographical distribution of phthisis it is found that density of population stands out most prominently and constantly in connection with a high death-rate from phthisis. The reasons for this would appear to be chiefly increased danger of infection and increased impurity of air. The fact that the death-rate has been lowered in cities after sanitary improvements have been carried out confirms this opinion. The wide distribution of phthisis, even in localities where the climate itself is inimical to the disease, shows that man's neglect of sanitation can render futile all natural safeguards.

Unhealthy Occupation. The next important cause seems to be unhealthy occupation, particularly if attended with inhalation of irritating dust or fibre, as in knife-grinding and wool-sorting. With few exceptions indoor occupations are, for obvious reasons (such as confinement, impure air, etc.), more injurious than outdoor ones; but among indoor occupations, where other conditions are similar, those which necessitate a cramped position, thus interfering with the free expansion of the chest, increase the percentage of phthisis.

Dampness of soil probably ranks next as a predisposing cause of phthisis. Observations made by various specialists in different countries, beginning with the work of Bowditch in New England, and followed by that of Buchanan in Great Britain and Pepper in Pennsylvania, have established its injurious influence.

Variability of temperature, with moisture, should perhaps follow next, especially when the temperature is low; for instance, in a damp climate variability is always greater in winter than in summer. Old and New England are notable examples, being countries where consumptives can often reside during the summer without harm, whereas it is always more or less dangerous for them to remain in such climates through the winter. The death-rate from phthisis is also much higher in winter; but, as there are so many other causes operating to bring about this result, the evidence of mortality is not so important as that of morbidity.

Variability of temperature, with dryness, on the other hand, seems to lessen the death-rate from phthisis both in warm and cold weather; for while the death-rate is higher in winter than in summer in dry, variable climates as well as in moist, variable ones, yet the winter death-rate of the dry is much less than that of the moist climate. (The probable causes of the higher winter-rate will be discussed else-

where.) In deserts and on open plains and mountain-sides variability
of temperature with dryness is exhibited in the change of temperature from day to night and from sunlight to shade, and, in most of
them, in much wind; yet it is in these places that phthisis is rare,
the death-rate low, and, among the consumptives resorting to them,
the improvement is greatest. Further, in these climates the majority
of patients improve more in the dry, variable cold of winter than in
the dry, variable warmth of summer. Dr. Hermann Weber said
in his Letsommian Lectures: " Many people still cling to the idea
that cold is injurious and warmth curative in phthisis, but this idea
is quite incorrect. Another idea that equable climates are the best
in the treatment of phthisis should likewise be much restricted. The
most important point of all good climates for phthisis is purity of
air. This is to be found, first, in elevated regions, second in the
desert, and third on the sea "

Mobility of Atmosphere. It is extremely difficult to general-
ize upon the influence of this meteorological factor on the development or progress of phthisis. As wind is usually associated with
increase of heat and cold, dryness or dampness, its effects must
be estimated in connection with them. Thus, a cold wind is
more severe than still cold; an extremely hot, dry wind, such as
the sirocco or khamsin, is more irritating and trying than still heat
of the same temperature; and a cold, damp wind is more depressing
and dangerous than still, damp cold. On the other hand, with
moderate warmth, either damp or dry, a breeze, though not directly
stimulating, is refreshing on account of its causing increased evaporation from the skin, even when the actual temperature of the air is
not lowered. This is shown by the relief afforded by the use of
punkahs and fans. With moderate cold a breeze is usually stimulating whether it be damp or dry. It may be said that, given certain conditions of the air with respect to temperature and humidity,
a wind which is not greatly different from the atmosphere in these
respects increases the previous stimulating or depressing qualities of
the air of the place in which it blows. As a matter of fact, the
temperature and humidity of the wind are usually different from
those of the atmosphere of the place to which it comes, and this
difference and the difference in the conditions of the evaporation
from the lungs and skin to which it gives rise form the dominant
characteristics of a climate.

Wind may be said, broadly speaking, to make a bad climate worse and a good one better with these qualifications: that even in a bad climate the wind commonly acts beneficially as a scavenger and purifier, and as a whip to sluggishness, and in hot weather it is more or less refreshing. While it may perform these same functions with like advantage in a good climate, it may also blow too strongly and stir up dust and restrict outdoor life, or it may prove too stimulating and so become irritating. When we find many consumptives doing equally as well in good climates that are windy as in good climates that are calm we cannot say that wind, *per se*, is good or bad. We may say, however, that wind increases or decreases the climatic characteristics of a resort, and sometimes transports to it, for the time being, a totally different climate.

Consumption is neither more nor less prevalent in a place simply because it is windy, nor are consumptives as a class made better or worse by this element alone. It is beneficial or detrimental according to its temperature and humidity and the patient's condition—that is, according to his need of stimulation or sedation.

Cold, moist wind sometimes soothes, but more often depresses the patient, while it aggravates catarrhal affections if they be of a relaxed type.

Cold, dry air simply stimulates or else irritates the patient, and hence it improves relaxed catarrhal conditions but makes those which are inflammatory worse.

Warm, moist wind lessens irritability, and is either soothing or depressing.

Warm, dry wind acts as a tonic or increases irritability.

Types of Climate as Distinguished by their Temperature and Humidity. In considering the influence of temperature as a cause of the prevalence of phthisis the degree of heat or cold is important. Taking extreme climates, the polar regions are certainly freer from phthisis than the tropics, and transported consumptives (at least those who, possessing a good physique, are in an early stage of the disease) do better in arctic than in tropical climates; while in resident consumptives the disease, as has been stated, is much more rapid and fatal in tropical than in arctic regions. In the European arctic zone phthisis is very rare; but this is not true to the same extent in that of the Western hemisphere, apparently because the traders have done more to introduce among the natives rum, syphilis, and the evils which too often accompany civilization. How far the difference in absolute humidity is a cause cannot be clearly determined, though it is thought by the best authorities to be of much importance; but as the evidence strongly shows that phthisis develops and flourishes less in dry than in damp air, this is a reasonable explanation of the fact that tropical heat is worse than arctic cold for phthisis. As has been shown in the chapter on Physics, the capacity of cold air for holding moisture is extremely limited. Therefore, the absolute humidity is necessarily low in arctic climates, while as the capacity of the air for holding moisture increases rapidly with increased temperature, it follows that in tropical climates the absolute humidity is necessarily high—as a matter of fact, it is usually very high indeed.

With regard to moderately cool and moderately warm climates, it would appear that phthisis is more prevalent and progressive in moderately cool, moist climates than in moderately warm, moist climates. As an illustration, phthisis is more prevalent in England than in Italy, in Germany than on the Riviera, in New England than in Southern California.[1]

In contrasting moderately cool, dry and moderately warm, dry climates it is not easy to say that one is better than the other, since in both phthisis is infrequent, rarely progressive, and generally tends to recovery. Climates which are moderately cool and decidedly dry can be said to exist in the United States only at elevations over 4000 feet and above latitude 35°, and climates which are moderately cool but only relatively dry may be found at elevations between 1000 feet and 4000 feet. Of the former class Colorado, during the summer and on most winter-days, is an example, although an opinion based upon the rigor of its winter-night and its shade-temperatures would place it among cold, dry climates. Of the latter class—*i.e.*, moderately cool, relatively dry climates— Bethlehem, in the White Mountains, is an average example; but it should be remembered that, during the winter season, such climates as this are *cold*, and not at all dry. Moderately warm, dry winter climates are found in Egypt, at varying elevations in parts of New and old Mexico, and in Arizona.

Depressed Climates. Increased barometric pressure, as a thera-

---

[1] Examples of moderately cool, moist climates are found in the summer on the North Atlantic coast of the United States, in the woods of Maine, the forests of the Adirondacks, and in the British Isles. In the winter season such places as the coast-resorts of Southern California, the Riviera, Thomasville, parts of Florida, and Madeira, while they vary in degree, are all examples of moderately warm, moist climates.

peutic agent, enters very little into the domain of climatology, and is chiefly of concern in contrast to its opposite—decreased barometric pressure. There are, however, some few places where, owing to depression of the earth's surface below sea-level, the influence of increased barometric pressure is felt, and these will be discussed under the head of depressed climates as distinguished from elevated climates.

Under this head are classed the climates of those places which are below the level of the sea. They exhibit more or less of the phenomena of increased barometric pressure, the physical and physiological effects of which have been discussed in the chapters on Physics and Physiology.

Places of any considerable depth below sea-level are not numerous, and none of them have been fully and scientifically studied with respect to their physiologic and therapeutic influence, but it may be said that phthisis is not prevalent in any of them. The surface of the Dead Sea in Palestine, according to Lieutenant Dale, is 1316.7 feet below the level of the Mediterranean, while Lieutenant Symonds found the greatest depth of its waters was 1350 feet. This makes the bottom of the Dead Sea 2666.7 feet below the surface of the Mediterranean. It is, therefore, so far as is known, the deepest fissure in the earth's surface. It has been stated that residence upon its shores has brought relief to chronic bronchitics and asthmatics, and that a sanitarium is to be erected for their accommodation. Lake Assai, in the Afar country of Africa, is 760 feet below sea-level, and there are said to be other depressions or sinks in that region averaging about 600 feet in depth. In the Libyan, Sahara, and Algerian deserts are several oases of less depth. The Caspian Sea is 85 feet below sea-level. No records of the therapeutic use of any of these places are obtainable. On the North American continent the most remarkable and important depressed climate is found in southeastern California, in the Colorado Desert. It is located in Riverside County, about 100 miles from Los Angeles, is known as the Sink of San Felipe or the Conchilla Valley, and is crossed by the Southern Pacific Railroad. It is about 130 miles long and 30 miles wide, and is 360 feet below sea-level in its deepest part. Numerous palm-trees, some reaching the height of eighty feet, grow on its northern margin, close under the mountains. Dr. Walter Lindley, in an interesting account of Indio, which is a health-resort standing some 50 feet below sea-level on the northern edge of this sink, writes as follows. He begins by referring to Salton, which

is at the southern end of the sink on the line of the Southern Pacific
Railroad, and about 270 feet below sea-level :

"At Salton, on the Southern Pacific Railroad, the surface of the
earth, for nearly ten miles square, is covered with a crust of salt
from four inches to a foot thick. I stopped there in midsummer
and went out on this great white field about noon. The mercury
indicated about 105° F. The workmen out in this peculiar har-
vest-field were as cheerful as any set of men I ever saw, and there
was far less exhibition of suffering from heat than is to be seen
ordinarily in July in the wheat-fields of the Mississippi Valley.
The low relative humidity explains the total absence of sunstroke
here. The atmosphere in this region, adulterated by the chlorine
gas emanating from the salt-beds, must be nearly aseptic. There
are extensive mills here for grinding the salt. It is not put through
any system of purification, but, after grinding, proves to be excellent for table-use. Several hundred tons are thus prepared every
month and shipped away.

"A few miles east of here are the famous mud-volcanoes, which
are equal in wonder to the geysers of this State. Owing to the
treacherous state of the ground around them they have never been
thoroughly examined. Professor Hanks, the State Mineralogist,
undertook it; but breaking through the crust he was so severely
burned that he was compelled to abandon his investigations. . .
Here is an extensive, almost unexplored field for some adventurous
scientist. . . .

" In this valley live about four hundred of the Cochnilla Indians.
This is an interesting tribe. Dr. Stephen Bowers, in a paper read
before the Ventura County Society of Natural History, March 5,
1888, said that he believed them to be of Aztec origin. They are
sun- and fire-worshippers, and believe in the transmigration of souls,
and that their departed friends sometimes enter into coyotes, and
thus linger about their former habitation. They practise cremation.
The ethnologist can, by gaining their confidence, get much interesting information from these very peaceable Indians. . . .

" I found at Salton and Indio asthmatics, rheumatics, and consumptives, all of whom report wonderful recoveries. These asthmatics and consumptives claim that the further they get below sea-
level and the drier the atmosphere the easier they breathe. The
rheumatics claim that the heat and dryness improve the circulation,
and thus relieve them.

"My stay was not long enough to make any trustworthy observations, but it occurred to me that, aside from dryness—mean annual relative humidity certainly not over 25 per cent.—and equability, there was considerable atmospheric pressure at a point three hundred and fifty feet below sea-level, and that we had here moderately compressed air on a large scale. In a recent paper on the use of the pneumatic cabinet the author in many cases in practice shows that compressed air relieves asthmatics and cases of phthisis. He says that compressed air will gradually force its way into every part of the lung, in order that the pressure may be the same on the inside as on the outside. While the proportion of oxygen is, of course, not increased, yet there is an increased quantity in a given space, and we really have the oxygen treatment here on an extensive scale.

"The physician may say that at from two hundred to three hundred and sixty feet below sea-level the pressure would not be as much as in the cabinet. That is true, but the patient goes into the cabinet for, say half an hour, three or four times a week ; while if he is at a point like Salton, he is breathing this moderately compressed air all the time, day and night. This is simply on the principle of the pneumatic chamber of Tabarie, the first one ever employed. This is the method recommended by Dr. A. H. Smith. He refers to the therapeutic value of the increased amount of oxygen inhaled. He says compressed air is useful in catarrh of the mucous membrane, in acute and subacute inflammation of the respiratory mucous membrane, in restoring the permeability of air-tubes, occluded by exudation or otherwise, in asthma, in pulmonary hemorrhage, in pleuritic effusion, in simple anæmia, in inveterate cases of psoriasis and ichthyosis, and in the various forms and stages of phthisis. He does not recommend it in pulmonary emphysema. Dr. Smith says compressed air should be used promptly and perseveringly on the earliest recognizable signs of apical catarrh in those predisposed to chest disease. He also especially recommends it as an alterative.

"If a phthisical or asthmatic patient of considerable vigor intends coming to Southern California, his physician might be justified in suggesting that, except during the summer months, he stop at Indio and from there test the climate of this basin."[1]

During the summer the intense heat makes residence in the

---

[1] Indio. Walter Lindley, M.D., in New York Medical Record, September 1, 1888.

desert impossible for the delicate, and is generally a drawback to it even in winter, in spite of the very low humidity. The climate and resources of Indio are mentioned in Section III. I have had some slight experience with patients in this region, and believe that it has a place in climatic therapeutics; but I am not yet prepared to regard the matter as anything but experimental. The observations of physicians who have used compressed air give some indications for the theoretical use of these depressed desert climates; but the moderate increase of barometric pressure and decided increase of dry heat qualify considerably any deductions that may be drawn from them. While the accommodations at Indio are sufficiently comfortable, luxury and variety of any kind are absent, and while these deficiencies may or may not be beneficial to the patient, it is necessary to take them into consideration before sending him there.

Compressed Air. Upon the therapeutic uses of compressed air Dr. Andrew H. Smith writes as follows : " Compressed air is used remedially in two ways: the first is by placing the patient in a suitably constructed chamber and condensing the air about him; the second is by causing the patient to breathe from a vessel containing air under pressure.

" By the first of these methods, to which alone our attention will be directed, we have the conditions reproduced which we have already considered in connection with caisson-work, except that the pressure employed is very much less—usually not more than ten pounds to the square inch. The physiological results are such as we have already studied, but on a scale reduced in proportion to the lessened pressure.

" Pathological results ought not to be encountered so long as proper care is exercised in the management of the apparatus and the selection of cases.

" It is the united testimony of many observers that the therapeutical effects are, under due limitations, of the greatest practical value.

" The treatment by compressed air in this form is applicable to only a limited circle of diseases. It is in place only when equally good results cannot be obtained by more convenient and more accessible methods.

" The general indications for this form of treatment are derived from its mode of action, which is twofold : 1. As simply a mechanical agent by virtue of the pressure upon the several tissues and organs. 2. As exciting certain chemico-physiological processes by greater activity.

"As a mechanical agent it is capable of lessening hyperæmia in situations accessible to direct pressure.

"As a consequence of this it lessens hypersecretion, and is useful in catarrh of the mucous membranes. It also promotes the reabsorption of inflammatory exudations.

" By increasing the supply of blood to those parts less accessible to direct pressure, it is useful in anæmia of the brain and cord, and can be employed to stimulate the functions of the deep-seated glandular organs, as the liver and kidneys.

" It tends to expand the pulmonary air-vesicles, increasing the vital capacity, and may be used to restore the permeability of air-tubes occluded by exudation or otherwise.

" In its chemico-physiological action the indications for the use of compressed air in this form are to be found in the increase which it occasions in the amount of oxygen taken up in the lungs, and the improved condition of the blood which results. This is shown in the more rapid oxidation of the tissues, in the greater activity of the vital processes, and in increased muscular power. .

"This use of compressed air is contraindicated in weakness of the heart from degeneration of the muscular fibre; in renal disease accompanied by active or passive congestion ; in diseases of the spinal cord of which congestion is a leading feature; in hyperæmia of the alimentary canal, of the ovaries, and of the uterus ; and in hyperpyrexia especially from pulmonary disease (Oertel)."[1]

Dr. Smith goes on to state, mostly on the authority of Oertel and Simonhof, giving apparently no clinical experiences of his own, his belief in the beneficial influence of compressed air in such conditions as acute and subacute inflammation of the respiratory mucous membrane and the upper air-passages, provided the inflammation is not of long standing. He also recommends it in pulmonary emphysema, bronchial asthma, and certain forms of phthisis, particularly because of the improvement that it causes in the general nutrition. He writes that in chronic phthisical parenchymatous inflammation of the lungs " the results are far less favorable, and only palliation can be usually looked for, fever being an unfavorable indication, while hæmoptysis is not necessarily so ; but it is contraindicated when large cavities

[1] The Physiological, Pathological, and Therapeutic Effects of Compressed Air. Andrew H. Smith, M.D.

are present. Pleuritic effusions are usually much benefited and
sometimes empyemas." In organic cardiac disease opinions differ,
Pravaz and Devay being opposed to its use, while most other writers
commend it.

For obesity, anæmia and chlorosis, lithæmia, and in certain hyper-
æmic conditions it is stated to be useful. Smith concludes with these
words : " Finally, as a general *alterative* this method, combining as
it does a vital effect upon the constitution of the blood and a mechanical effect upon its distribution, offers promise of benefit in cases not
amenable to ordinary measures."

Dr. Theodore Williams discussed the treatment at some length in
his lectures on the compressed air-bath and its uses in the treatment
of disease.[1] In his recent work on *Aërotherapeutics* he gives the
following condensed statements of the effects of compressed air in
lung-diseases : " We will now consider the effects of compressed air
on various lung-diseases, and I may state that my conclusions in
the present instance are based on cases from my private note-books
and on sixty-six patients suffering from various forms of lung-
disease under my care at the Brompton Hospital treated in these
baths, most carefully tabulated by my late house-physician, Dr.
Horrocks.

" **Asthma.** Of bronchial asthma there were 15 cases—10 males
and 5 females. In 7 asthma was largely complicated with emphysema. The average number of baths taken was from 12 to 15.
Of these patients 12 improved and 3 did not improve. Out of
11 whose weights were taken 9 gained weight—on an average $4\frac{3}{5}$
pounds—and 2 lost. The measurement of the circumference of the
thorax at various levels was made in 7 patients before and after the
baths, and the circumference increased in 4 and diminished in 3.
The spirometer showed an increase of 25 to 33 per cent. in the 3
cases in which this test was applied. The principal effect of the
baths on asthma appears to be sedative to the pulmonary plexuses
of nerves and to the pneumogastric. The attacks are rendered less
severe, and after a course of twenty or thirty baths the intervals
between the attacks become much longer. I do not remember one
case where a complete cure was effected, but I recollect several
where the patient remained free for months, and, in one instance,
for years from asthma. The effect on the paroxysm is immediate

---

[1] Smith, Elder & Co., London.

and wonderfully efficacious—in fact, more so than any medicines, and many asthmatics have expressed to me the wish that they could live in the bath and thus be freed from their sufferings. The effect on the emphysema accompanying it is to reduce it, as percussion and auscultation show.

"Chronic Bronchitis and Emphysema. In chronic bronchitis and emphysema the effect is satisfactory; the cough diminishes and the expectoration is lessened; weight is gained; breathing is easier; but the great feature is the reduction of the emphysema. Examination of the chest shows diminution of the hyper-resonance and a return of the various displaced organs to their normal positions, and cyrtometric measurements give a reduction of the chest-circumference at different levels, a reduction varying from $\frac{1}{2}$ to $1\frac{1}{2}$ inches. This indicates that a great deal of emphysema even in adults is of a temporary nature, produced often by severe paroxysms of coughing or dyspnœa, and capable of reduction if respiration be rendered more easy, as in a compressed-air atmosphere. In my 33 cases of chronic bronchitis and emphysema, who had on an average eighteen baths, 15 were measured cyrtometrically, and of these, 11 were found to have decreased in circumference and 4 to have increased; 26 of the patients improved generally, 5 did not improve, and 2 died, 1 of heart-failure from cardiac dilatation after five baths, and the other from an attack of capillary bronchitis after having greatly improved from nine baths, when he contracted fresh bronchitis and died later.

"Phthisis. My experience of compressed air in phthisis is not altogether favorable. In 9 of these cases I submitted to the bath there were gain of weight and some diminution of cough and expectoration, and apparently the respiration became freer in the unaffected portions of the lungs; but in 2 cases the bath appeared to bring on hæmoptysis, and in 4 patients hæmoptysis came on during the treatment, though it could not be distinctly connected with it. Beyond the opening up or aëration of portions of the lung which had not been brought into play for some time, I could see none of the improvement resulting from compressed air which is so loudly proclaimed by Oertel and Simonhoff, nor could I discover that it facilitated the absorption of lung-consolidations or infiltration, though cases of this class were submitted to the bath; or, lastly, that it promoted, as Oertel states, the absorption of serous exudation in acute pleurisy and tended to expand the compressed lung. My experience is that it exercises no effect in expanding a lung which has been compressed with fluid when the fluid has been removed, and, even after a course of air-baths steadily persevered in, the fluid will reaccumulate and will make itself known by indubitable physical signs and by the diminishing amount of expiratory power, as evidenced by the spirometer."[1]

Caisson-disease. As explaining most satisfactorily the pathology of the caisson-disease, the article by Dr. Howard Van Rensslaer[2] should be consulted.

In France experimental investigation has for some time been directed to the discovery of the cause of the symptoms characterizing this affection. The experiments of M. Hersent, a French engineer, are particularly interesting, and through them he has arrived at the opinion that the loss of life among workmen engaged in the construction of submarine masonry, etc., is due to quick changes in the degree of compression. In his experiments dogs were, with safety, exposed to a pressure of from seventy-five to eighty pounds to the square inch for five hours; but it was necessary that the increase of pressure should be gradual, occupying at least twenty-five minutes, and its decrease was a much slower process, lasting for ninety minutes. A uniform temperature was maintained throughout. These tests were afterward applied with still greater caution to workmen, and produced merely a degree of languor, a tingling of the surface of the skin, and lumbago. M. Hersent thus established the fact that workmen can successfully pursue their avocations for four consecutive hours, or even longer, exposed to a compression of over seventy-five pounds, provided that the temperature be uniform and that the change in pressure be gradual. This amount of compression is the same as that felt at a depth of 150 to 175 feet below sea-level.

Elevated Climates. The physical and physiological effects of *diminished barometric pressure* have been discussed in preceding chapters, and we will now further consider it only in connection with elevated climates, as its effects are so intimately associated with the other elements of high altitude. It was stated previously that experiments have proved that the most important effect of diminished barometric pressure was to increase the amount of hæmoglobin and red corpuscles in the blood. Regnard established this truth, as will be detailed later. This hæmatogenic effect undoubt-

---

[1] Aërotherapeutics. C. T. Williams, M.D.   [2] Medical Record, August 8, 1896.

edly explains two facts—first, that phthisis is rarer in elevated climates than in others even when due allowance is made for sparsity of population, variety of occupation, and for the qualities ascribable to altitude ; secondly, that consumptives, as a class, are more benefited by high altitudes than any others, as proved by statistics.   Combining these two facts in relation to the blood-changes referred to, we are, I believe, justified in stating that diminished barometric pressure is the chief cause both of the rarity of phthisis and of the exceptionally high average of improvement in high altitudes.   The special reason for the infrequency of phthisis probably is that the increased germicidal power of the blood makes it more difficult for the bacilli to effect a lodgement in the tissues of the inhabitants of elevated regions.   In suitable cases of consumption (and by suitable cases I mean chiefly those tending to the pure tuberculous type) the improvement is apparently due, primarily, to the checking of the further growth of the bacilli by the germicidal quality of the blood; secondarily, to the counteracting by the same quality of the effect of the septic conditions which usually ensue; and, thirdly, to the correction, through the improved characteristics of the blood, of the anæmia and cardiac weakness which almost invariably exist in tuberculous cases as well as in all forms of phthisis.   Thus the tuberculosis which cannot be eliminated is limited, and the general powers of resistance of the patient are increased.

Before proceeding to give at greater length the evidence as to the blood-changes produced by altitude, it will be well to quote at length the opinions of Hirsch and other authorities upon the comparative immunity from phthisis to be found in elevated regions, and its apparent causes.

"The observations published by Archibald Smith and Tschudi as to the extreme rarity of phthisis on the high plateaus of the Andes in Peru, and as to the good effects upon the phthisical of a residence thereon, were the first statements to direct general attention to the comparative immunity from consumption of regions at a great elevation.   Further inquiries in the same direction have confirmed the general fact, but they have in part also given color to an opposite conclusion, so that the question may be said to be still *sub judice* for those who would decide it absolutely and without regard to accessory circumstances.

"It is not to be denied that phthisis does occur at the highest inhabited spots on the globe, and that it is rare in many places situated on low plains; none the less it is an incontestable fact that consumption is, *cæteris paribus,* much less frequently met with at high-lying places than in those at a lower elevation or on the sea-level. Not only so, but the number of cases stands in some kind of definite proportion to the degree of elevation, while the exceptions to the general rule find satisfactory explanation in other etiological factors coming into play at the same time.

"The rarity of phthisis at high elevations is well shown in the returns of sickness from that most extensive of the earth's mountain-chains which runs along the whole Pacific coast of the Western Hemisphere. For the Rocky Mountains of North America we have evidence of the fact from a number of places in the territories situated toward the southern end of the range, such as New Mexico, Arizona, Colorado, and also Utah. In like manner all the authorities speak of the rarity of the disease on the plateaus and mountain-slopes of Mexico, Guatemala, Salvador, Costa Rica, and Panama (for example, on the Cordilleras of Veragua and Chirigui). From Bogotá, in New Granada, Holden writes that he did not see one consumptive person in the hospitals of the town during a prolonged residence there. Referring to Quito, Ecuador, Gayraud and Domec say: 'Notre expérience personelle nous permet d'affirmer, que la phthisie y est tellement rare qu'elle n'y existe pas, au moins comme maladie prenant naissance dans le pays lui-meme. . . . Le fait est donc pour nous indubitable, on ne devient pas phthisique a Quito.'"

For the Peruvian Andes we have the statements of Smith and Tschudi, already mentioned. During a year's stay on the Cerro Pasco the former saw only one case of consumption, and that was in a woman who had come from Europe. "There is no doubt," says Andrew, "that as regards altitude the prevalence of phthisis at considerable heights, although instances of it do exist, is exceptional : and Burkhart mentions that he had not seen a single case of phthisis during a period of three months among the Europeans occupied at the mines in Mexico.

"In those parts of the Argentine Republic that are within the limits of the Andes the influence of high elevations upon the rarity of phthisis is observable as far down as Salta ; it is still more obvious in the elevated valleys on the western side as well as on the Bolivia plateau at Chuquisaca, Cochabamba, Potosí, and other places. In the mountainous parts of Guiana also consumption is almost unknown.

" In the Eastern Hemisphere this immunity from phthisis is shown most decidedly on the plateau of Armenia, where the disease is found almost solely among those who have come from less elevated places ; also on the tableland of Persia, where it is extremely rare, and among the natives of the country almost unknown ; on the northern and southern slopes of the Himalayas, at the elevated points of the Western Ghats, on the Nilghiri Hills, on Mount Abu (4000 feet) in the Aravalli range, and in Nearer India ; on the plateau of Abyssinia and on those of southern Africa.

" In Europe a certain rarity of phthisis begins to be noticeable even at comparatively small elevations, as in the Iser range, on the northern spurs of the Carpathians, in Upper Silesia, on the elevated plain of Thuringia, in the Upper Hartz, and in the Spessart."

Writing of Upper Silesia, Virchow says : "Although I have seen an exceptionally large number of sick persons of the poorer class both in town and country at their homes and in hospitals, yet there has not come under my notice a single case of phthisis ; and the statements of the medical men bear out the notion that the disease is rare. In the Upper Hartz consumption is so unusual that Brockmann, during a practice of many years and extending to eighty thousand sick persons, found only twenty-three phthisical patients, of whom only fourteen had been born in the Upper Hartz ; in the lower valleys the malady is more common, but the high plateau is almost exempt." In the Spessart, according to Virchow, phthisis is not so rare, although in the lower villages he met with only an occasional case, and the registers of deaths rarely contained the entry of consumption or decline. I shall add here the interesting note by Gross that consumption is almost unknown in Briançon (Hautes-Alpes), the most elevated town in Europe (1306 metres, or 4285 feet), although the place is a small fortress with a number of industries and a good deal of filth.

Statistical inquiries, such as have been made in Saxony, Baden, and Switzerland, in regard to the prevalence of phthisis in the elevated regions as compared with low-lying districts close at hand (due allowance being made for any differences in the mode of life), have confirmed the law of immunity of the more elevated places from phthisis which had been deduced from the study of the higher elevations by themselves. The following is Merbach's table for Saxony, based on a period of three years, from 1873 to 1875, and including only those towns which have more than five thousand inhabitants, and only patients whose ages range from fourteen to sixty years :

Altitude in metres (3¼ feet).		Deaths from phthisis within the limit of age.
100 to 200 . . .	.	4.9 in 1000
200 " 300 . . .	.	3.3 " 1000
300 " 400	.	3.2 " 1000
400 " 500	.	3.5 " 1000
550 " 650	.	3.3 " 1000

Merbach concludes as follows : " There is certainly nothing shown here of any marked influence due to the elevation of the various localities, or of such an influence as would cause the number of deaths from phthisis to decrease *pari passu* with the increase in elevation. A result of that sort was indeed not to be looked for, inasmuch as the several towns are subject to other influences—and some of them noxious ones, such as the occupation of the inhabitants, the density of the population, etc.—which are capable of neutralizing the effect of an elevated location. At the same time, even in instances before us, the good effects (otherwise sufficiently proved) of a high situation upon the prevalence of consumption can hardly be overlooked whenever we begin to compare the villages in lowest situation with those in the highest. . . . The contrast comes out with special clearness when the averages calculated for towns situated at one and the same level are compared together."

Corval has worked out this relationship from the Baden bills of mortality over a period of four years (1869–'72), including in his total all those cases in which the cause of death was given as tuberculosis, " chronic pneumonia," or "phthisis." He distinguishes six groups of localities according to elevation :

TABLE OF MORTALITY FROM PHTHISIS IN BADEN, ACCORDING TO ELEVATION.

	Elevation in feet.	No. of towns or villages.	Population, aver. of 4 yrs.	Deaths from phthisis per 1000.
I.	300–1000	750	933,773	3.36
II.	1000–1500	337	224,210	2.75
III.	1500–2000	160	81,066	2.60
IV.	2000–3000	190	104,289	2.75
V.	2500–3000	97	59,155	2.33
VI.	Above 3000	47	20,367	2.17

In order to ascertain what effect is produced upon the death-rate from phthisis by density of population, industrial pursuits, and other conditions peculiar to towns, we may make a calculation of the mortality according to the size of every town or village in Baden, using Corval's figures.   We shall find that it is 3.12 per 1000 inhabitants for the whole of Baden, 3.00 for villages of 3000 and under, 3.49 for towns from 3000 to 10,000, and 4.56 for towns with more than 10,000 inhabitants.   If, now, we arrange the places that are respectively over and under 3000 population in two columns, classifying them in six groups according to elevation, we shall get the following table of the death-rate from phthisis:

Altitude-groups.	Under 3000 population.	Over 3000 population.
I.	3.11	4.05
II.	2.73	3.08
III.	2.49	4.99
IV.	2.71	4.72
V.	2.29	3.06
VI.	2.17	

In the series with less than 3000 inhabitants the favorable influence of increasing elevation is quite obvious; but in the second column of death-rates it will be seen that the benefit is, in some instances, neutralized by detrimental factors belonging to the social and industrial life of the larger centres or the towns.   Still from the facts, such as they are, we have adopted Corval's conclusions, " that cases of phthisis decrease with increasing elevation; or, in other words, in mere increase of altitude we may discover one of the most important factors in checking the development of consumption."

Müller's inquiries into the effect of elevation upon the prevalence of phthisis in Switzerland have led him to the same conclusion, although the results, as he is careful to explain, can only be said to be approximately correct, for the reason that the data at his service were not free from a good many omissions and errors.   He distinguished three groups of places: 1. Those in which from 43 to 63 per cent. of the inhabitants follow some industrial occupation (cantons of Outer Appenzell, Glarus, Neufchâtel, town and country divisions of Bâsle, and Geneva).   2. Where the industrial part of the population reaches from 31 to 43 per cent. (cantons of Zurich, St. Gall, Thurgau, Zug, Inner Appenzell, Aargau, Schaffhausen, Solothurn, Bern, Ticino).   3. The agricultural cantons, where the industrial population is only from 13 to 26 per cent. of the whole (Lucerne, Schwyz, Unterwalden, Vaud, Freiburg, Grisons, Uri,

Vallais).   Grouping the places in each of these divisions according
to their elevation within a limit of 200 to 1800 metres (650 to 6000
feet), we get the following table of death-rates :

TABLE OF DEATH-RATES FROM PHTHISIS IN SWISS TOWNS AND
VILLAGES.

Elevation (in metres).	Industrial Cantons.	Mixed Cantons.	Agricultural Cantons.	Average.
200– 500	2.7	1.85	1.4	2.15
500– 700	3.0	4.55	1.2	1.9
700– 900	1.35	1.7	0.7	1.0
900–1100	1.5	1.9	1.9	1.2
1100–1300	2.3	2.3	0.7	1.9
1300–1500	...	1.4	0.6	0.8
1500–1800	...	1.3	0.7	1.0
Average,	2.55	1.7	1.1	1.86

Müller concludes from these facts " that in Switzerland consumption can be shown to decrease as we ascend; that the malady does
occur, although rarely, at the highest inhabited spots; that the
lowest localities have on the average about twice as many consumptives as the highest, and very much more than that if the cases
where the phthisis had been acquired elsewhere be subtracted; and
that the decrease of phthisis with ascending elevation is, however,
neither constant nor proportional; and that the irregularities and
fluctuations which are noticeable are owing mostly to the position
in the social scale, inasmuch as the industrial group of places shows
the irregularities most, and the mixed groups, on the whole, a regular decrease with height; while the agricultural groups touch their
lowest death-rates at a comparatively small elevation.

" What the minimum of elevation is that a locality must have
before it feels the good effects of altitude on the prevalence of consumption is a question that cannot be answered with certainty from
the facts before us.   Gastoldi puts it at 600 to 1000 metres (2000 to
3300 feet).   It seems to me, however, that a notable decrease in the
disease can be shown to occur at as small an elevation as 400 or 500
metres (1500 feet), provided other circumstances are favorable.
But any immunity from phthisis due to the height of the place
does not come out decidedly until we go to elevations so great as to
be uninhabitable in temperate climates like that of Europe.   We
must go to the equatorial parts of the world to study the sanitary
effects of altitude ranging from 2000 to 3000 metres (6000 to 10,000

feet); and, inasmuch as the question is one of populous places and the seats of industry, we must take the large towns on the plateaus of the Andes in Central and South America, such as Puebla, with 80,000 inhabitants, and at an elevation of 2300 metres, or 7500 feet; Mexico (320,000 inhabitants, 7500 feet); Quito (60,000 inhabitants, 9300 feet); Bogotá (40,000 inhabitants, 8500 feet); Chuquisaca (25,000 inhabitants, 9800 feet); Cochabamba (40,000 inhabitants, 11,000 feet); and Potosí (20,000 inhabitants, 13,000 feet).

"Causes of Rarity. In all of these, which are to some extent industrial towns, or at any rate much occupied with trade and commerce, and by no means models of good sanitation, consumption, according to the unanimous testimony, is either rare or among the natives it does not occur at all. And that is a proof that the influences which go with a very considerable altitude have the power to overcome those detrimental things which arise from a bad kind of hygiene and social life, in so far as these tend to produce consumption.

"Opinions differ as to the nature of the influence of altitude. Some trace the beneficial effects to the freedom of the air from decomposition-products—dust and the like; others to the dryness of the air and soil; but both of these opinions seem to me to be overthrown by the details given above, as well as by the fact that immunity from consumption is found equally on dry and on damp plains, or in mountain-valleys abounding in lakes and pools, provided only that they possess a considerable elevation. The only explanation that I can offer, and one to which I shall hold until something more satisfactory presents itself, is that people who have been brought up at great elevations have been always under the necessity of taking frequent (or perhaps deep) inspirations as a consequence of breathing rarefied atmosphere—that they are continually practising a kind of pulmonary gymnastics, from which there proceed a vigorous development of the breathing-organs and a greater power of resistance on their part to noxious influence from without."

"After looking at the bustle of traffic in towns like Bogotá, Micuipampa, Potosí, at elevations of 8000 to 12,000 feet," says Boussaingault; "after witnessing the strength and marvellous skill of the toreadors in the bull-fights at Quito, 9000 feet above the sea-level; after seeing young and delicate girls dancing a whole

night at places almost as high as Mont Blanc, on which the celebrated Saussure had hardly strength enough to use his instruments
of observation, and his hardy guides fell down in a swoon as they
proceeded to dig a hole in the snow; when we remember, finally,
that the famous battle of Pichincha was won almost in the altitude
of Monte Rosa—I think that you will agree with me that man can
become adapted to breathing the rarefied air of the very highest
mountains. I will readily grant that many of the accounts of
embarrassed breathing experienced by the natives of the plains on
ascending very high mountains are exaggerated; and I must con-
fess that in my own case, after resting a short time at elevations of
ten thousand feet and upward, I was conscious of no considerable
want of breath, or did not become aware, at least, of any need for
quicker and deeper inspirations. At the same time it is not to be
denied that the atmosphere at elevations of ten thousand feet, especially in a warm climate, is rarefied to the extent of more than one-
third of its volume at the sea-level. The quantity of oxygen contained in it is therefore considerably diminished, and a man must
take in a larger quantity of air in a given time or must inspire
oftener than on the plains so as to cover his requirements for oxygen. To that assumption no well-grounded objection can be raised,
whether from the side of physics or of physiology; and there is
equally little reason why we should not assume that those who have
been born and have lived all their lives under such circumstances
will have had their breathing-organs powerfully developed. I do
not hesitate, therefore, to believe that the reason of the immunity
from phthisis enjoyed by the residents of elevated places is the influence which a continuous residence in a rarefied atmosphere exercises over them.

"On the other side, we may thus further explain the exemption
from phthisis of many parts of the world by reason of their favorable weather-conditions and the consequent rarity of all pulmonary
affections therein. The immunity from consumption enjoyed by
natives of elevated regions seems to me to be referable to a peculiarly strong development of their breathing-organs and a corresponding power of resistance in them to noxious influences from
without. It is proved that this is now not all an affair of 'purity
of the atmosphere,' and as some have supposed by the fact that
the state of hygiene in the towns of Ecuador, Bolivia, and Peru,
situated at a great elevation, is by no means distinguished for its

excellence, for cleanliness in the houses or streets, adequate ventilation of the rooms, and the like."

The Physiological Effects of Altitude upon Phthisis. It will be noted that, at the time Hirsch wrote, the effect of diminished barometric pressure upon the blood was unknown, and he ascribes freedom from phthisis in high altitudes to the powerful development of the lungs caused by breathing the rarefied air and their consequent heightened power of resistance to noxious influences from without, and he says that the purity of the atmosphere is not the only cause.

In regard to the differences between high and low climates and the causes and results of such differences, perhaps a general, short view of the whole matter is best given in the words used by the author in addressing the American Public Health Association, in October, 1895:

Looking, first, at the physical facts, we find there is one element only, but a most important one, which a high climate possesses and a low one does not, viz., diminished barometric pressure. While possessing also in an unusual degree and combination abundance of sunshine and pure, cool, dry air, it shares these more or less with other climates. Much speculation and various theories have been indulged in to explain the effects of reduced barometric pressure. When I first came to live in Colorado Springs, now some twenty-one years since, the solution to the problem was hindered by wrong conclusions drawn by the great traveller and scientist, Von Humboldt. Since then physiologists have been at work and have given us a sure foundation upon which to lay, in their proper order, the facts gathered from observation.

With diminished barometric pressure we have, of course, less weight of atmosphere pressing upon the surface of the body, and this was long thought to be important in accounting for the phenomena of mountain-sickness. It has since been shown, however, that as the pressure from the air within the body is equally diminished, and the laws regulating the diffusion of gases equalize the pressure throughout the body, there is no direct practical effect from diminished barometric pressure *per se*, or at least not more than a passing one. So, because with the lessened amount of air in each cubic inch of atmosphere there must also be less oxygen, the conclusion was wrongly drawn that it was the diminution in the amount of oxygen which accounted for the shortness of breath and other phenomena of mountain-sickness.

As the actual amount of oxygen, even at the greatest heights to which man has penetrated, is always in excess of that required in order that the blood may carry on its functions, this theory is not satisfactory. It has been proved that the deficiency is in the *proportion* of oxygen in each cubic foot of air inhaled, and not in its actual amount. In short, the special effects of altitude are directly produced, not through the influence upon the lungs and heart of the reduced pressure of the atmosphere nor of the reduced amount of oxygen, but by the *reduction of oxygen-pressure*. When air is taken into the lungs a certain proportion of the total amount of oxygen contained in this air is absorbed by the hæmoglobin, which is that element of the blood contained in the red corpuscles, and is the direct receiver and carrier of oxygen to the system. As each drop of hæmoglobin can only take up oxygen in a certain proportion to the oxygen-pressure of the air, it therefore follows that when this oxygen-pressure is lowered, as it is in elevated climates, the newcomer is rendered uncomfortable because his blood is not sufficiently oxygenated by each ordinary inspiration. To remedy this he breathes faster, so as to take in more air and more oxygen in a given space of time; the heart has therefore to pump blood more frequently into the lungs, and his pulse beats faster. These are, however, only the immediate effects; in time the chest becomes expanded, so that more air is taken in at each breath; the heart is increased in size, its cavities hold more blood, and its walls are thickened so that the muscle has more force to pump the increased amount of blood.

Blood-changes Produced by Altitude. The natural method of compensation for the deficiency in oxygen-pressure does not stop here, however, and a still more remarkable change takes place, in which lies the special therapeutic value of high climates for appropriate cases. This change is an increase in the amount of red corpuscles and hæmoglobin, whereby the oxygen-absorbing power of the blood is largely increased, and in a definite ratio to the elevation. Some of the experiments which have established these facts will be briefly referred to.

Paul Bert, on examining the blood of the llama (the Peruvian mountain-sheep), found that for every 100 cubic centimetres of blood it absorbed an average of 20 cubic centimetres of oxygen, while the sheep of the plains did not absorb more than an average of 15 cubic centimetres of oxygen; Viault, Muntz, and others repeated and confirmed these experiments.

That the increase in the number of red corpuscles and hæmoglobin is due to the diminished atmospheric pressure, and not to the
other accompanying conditions, was proved by Regnard, who at sea-
level placed a rabbit under a bell-glass and reduced the atmospheric
pressure until it was equivalent to an altitude of 9500 feet.   Two
bell-glasses were used, so that the rabbit could be changed from
one to the other when cleaning and disinfecting became necessary.
The rabbit was kept in this atmosphere for a month, and then
came out fat and healthy.   Examination of its blood showed that
it absorbed 21 cubic centimetres of oxygen, while under normal
conditions its blood and that of its fellows at sea-level could only
absorb 17 cubic centimetres of oxygen.   These changes in the respiratory capacity were completed in about four weeks.

At the Congress für innere Medicin, held at Wiesbaden, April,
1893, Dr. Egger read a valuable paper upon "The Blood-changes
in High Mountains."   He began by referring to the experiments of
Viault made at Merococha, Peru (altitude, 14,275 feet), upon six persons, by which he found that after they had ascended above sea-level
and remained three weeks the number of red corpuscles had increased
from five millions to eight millions.   He then passed on to detail his
own experiments, made during the summer of 1891 at Arosa (altitude, 5500 feet), upon twenty-seven persons, twenty-one men and six
women.   In every case there was a marked increase of blood-cor-
puscles, the average increase being 16 per cent., and it was greater
in tuberculous than in sound persons.   The diet and the general
conditions of daily life were the same which had been habitual to the
patients at home.   Like experiments upon nine rabbits showed 17 per
cent. increase.   The increase was permanent, as the blood was retested
in several cases some months later, and, further, the blood of ten
natives showed the same average in the number of red corpuscles,
viz., seven millions.   Experiments upon six rabbits also demonstrated the fact that the increase was not in the capillaries or cutaneous vessels alone, but that it was universal, being found also in the
carotid and femoral arteries.

Egger proved further that the increase was not due to the evaporation of the serum in the drier air of the high ground and a consequent
increase in blood-density and in the relative number of corpuscles,
but was an actual increase.   He established this by withdrawing
blood from two rabbits on low ground, at Bâsle, treating it with
the hæmatocrit, and performing the same experiment three weeks

later, upon the same rabbits after transferring them to the high ground of Arosa. He found that the percentage of solid constituents in the blood examined was practically the same on high as on low ground. In five persons, three men and two women, resident in Bâsle, who were suffering from marked oligocythæmia—*i.e.*, scarcity of red corpuscles—there was, after a few weeks' change to Arosa, an increase in this respect from five millions to six millions, and their disease was cured. He found also that in healthy persons who returned from Arosa to low ground the number of red corpuscles became normal again in three weeks, and in the case of those who were anæmic that, while they lost their increase, the number of corpuscles did not again fall below the normal.

He also made experiments for hæmoglobin, using Fleischl's hæmoglobinometer modified somewhat by Miescher. In eleven persons, a month after their transferrence to Arosa from sea-level, he found, without exception, 16.35 per cent. increase of hæmoglobin, and the same was shown in three rabbits. He noted, however, that the hæmoglobin increased much more slowly than the corpuscles. As this is exactly what takes place after a hemorrhage, when there is a rapid increase of red corpuscles followed by a gradual increase of hæmoglobin, he believes it to be an additional proof that altitude increases the actual amount of both corpuscles and hæmoglobin, and that it is no mere relative, apparent increase, but a real one.

Muntz, by experiments upon rabbits at sea-level on the Pic du Midi (9000 feet), found that the specific gravity of the blood was increased by altitude, as was also the amount of iron. Viault, in the year 1890, conducted similar experiments (also on the Pic du Midi), the results of which were confirmatory of those made by Muntz.

Egger, in closing, said that the one important factor in the causation of these changes is the diminished oxygen-pressure, and that the crucial experiment was made by Regnard when he proved that if barometric pressure was decreased, the blood's power of absorption of oxygen was increased.

At the same congress Drs. Koeppe and Wolf also reported upon the blood-changes produced by altitude. Their experiments were made between Leipsic, near sea level, and Reiboldsgrün (elevation, 2200 feet). They found, in experiments upon healthy and sick persons, especially among the latter, a marked increase in the number of red corpuscles, a marked reduction of hæmoglobin, " a striking behavior in the proportion of blood-plates, and a unique appearance in them." These changes remained constant after the eighth or ninth day. On comparing notes with those who had experimented at different altitudes their conclusion that the number of corpuscles increased in proportion to the altitude was corroborated. They stated, further, that the size of the new corpuscles is smaller than that of the old, and that they are not so rich in hæmoglobin. These physicians consider that the facts justify the belief that altitude increases tissue-change, and is of special value to consumptives. Egger noticed that while the corpuscles increased 19.72 per cent. during the first twelve days' residence on high ground the hæmoglobin increased only 7.23 per cent.; but in the next twenty-four days, while the corpuscles increased only 4 per cent. more, the hæmoglobin increased 8 per cent. The fact that Koeppe's cases were observed for an average of but fourteen days, while Egger's were observed for an average of thirty-six days, partially explains Koeppe's evident error in stating that the hæmoglobin was not increased. His experiments were not so elaborate nor so precise as those of Egger, and, with respect to the increase of hæmoglobin, his conclusions are opposed to those of all other observers and conflict with reasonable deduction.[1]

We see, then, that the following facts are established, namely, that the peculiar effects of high altitudes are to increase the number of red corpuscles in the blood and also the amount of hæmoglobin, as well as the power of oxygen-absorption. The chest-expansion is also increased, this fact being admitted by all observers, who have also remarked frequently upon the almost emphysematous character of the breathing (especially at the apices of the lungs) of native-born children and old residents.

With regard to the assumed enlargement of the heart-cavities and heart-muscle no sufficiently precise or extensive experiments have as yet been made to prove what logical deduction and common belief indorse, viz., physiological cardiac hypertrophy. The remarkable power of horses raised in high altitudes to endure prolonged fatigue and great speed, both in their native uplands and also when transferred to the sea-level, is a matter of common observation and belief. The superior endurance of mountaineers is also well known. This has been especially noted by Hirsch in accounting for the endurance

---

[1] The more recent experiments of Lopez in Mexico confirm the statements of Egger, as do also those of Fraenkel, Gebhard, and Gravitz.

of natives resident in the high Andes. Further, the results of the treatment of heart-disease by the strengthening effect of climbing and by residence in high altitudes, as recommended by Oertel, tend to confirm the belief that the normal heart (and sometimes the abnormal heart also) is increased both in strength and capacity by altitude.

We have, then, as the especial physiologic effects of altitude, acting through barometric pressure, greater lung-capacity, as shown by the increased chest-expansion and more forcible breath-sounds, a stronger and larger heart, and an increased ability of the blood to absorb oxygen. Therefore, we have good reason to believe that there is developed a greater power of resistance to the lodgement of germs within the body through the increased germicidal character of the more highly oxygenated blood, as well as through the more perfect working of the lungs and heart. We may also assume that there is an equalizing of the circulation throughout the body, and that the tendency to local stagnation—that is, to chronic congestion—is thereby lessened. On the other hand, the more vigorous circulation, if by any accident it should be dammed up at any point, would cause a more violent local inflammation and a greater general fever and disturbance. This is shown in acute pneumonia, a disease more severe on high ground than at sea-level under otherwise similar conditions.

In connection with these changes in the blood and its circulation and in the respiration, and occurring at the same time, certain peculiar effects upon the nervous system have been noted; but as this phase of the question has not yet been fully developed, and as it does not affect the main points of the discussion, we will not here do more than refer to it, although we know that it must largely modify the influence of altitude.

General Therapeutic Effects. Turning now to consider briefly how the results of observations on the effects of altitude upon disease agree with the physiological facts, proved and inferred, we find that better results, as a rule, are obtained at high than at low altitudes in all diseases of which anæmia or any other deficiency of the blood is the most important feature, and also in chronic germ-diseases, such as tuberculosis, provided, of course, that in either case the accompanying conditions are not of a nature to be seriously aggravated by altitude.

Thus it is true that cases of pure anæmia, or those in which the anæmia is the controlling factor (as in most cases of neurasthenia and cases of pure or mixed tuberculosis), without grave complications from catarrhal, inflammatory, irritable nervous, or cardiac conditions, are, as a rule, most markedly benefited by residence in a high climate. As might be expected from the equalization of the circulation before referred to, the tendency to hemorrhage is lessened ; but when it occurs the bleeding is often more profuse. Many chronic heart-affections and aneurisms are benefited,[1] while, on the other hand, to cases where the disease is too far advanced to admit of improvement there is great danger in high climates.

Returning to the question of the assumed increase of the germicidal power of the blood, it should be stated that physiologists have not as yet proved it by experiments; but certain clinical observations strengthen a belief in its existence. For instance, when typhoid fever, scarlet fever, measles, etc., prevail at high altitudes they are much more frequently of a mild type than in low climates, and the symptoms are often so slight as to be unrecognized, showing, doubtless, that the germs are scanty and well resisted. But along with these mild cases there are seen now and again some of the most severe types; indicating apparently that when the attack has been caused by an unusual quantity of germs even the especially strong germicidal power of the blood is overcome and the disease progresses uncontrolled. This occurs in the outset, at least ; though, if death does not ensue, the increased power of resistance returns and the convalescence is more rapid and complete than is usual at sea-level. This increased germicidal quality of the blood is shown, I believe, in many ways, as, for instance, the comparative difficulty in making successful vaccinations at high altitudes and the rarity of hydrophobia.

I will now briefly refer to the effect produced by some of the elements of elevated climates, other than diminished barometric pressure, which they share in common with certain of the climates of low altitu. e. While there may or may not be less precipitation, there is almos invariably a drier atmosphere, as shown in the smaller degree of hun 'di.y, both relative and absolute, and there is usually a dry, well-drai. .. soil. This quality of dryness lessens the effect of the temperature upon the body by increasing the evaporation from the skin and lungs, and consequently such affections as sunstroke are

---

[1] Ar...i   S.nith's observations in the Andes confirm these statements.

rare. In nervous disorders, while neuralgia accompanied by anaemia is not common, the various myalgias and dry pleurisies are frequent, and inflammatory and irritable conditions and disturbances of nervous equilibrium are not unusual. As might be expected, the dryness improves the action of the skin, mucous membranes, bowels, liver, etc., in one class of individuals while it disturbs it in another. The dry air further increases the nerve-tension of the individual and also the electric phenomena of the atmosphere.

In summer, warm, dry air, by increasing evaporation, lessens the injurious effect of heat upon the body; and in winter, on the other hand, the cold, although it chills the surface of the body, does not materially affect the heat maintained throughout the system, because dry air is a non-conductor.

As in dry air little vapor is interposed between the body and the sun's rays, the latter have more power, both as to heat and light, than in most climates of low altitude, and therefore the vivifying effects upon animal life are strongly exhibited; but the air itself, which is unusually cold in the shade and at night, because there is little vapor to retain the warmth, mitigates the injurious influence of too great a degree of heat.

Effects upon Sanitation. The increased sunlight and sun-heat, with the increased electric tension and consequent increased ozone, insure a normally pure air and a readier destruction of germs which are conveyed into it from the bodies of sick persons or animals or from garbage or filth. Thus it is found that the tubercle-bacilli in sputum are more quickly destroyed, and infection more rarely follows even when the conditions are otherwise favorable; but it must be remembered that out of the body as in it, if for any reason the number of germs is excessive, the barriers due to altitude are swept away, and infection will be as sure and fatal as in low countries. Therefore, while it is reasonable to believe from the facts and logical assumptions that both the body and the air are better protected against injurious germs at an altitude, and that these germs are feebler and less abundant, yet when the safety-line is once passed the pestilence will walk and the body wither as surely on the mountain as the plain. The mountaineer may trifle longer with the laws of health and sanitation than the plainsman; but if he does not mend his ways, the day of reckoning will surely come.

In all climates, whether at high or low altitudes, a law of health must be obeyed.   To preserve the body in a state of resistance and keep the home sanitary, both must be clean, and pure air and sunshine must be freely admitted.   Where these precautions are taken no other germicides are needed nor will infection be rife.   During the twenty-two years of my practice in Colorado Springs my colleagues and myself together have only been able to collect some twenty cases of phthisis which had originated in the town (population about 22,000), and this in spite of the fact that the hygienic conditions of some of the poorer parts of the city are very indifferent.[1]

[1] On this point see article by C. F. Gardiner, M.D., in the American Journal of the Medical Sciences.

CHAPTER VII.

FORMS OF PHTHISIS AS INFLUENCED BY CLIMATE.

In order to comprehend fully the effect of climate upon phthisis it is necessary to remember not only that consumption is a germ-disease, but also that it is more or less connected with inflammatory and catarrhal states of the respiratory tract, and that it is usually complicated by anæmia.

For our purpose we will find it convenient to divide phthisis into three forms—the tuberculous, the pneumonic, and the catarrhal. While each form is accompanied by more or less inflammation or catarrh, and eventually becomes tuberculous, yet in their inception, and usually throughout their course, either the tuberculosis, the inflammation, or the catarrh is the most prominent feature. For this reason the *form* of phthisis has much to do with its relative prevalence in the various climates and also with the selection of the appropriate climate for a developed case.

Being a germ-disease, phthisis, as we might expect, develops less readily in a cold climate than in a warm one, provided it be dry; and when established its advance is more rapid in a hot climate.

In view of the fact that its origin is often catarrhal, we are not surprised that it prevails less and is oftener cured in climates which are dry rather than damp. Further, in view of its dual character as both a germ and a catarrhal disease, a cold, dry climate is rather better for it than a hot, dry climate, and a cold, moist climate worse than a hot, moist climate.

In order to bring out these points most strongly the extremes of climate have been spoken of, but in actual therapeutic application only moderate climates are used. Therefore, those referred to are the warm and cool, the dry, but not the excessively dry, and the moist, but not the extremely damp.

As a broad statement it may be said that the majority of purely tuberculous cases undoubtedly do best in cool, dry climates and next best in warm, dry ones; on the other hand, the inflammatory cases do best in warm, moist climates and next best in warm, dry ones; while the catarrhal cases do best in a warm, dry atmosphere and next best in a warm, moist one.

Elevated climates are far more beneficial than low ones for the majority of cases of phthisis, and particularly for the more purely tuberculous form, as they afford greater relief to the anæmia which is usually present.

The victims of the purer tuberculous type are found mostly among those in whom the lungs are poorly developed or imperfectly used. A cool, dry air is beneficial to this class of invalids, because the cold air is more bracing and stimulates to more exercise and to greater consequent lung-expansion and use; also to larger appetite and an improvement in the quality of the blood. While the dryness prevents the depressing effects of the cold and causes more evaporation from the lungs and a quickening of tardy circulation, it also increases the nerve-tension, and therefore makes respiration and all other functions more active. The dry, cool air also tends to relieve chronic congestion, particularly of the lungs, and to cure anæmia. Moreover, in such a climate exercise is enjoyable because of the coolness and because the opportunities for taking it in the outer air are frequent and pleasant, there being a scarcity of precipitation and cloudiness and an abundance of sunshine.

On the other hand, the inflammatory or pneumonic type, which is generally found in persons of an irritable circulation and nervous system and of a full habit, is in danger, in cool, dry air, of an over-stimulation of the lungs, and the increased evaporation may cause the tissues to become charged with salts, which form an additional source of irritation. Then, again, in this type the circulation and nervous system cannot stand the unusual stimulation, and there is great danger of a fresh pneumonia being developed, so that a warm, moist climate, where the effects are of a sedative nature, is preferable.

The catarrhal cases, on the contrary, usually do better in a warm, dry air, because the dryness tends to lessen the secretion from the mucous membrane, and, combined with the warmth, acts as a moderate stimulant, mitigating at the same time the danger of catching cold.

The Individual in Climatotherapy.

In considering the personal equation in the question of climatic change for an invalid, and particularly for a consumptive, I shall avail myself largely of the studies I made upon this subject and presented before the Colorado State Medical Society,[1] and also of an address upon temperament given before the Denver Medical Association.[2]

The question which I desire to bring to the attention of the reader is the influence which the conduct and temperament of the consumptive exert upon the progress of his disease. We all of us doubtless believe that the degree of prudence and intelligence shown by the invalid in regulating his life greatly modifies the result; and, further, that his general physique and his temperament are important elements in determining improvement or deterioration. Nothing, however, has been done, so far as I can ascertain, to demonstrate by statistics the proportionate influence of wisdom and unwisdom, of the quality of the physique, and of the special kind of temperament. The difficulty in correctly classifying each case and the difference among observers both as to definition and in ability to read the signs of each quality, have doubtless tended to discourage attempts in this direction; and at the best it must be admitted that, in dealing with many cases, a large margin must be allowed in the exact figures given for the precise influence of such factors as conduct and temperament. It is only when the main points are agreed upon, the material plentiful, the observers many, and their opportunities of judging of these personal matters frequent and lasting through a good portion of the period of illness that such facts can be arrived at. It is not the consultant, ignorant of the patient's daily history and seeing him but once or twice, who can determine how far over-exercise or sloth, apathy or worry, lack of physical resistance or excess of reaction has been displayed during the months or years of the sickness, so as to bring the patient correctly under the special denominations used in the classifications which follow; it is for the physician in charge to furnish the facts upon which such matters can be determined.

Admitting that it is possible to produce a rough scheme of classification out of elements so indefinite and complex, and then to demonstrate the influence of each class, it may be asked, *Cui bono?* The answer is that although we are probably correct in our general common-sense views of these subjects, and although we usually influence our cases so as to mitigate the various evils arising from the different elements, yet, if it be well to believe rightly, it is still better to know correctly, even though in each case the consequent treat-

ment be the same. Further, our patients will hearken to us the
more when we can speak to them to the effect that the percentage of
improvement is much greater among wise invalids than among the
foolish, who perish through their own folly in the proportion shown
by the statistics of so many thousand cases. The physician, also,
if he knows the relative percentage of danger lurking in a special
physique or temperament, is forewarned and so forearmed, and can
shape his treatment and his prognosis with greater accuracy.

With these objects in view I have further analyzed the 141 cases
of phthisis treated by me in Colorado which I reported to the
American Climatological Association.[1] The cases are too few in
number to enable us to take the results as absolute proof of any of
the points inquired into, and these statistics can be received only
as foreshadowing in a greater or less degree the nature of the truths
to be verified and elaborated by longer and ampler investigations.

Temperament. Looking carefully through the original notes
of my 141 cases, and recalling mental pictures of each individual,
their mental and physical peculiarities, and the tendencies exhibited
in the reaction of each individual to his surroundings, I find it comparatively easy to range them under the several temperaments which
I have described in the paper on this subject before referred to. In
order to explain my views, I quote the following extracts :

The ancient writers upon our art endeavored to explain these different underlying forces as due to certain humors ; the history and
description of their views are too well known for me to recapitulate
them here. We moderns have accepted, and must still accept, much
of their nomenclature ; but we have rejected their explanations of
the causes of the several temperaments without seriously troubling
ourselves to find new ones. A writer,[2] in reviewing two essays of
Hellwig upon temperament, says :

" Physicians learn consciously or unconsciously to recognize temperamental differences, and suit both manner and medicine to the
fact." He further goes on to remark that the best definition has
been given by Müller, who essentially describes temperament as
" the reaction of the individual to his environment." In the same
article is presented Hellwig's tabular definition of the term founded
on the view that it is the varying strength of the reception of an
impression, and of the reaction of the individual to it, that characterizes temperament.

---

[1] Transactions, 1890.   [2] The Medical Record, August 4, 1888.

HELLWIG'S TABLE.

Temperament.	Reception.	Reaction.
Choleric,	Strong,	Strong,
Sanguineous,	Strong,	Feeble,
Melancholic,	Feeble,	Strong,
Lymphatic.	Feeble.	Feeble.

What is the essential quality of living matter? Its power of renewal—that is, nutrition. When a portion of elementary living matter which we term protoplasm becomes separate and individual, as in an amœba, what is the essential quality of its individuality? It is its capacity to receive an impression from and its power to react to its environment. This quality is exercised through nerve-force. It is true that we cannot detect nerve-structure, as we know it, in the dawning life of the individual; but though the localized and visible machinery, which we term nerve-tissue, is not apparent, the real, essential element of nerve-force is undoubtedly diffused through the general mass of the individualized protoplasm, conferring on it the capacity to receive impressions, at least in an elementary manner.

The first reaction of the separate piece of protoplasm to the reception of an impression received from its environment would appear to be the formation of a cell-wall, showing that it reacts to external pressure by hardening itself superficially. Thus it defines its individuality and protects itself in the exercise of its essential function of nutrition, which consists of the importing of raw material for food and converting it into the structure of the individual. The first evidence of the existence of a nervous system is in receiving and reacting to impressions made from without; the passing food is drawn in when reflex action is developed by the impression received from without.

Thus we see that a living individual has two essential qualities —nutrition, whereby it lives, and innervation, whereby it individualizes itself—each essential to the other.

The evolution of nutrition is briefly thus: simple absorption and assimilation of food by the whole mass of protoplasm and the general excretion of its waste; then the localization of digestion in a stomach; next the carrying of the digested nutriment to remote parts by lymphatic vessels; then this circulatory process elaborated into a vascular system, with its heart or pump. Then a portion of the clear, white lymph gradually changed into red blood, then the chemical changes producing bodily heat. Thus the circulatory system of nutrition passes from a lymphatic, cold-blooded state to the warm, red-blooded form seen in the mammal.

The nervous system develops first in the sympathetic form, next the motor, then the sensory, up to its highest elaboration in the brain of man, with its power of receiving impressions without bodily contact by means of thought.

Through innervation comes the power to receive impressions made upon the individual.

Through nutrition comes the power of reacting to such impressions, the latter being exhibited immediately through its circulatory system, which in man in its most important form, with respect to the power of reaction, is sanguineous.

The essential difference in reception is in speed, and it is, therefore, classed in individuals as quick or slow. Quick reception may be better characterized as " nervous," and slow as " phlegmatic."

The essential difference in reaction is in strength; it may be classified as strong or weak. Strong reaction may be called " sanguineous"; weak, " lymphatic".

Temperaments should be primarily divided into groups according to the receptive powers, those in which the reception is quick being denominated "nervous," and those in which it is slow, " phlegmatic"; according to the power of reaction, they would fall under the heads "sanguineous," or those in which reaction is strong, and "lymphatic," or those in which it is weak.

An individual born with a certain temperament can undoubtedly modify it considerably by force of will and education. Circumstance or disease will also modify and temporarily or permanently change the relative force of its phenomena. Change of climate often exaggerates or diminishes certain of its manifestations. The impression made upon temperament by disease is what chiefly concerns us as physicians. The reception of the impression made by the invasion of the body by disease is quick or slow, excited or calm, according as the individual is of the nervous or phlegmatic temperament, and reaction is strong or weak as he is of the sanguineous or lymphatic temperament. Knowing the temperamental type of a patient, we can explain and allow for many of the incongruities of pulse, temperature, and nervous phenomena that we meet with.

How are we to diagnose the temperament? Is the individual plethoric or anæmic in appearance, finely chiselled in feature and small-boned, or coarse in outline and large-boned? Is he mentally quick or slow in conversation and nervous or phlegmatic under our examination? Is his view of his case exaggerated in its despondency or cheerfulness? Does his history show a tendency to inflammation or to passive congestion? Is he inclined to fever? Does he react quickly to cold? Are his feet usually warm? These, suggestively, are some of the observations and questions which will give us the material for classifying a patient's temperament.

The old classification of temperaments into hot and cold suggests the sanguineous or hot and full-blooded, the lymphatic or cold and thin-blooded. The old forms of dry and moist are suggestive of the nervous and phlegmatic, high nerve-tension and dryness being necessarily allied, while moisture and low nerve-tension are equally inseparable.

It is perhaps best to speak of the peculiarities of reception as grouped under two kinds of temperament, viz., the nervous temperament, in which the reception is quick, the phlegmatic, in which the reception is slow; as regards the power of reaction, it may be classed under one of two heads, viz., strong physique and weak physique. On this basis I analyzed 141 cases of phthisis treated by me in Colorado, as shown in the following table:

TABLE SHOWING PROPORTION OF CURED AND BENEFITED IN THE DIFFERENT TEMPERAMENTS, IN ALL STAGES OF THE DISEASE.

Temperament.	Physique.	Cured, per cent.	Benefited, per cent.	Cured, per cent.	Benefited, per cent.
Phlegmatic,	Strong,	59	86 }	41	70
"	Weak,	13	47 }		
Nervous,	Strong,	45	68 }	37	60
"	Weak,	18	62 }		
Strong physique, both nervous and phlegmatic				49	84
Weak " " " " "				19	57

We find from this table that the greatest number cured and benefited is among those possessing a strong physique irrespective of the temperament; while, as to temperaments, the phlegmatic are more benefited than the nervous, providing they each possess a strong physique; but, on the other hand, when the physique is poor those with a nervous temperament show a greater power of resistance to the disease than the phlegmatic.

Wisdom and Unwisdom. To find out in a measure what amount of influence the prudence and common-sense of the individual exerted over the course of the disease, I went carefully over the records and marked down each person as wise or unwise, in the matter of taking care of himself, and then found that of the 141 cases 86, or 61 per cent., were wise, and 55, or 39 per cent., were unwise. I then proceeded to find out the percentage of wisdom in the various temperaments, and this is shown in the appended table:

TABLE SHOWING THE PROPORTION OF WISDOM AMONG THE VARIOUS TEMPERAMENTS.

		Wise.		Wise.
Phlegmatic	$\left\{\begin{array}{l}\text{Strong, 82 per cent.}\\\text{Weak, 53 " "}\end{array}\right\}$			76 per cent.
Nervous .	$\left\{\begin{array}{l}\text{Strong, 62 per cent.}\\\text{Weak, 57 " "}\end{array}\right\}$			60 " "
Strong, both nervous and phlegmatic	.	.	67 " "
Weak, " " " "	.	.	56 " "

From this it appears that the greatest amount of wisdom was among those of a phlegmatic temperament, and that there was more among the strong than the weak; therefore, those of a phlegmatic temperament and a strong physique were by far the most apt to show wisdom.

Finally, I found the percentages of cured and benefited among the wise and unwise in the first-stage cases only; then in the second and third stages combined; and, lastly, in all stages together, heading each table with the average percentage for the total 141 cases, so as to show the relative effect of wisdom upon the results:

TABLE SHOWING PROPORTION OF CURED AND BENEFITED AMONG WISE AND UNWISE IN THE DIFFERENT STAGES.

		Cured.	Benefited.
1st stage	Total, 141 cases,	58 per cent.	87 per cent.
	Wise cases,	68 " "	91 " "
	Unwise cases,	31 " "	75 " "
2d and 3d stages combined	Total, 141 cases,	14 per cent.	52 per cent.
	Wise cases	21 " "	59 " "
	Unwise cases,	2 " "	41 " "
All stages combined	Total, 141 cases,	33 per cent.	68 per cent.
	Wise cases,	42 " "	77 " "
	Unwise cases,	11 " "	51 " "

From these tables it would appear that among the wise the percentage of cures in all stages was a third more than the average

and nearly four times as many as among the unwise; and the
percentage of those benefited was similarly higher, though in a
less degree. In the first stage a like difference prevailed, but in
both the cured and benefited in a degree about 50 per cent. less ;
while, taking second and third stages and omitting the first-stage
cases, the difference was similar, but among the cures very much
more marked in this than in the other tables. This would indicate
that, although imprudence is a bad thing in an early stage, it
is far more serious in an advanced one, in which an act of folly
is often irremediable. The superiority of results among the prudent serves to explain in a measure the better reports obtained from
sanitariums than from open resorts, a fact which is demonstrated
by my tables in the article upon Climate, already referred to, for
doubtless the example and the discipline enforced in sanitariums
turn many of the would-be unwise into wise invalids.

Conclusions. If, as would appear from the comparison made
with the other reports of cases treated in climates of high altitude,
these 141 cases represent the average qualities of such cases, then
the truths indicated by these inquiries are that the qualities which
most aid the consumptive in recovery are, first, strength ; secondly,
wisdom; and, thirdly, equability of temperament.

*Therefore, the essentials of the general treatment of phthisis are to
preserve and strengthen the physique, to enforce prudence, and to
induce placidity.*

CHAPTER VIII.

RESULTS OF THE TREATMENT OF PHTHISIS BY CHANGE OF CLIMATE.

In order to ascertain as far as possible what are the definite effects of change of climate upon the progress of phthisis, I have searched the medical libraries and press as thoroughly as my opportunities allowed. The total number of cases which I found reported in such shape that they could be tabulated for the purpose of comparison was 7795. The number of separate reports was thirty-six and the number of individual observers twenty-four. Beside the above-mentioned reports, the statistical results of which will be found in the table on pages 132 and 133, there were twelve reports by seven additional observers; these are given in the foot-notes, but could not be introduced into the tables, and they complete the sum total of all reports with statistics which I could find.

The material from which the statistics are compiled was extracted mainly from the pages of the masterly and exhaustive treatise upon *The Diseases of the Lungs,* by the late Dr. Wilson Fox, while I am also indebted to Dr. Hermann Weber's *Climatotherapy;* to the late Dr. C. T. B. Williams's work upon *Pulmonary Consumption;* to Dr. C. Theodore Williams's *Aërotherapeutics;* to *Diseases of the Lungs,* by the late Dr. Walshe; to the late Dr. Austin Flint's work on *Phthisis;* to Dr. Charles Denison's *Rocky Mountain Health-resorts;* to the *Transactions of the American Climatological Association,* and various reports of hospitals and sanitariums.

In such a disease as phthisis it must be obvious that the material of which the cases are composed is too variable in quality and tendency to allow any close comparison of results. The most convenient and practical division of phthisis is into three stages— the first stage being that of infiltration, with more or less consolidation of the lung-tissue, but no softening; the second stage that in which softening has commenced, but no cavities can be detected; and the third stage that in which cavities are present. These divisions, based on the local state of the disease, are used by the majority of the reporters. The same terms, however, are frequently applied not so much to the local as to the general condition of the patients, for it is recognized that the disease may be in the first stage locally, and yet its extent or its accompanying serious symptoms be such that the condition of the patient as regards any prospect of cure must rank with the second or third stage of the disease. In short, while the tuberculosis may locally be limited or incipient, or may not have proceeded to the stage of softening, yet the phthisis, with its general disturbance of the health and waste of the system, may be far advanced. On the other hand, softening or cavities may be well marked and even extensive, while the phthisis is comparatively slight or non-progressive, so that the case may be in a distinctly higher class, viewed as material for climatic cure. Again, the three stages are sometimes used chiefly to define the length of the time during which the disease has existed; while it is perhaps true that the majority of first-stage cases are recent—that is, early or incipient—yet some cases remain in the first stage for years and some never pass beyond it, although they continue to show symptoms of pulmonary tuberculosis or phthisis. Also softening and even excavation may occur very early in the case, but if the patient show power of resistance, both local and general, the disease tends toward self-limitation, and so the case may offer more encouraging material for climatic relief than one where the disease is in the first stage or is of more recent origin.

Heredity, physique, temperament, and circumstance, as well as accompanying complications, modify considerably the quality of the case, under whatever stage the disease may be grouped; while the temperament, accuracy, reliability, education, and experience of the reporter also influence the value of any single set of statistics. It is therefore impossible to gauge the reports, one by another, and to make exact comparisons.

The value of the climate also varies with the particular season, and passing influences, such as previous epidemics of influenza, modify the general character of the phthisis exhibited by patients during particular periods of observation. Thus we see that it is impossible to make close comparisons of the influence of each variety of climate upon phthisis, either upon the disease as a whole or its various stages.

We have, however, in the relatively imperfect and heterogeneous statistics available and here presented, a fund of recorded observations formed by the laborious industry of men of character and experience, and it would be a grievous pity if, instead of putting this talent out at interest, we were to wrap it in a napkin and wait for the fulfilment of a dream, when all reports shall be uniform and all reporters exact and we shall be able fully to comprehend and classify cases of phthisis and climates in all their bewildering variations and intricacies.

Fortunately every shield has another side, and if we turn from that on whose face we fail to see order, " heaven's first law," we may clearly discern on the reverse side another law, *the law of averages*. In the use of this law the very widespread diversity of opinions, ability, opportunity, material, and season comes as an aid instead of a hindrance, and enables us to find, in the place of nice comparisons, governing principles, and so we gain some insight into the why, wherefore, and whither of these varied and sometimes incongruous statements. It is necessary therefore to arrange these reports in groups, in which are given a sufficient number of reports and reporters to equalize the variations and to afford a fair average; thus comparisons can be made which, it is reasonable to believe, indicate in a general way the relative values of the several varieties of climates in the treatment of phthisis.

The conclusions drawn from the statistics are confirmed or modified by what we may regard as reasonable deductions from our knowledge of the physiological and other influences of climate, and by other statistics which show the geographical distribution of the disease.

Statistics of Results of Change of Climate. In this spirit I have set out all the statistical information that could be obtained, which was sufficiently complete, in the tables, grouping each case according to its stage, because this is the best common ground for classification, although it is an imperfect one. It will be seen, however, that the separation has only been carried out in a portion of the original reports, and therefore the variety and number of reports in many of the groups are not sufficient to indicate clearly the general influence of the group in this particular stage.

The second and third stages were so frequently combined in the original reports that it seemed best to unite them in the table. In consequence of the different definitions of the stages and the scarcity of material under each, it is undoubtedly true that the results under the head of all stages are most reliable, for it is very unlikely that patients should be set down as consumptives unless they are so, whereas the exact stage reached by the disease is much more easily mistaken. By thus combining all stages a sufficient number of cases is obtained to produce an average quality all round.

With respect to results, the definitions of the terms " cured," " arrested," " partially arrested," " greatly improved," " improved," " stationary," etc., are so varied that I thought it best simply to group under " benefited " all the cases which the reporter did not classify as " stationary," " worse," or " died." Here, again, it is only the breadth and depth of the inquiry that save it from being useless, because the length of time during which the patient was under observation, and at the end of which the given result was reached, varied greatly. For instance, many reports coming from local physicians at health-resorts are made on the condition of the patient at the end of the season and at the sanitarium when the patient leaves; while in my own statistics I do not report upon any case which I have seen for the first time within two years of making the report of results. This allows a better opportunity of judging of the permanency of results. Most of my reports, of course, dated much further back. Again, some physicians attach most importance to the local improvement and some to the general, so that the material from several reporters must be combined in order to get a fair average in this respect.

Group I. *Home.Climates.* England can scarcely be said to illustrate change of climate, as nearly half the cases were still in the same city (for instance, most of those treated at Brompton Hospital), and in the other cases the climatic change was comparatively trifling. England is introduced, however, as a home climate, for the purpose of comparison, as showing the difference to a consumptive in remaining at home and in going to a different climate. England is the only country from which such reports were available; but home climates where the prevalence of phthisis is about the same would probably show figures approximating to these.

Group II. *Ocean-voyages.* This group is not so extensive as it should be, either in the number of cases reported or in the number of observers, particularly as the course and season of each voyage varied so much. Patients sent on voyages are generally carefully selected, and, according to Dr. Williams,[1] while they often improve in appearance and in general well-being, they frequently

---

[1] Aërotherapeutics.

show an advance in the local disease. Most of the cases here reported took a voyage from England to Australia or the Cape.

Group III. *Island Climates.* These embrace seven reports and 568 cases, of which all but 25 were from Madeira. The first-stage cases appear to have done remarkably well, though no doubt there are other sea islands less moist and relaxing which would give better reports for all stages.

Group IV. *Coast Climates* are represented by Riviera resorts alone. The number of cases is greater than in any other group (2328 cases), and they are given by four reporters. It is to be regretted that no reports can be given of the coast climates of Southern California, as they are very similar to the climate of the Riviera and have been extensively used by consumptives.

Group V. *Land Climates of Low Altitude* are represented by the Adirondacks, Aiken, Southern California, Asheville, Rome, and Pau, making a total of 789 cases by seven reporters. The value of the total percentage of these reports, viewed simply as an illustration of the influence of climate, is somewhat modified by the inclusion of the reports of sanitariums at Saranac, Sharon, and Asheville. At Saranac and Sharon the patients are selected, and this, coupled with better care and less danger from imprudence than falls to the lot of the average patient living in the same climate outside the sanitarium, probably raises the percentage of benefit over what it otherwise would be. The greater elevation of Saranac and Asheville over the open resorts may also aid in bringing about better results. Omitting the sanitarium reports from this group the percentage of benefit is only 58.

Group VI. *Desert Climates of Low Altitude.* Egypt and Syria are the only countries (in which a desert-air is the chief feature of the climate) from which reports could be obtained. The number of cases is small—154 in all—and the number of observers only three. From my personal experience in Egypt and similar climates the percentage would appear to me lower than it should be. These, though they come under the general head of land climates of low altitude, are, because of their greatly increased dryness and sunshine, put in a separate group.

Group VII. *Climates of High Altitude.* These are all well represented by reports of the cases of 2027 patients who resided in the Alps and 571 who were treated in Colorado, making a total of 2598. The Alps had five reporters and Colorado four. No reports could be obtained from other high climates. The Alpine percentage for all stages is 3 per cent. higher, which slight difference is apparently due not to any superiority of climate, but to a higher percentage of first-stage cases in the Alpine than in the Colorado reports (Alps 60 per cent. in the first stage, and Colorado only 40 per cent.).

It is probable that the cases were also better selected and the patients more prudent.

I may add that the percentage in Dr. Fisk's 100 and my 250 cases is lower than the average, because a longer time was allowed to elapse before the reports were made than is the case with the other observations given. However, the remarkable approximation in results obtained by most of the nine separate reporters would indicate the general correctness of the observations.

Table showing Reported Results of Climatic Change in 7795 Cases of Phthisis.

Climates.	All stages. No. of cases.	Benefited. No.	Per ct.	1st stage. No. of cases.	Benefited. No.	Per ct.	2d and 3d stages. No. of cases.	Benefited. No.	Per ct.
1. Home Climates.									
England:									
London. Williams[1]	700	210	30	235	98	42	465	112	26
Brompton Hosp., 1st Rep.[2]	535	150	28	187	75	50	348	75	22
Total	1235	360	29	422	173	41	813	187	23
2. Ocean Climates.									
Voyages. Williams[3]	65	45	70	41	30	73	24	15	63
" Flint[4]	13	7	54	...	...	...	...	...	...
" Weber[5]	19	9	47	...	...	...	...	...	...
" Faber[6]	26	5	19	9	...	...	17	...	...
Total	123	66	54	41	30	73	24	15	63
3. Island Climates.									
3a. Madeira. Williams[7]	63	33	53	...	...	...	...	...	...
" Lund[8]	100	54	54	48	37	77	52	17	31
" { Mittemaier } / { Goldschmidt }[9]	281	157	56	37	37	100	244	120	49
" Brompton Hos.[10]	20	9	45	5	0	0	15	9	60
" Renton[11]	79	30	39	35	31	88	44	0	0
Total	546	283	52	125	105	84	355	146	41
3b. Tenerife, St. Helena, West Indies. Williams[12]	7	2	28						
Corsica, Sicily, Malta, Corfu, Cyprus. Williams[13]	18	10	55		...				...
Total	25	12	48	...			...		...
Islands—total	568	295	52	125	105	84	355	146	41
4. Coast Climates.									
Riviera. Sparks[14]	1930	1196	62	...	...		...	...	...
" Bottini[15]	125	28	23	56	30	55	69	7	11
" Weber[16]	63	30	48	36	25	61	27	9	30
" Williams[17]	210	115	55	120	77	64	90	38	32
Total	2328	1369	59	212	132	62	186	54	29
5. Lowland Climates.									
5a. Open Resorts:									
Aiken. Geddings[18]	69	42	61	...	...	...	...	...	...
South Calif. Johnson[19]	6	4	67	4	3	75	2	1	50
Rome. Williams[20]	18	10	55	...	...	...	...	...	...
Pau. Williams[21]	43	21	52	...	...	...	...	...	...
Total	136	77	57	4	3	75	2	1	50

Climates.	All stages.			1st stage.			2d and 3d stages.		
	No. of cases.	Benefited. No.	Per ct.	No. of cases.	Benefited No.	Per ct.	No. of cases.	Benefited. No.	Per ct.
5b. Sanitariums:									
Sharon, Mass Bowditch[22]	40	23	58	22	19	79	18	4	22
Adirondacks Trudeau[23]	559	358	64	87	82	91	472	276	58
Asheville,N.C. Von Ruck[24]	54	34	65	6	6	100	48	28	58
Total	653	415	64	115	107	93	538	308	57
Lowlands—total	789	492	62	119	110	92	540	309	57
6. Low Desert Climates.									
Egypt. Sandwith[25]	104	72	69	...	...		...		
Egypt, Syria. Williams[26]	26	16	61	...	...		...	...	...
Egypt, Syria. Weber[27]	24	12	50	6	4	67	18	8	44
Total	154	100	65	6	4	67	18	8	44
Low Desert—total	943	592	63	125	114	91	558	317	57
7. High Climates.									
7a. Alps. Weber[28]	106	80	75	70	64	91	36	16	44
Davos. Spengler[29]	312	269	75	...	...	...	...	...	...
" Williams[30]	247	202	82	159	138	87	88	64	73
" Allbutt[31]	62	47	76	22	21	95	40	26	65
" Ruedi[32]	1270	953	75	...	...	...	...	...	...
Total	2027	1551	77	251	223	89	161	106	65
7b. Colorado. Denison[33]	202	162	80	75	69	92	127	93	73
" Fisk[34]	100	67	67	42	39	91	58	28	48
" Johnson[35]	19	15	79	9	7	78	10	8	80
" Solly[36]	250	176	70	106	92	87	144	84	58
Total	571	420	74	232	207	89	339	213	63
Altitudes—total	2598	1971	76	483	430	89	503	319	63

[1] C. B. Williams and C. T. Williams: Med.-Chir. Trans., 1871, vol. liv., table p. 108; also Trans., 1872, vol. lv., table p. 241. Dr. Wilson Fox extracted from these two reports all the cases that had remained in England of which the results were known—700 cases. See p. 902 Treatise on Disease of the Lungs, Wilson Fox, M.D. Churchill, London.

[2] Brompton Hospital, first report. This is the only report which is divided into stages; the other report given by Wilson Fox could not be used in this table.

[3] Voyages: (C. T. Williams) see p. 907, Wilson Fox.

[4] Voyages: (Austin Flint) Flint on Consumption.

[5] Voyages: (Hermann Weber) Braun's Curative Effects of Baths and Waters, p. 566.

[6] Voyages: (Faber) Practitioner, 1877, vol. xix. p. 275.

[7] Madeira: (C. T. Williams) Med.-Chir. Trans., vol. lv.

[8] Madeira: (Lund) from White's Madeira; also British Medical Association Journal, 1883; also Theodore Williams's Influence of Climate in Pulmonary Consumption. "It must be noted that, of the arrests of the first stage, relapses occurred in two. In the second stage, of five

cases of 'arrest,' relapses, followed by subsequent arrest, occurred in two ; two were much ameliorated, and in two more, though the disease progressed to the third stage, it was then arrested, and the general health remained good. I have placed the four last named among 'marked improvement.' In the third stage it is stated of three cases that on leaving after one winter their only symptoms were moderate cough and expectoration. I have placed these among 'relief or slight improvement.'" Note 17, p. 937, Wilson Fox.

[9] Madeira : (Mittemaier and Goldschmidt.) "Madeira, seine Bedeutung als Heilungsort. 37 cases, first stage, all recovered ; 241 second and third stages. 83 recovered, 2 relapsed, 49 were living with a mean of fourteen years subsequently, 14 lost sight of, 8 died later of other diseases, having lived for a mean (?) of fifteen years ; 12 died later of phthisis, having lived for a mean of ten and a half years. These all, with the exception of two relapses, are reckoned as 81 cures ; 39 died after living variable periods up to fourteen years. These are reckoned as 'slight improvement.' but are also included in the total of 173 deaths, which includes also the 12 before referred to among the cures." Note 18, p. 903, Wilson Fox.

[10] Madeira : (Brompton Hospital report, 1866) 20 patients were especially selected and sent out for the winter, but on their return the results were considered so discouraging that the experiment was not repeated.

[11] Madeira : (Rentou) Edinburgh Med. Surg. Journ., 1827.

[12] Teneriffe, St. Helena, West Indies : (Williams) Med.-Chir. Trans., vol. lv.

[13] Corsica, Sicily, Malta, Corfu, and Cyprus : (Williams) Med.-Chir. Trans., vol. lv.

[14] Riviera : (Sparks) note 10, p. 902, Wilson Fox.

[15] Riviera : (Bottini) Walshe, Diseases of the Lungs, see p. 619. Bottini gives 21 arrests in the first stage, and 55.3 per cent. improved ; taking the mean between them, as in Williams's cases, it yields 26 cases improved in both local and general condition.

[16] Riviera : (H. Weber) Klimato-therapié, Ziemssen's Handb. Allg. Therapie.

[17] Riviera : (C. T. Williams) Aërotherapeutics, p. 47. Williams gives a table on page 51 of Aërotherapeutics, with the number of first stages 123, but 3 were lost sight of, leaving 120. He states that these improved in the proportion of 41 to 30 per cent. more than the cavity cases, but the table and text do not make it clear, and there are some typographical errors. I have taken the highest number of improved possible, viz, 77 cases.

[18] Aiken : (Geddings) Trans. Amer. Climatol. Assoc.

[19] Southern California : (H. A. Johnson) Trans. Amer. Climatol. Assoc., 1891 ; see, also, Hare's System Therap., vol. i. p, 427.

[20] Rome : (C. T. Williams) Med.-Chir. Trans., vol. lv.

[21] Pau : (C. T. Williams) Med.-Chir. Trans., vol. lv.

[22] Sharon : (V. Y. Bowditch) Med.-Chir. Trans., Amer. Climatol. Assoc., 1893.

[23] Adirondacks : (E. L. Trudeau) from ten annual reports, the last being for 1894, Adirondack Cottage Sanitarium. The only reports obtained of first stage were for the years 1890-'92-'93-'94. Those who remained less than three months are not included, as they are in the all-stage tables, because they are not separated in Dr. Trudeau's report. He recently writes to me that the cases are selected before admission, and again picked out after three months' trial, only hopeful cases being retained.

[24] Asheville : (Karl von Ruck) Medical News, September 16, 1893.

[25] Egypt : (F. M. Sandwith) Egypt as a Health-resort.

[26] Egypt and Syria : (C. T. Williams) Aërotherapeutics, p. 28.

[27] Egypt and Syria : (H. Weber) Klimato-therapie.

[28] Alps : (H. Weber) Klimato-therapie.

[29] Davos : (Spengler) Wilson Fox, note 13, p. 903.

[30] Davos : (C. T. Williams) Aërotherapeutics. p. 111.

[31] Davos : (Clifford Allbutt) Lancet, 1878, vol. ii., and 1879, vol. i., and also Wilson Fox, note 15, p. 903.

[32] Davos : (Carl Ruedi) Dr. Ruedi very kindly gave me extracts from his case-books. He had noted his cases at the end of each season as improved or not ; he had not divided the cases into separate individual cases, but reported the same persons as often as they passed a season at Davos. Dr. Ruedi's report covered twelve seasons, with 1270 cases (an unknown number being repeated), 953 cases of these improved, thus making the benefited 75 per cent.

[33] Colorado : (C. Denison) Rocky Mountain Health-resorts. Houghton, Mifflin & Co., Boston

[34] Colorado : (S. Fisk) Trans. Amer. Climatol. Assoc., 1891.

[35] Colorado : (H. A. Johnson) Trans. Amer. Climatol. Assoc., 1890.

[36] Colorado : (S. E. Solly) Trans. Amer. Climatol. Assoc., 1890.

Reports not used in the Tables.

1. The second Brompton report of 6001 cases is referred to by Dr. Wilson Fox, but, as it is not divided into stages and does not make evident the total percentage of improvement, I have not used it. The extent of improvement was apparently only 15 per cent. (Wilson Fox p. 902).

2. Pollock's Brompton report of 641 cases, improvement 25,6 per cent., was not used, as apparently some of the cases appear in the last Brompton report (Fox, p. 902).

3. Williams reported 235 cases who had some climatic change, but as it is not defined I omitted these. The improvement was 38.7 per cent.

4. Champouillon reported 78 cases who tried various changes of climate. This is omitted for the same reason, and because the percentage of improvement is not given (Fox, p. 902).

5. Hameau reported 110 patients treated at Arcachon; improved, 13.6 per cent. This was not used, as Arcachon could not be ranged with the Riviera group (Fox, p. 902).

6. Leeson sent 29 patients to the districts of Buenos Ayres; improved, 48.2 per cent. These were omitted, as they could not be put in any group (p. 904).

7. Leeson sent 23 patients to Montreux; improved, 48 per cent. Omitted for the same reason.

8. Dettweiler reported 1022 patients treated at Falkenstein Sanitarium; improved, 25.7 per cent. The report does not show the stages, and I thought it fairer to omit it as the percentage seems too low (Fox, p. 904).

9. Thaon reported 151 cases treated at Nice; improved, 55 per cent. Omitted because apparently included in Spark's report (Fox, p. 904).

10. Williams reported "10 cases treated in India generally." Much too indefinite climatically to be used. The improvement was 90 per cent. He reported four treated in New Zealand, 75 per cent. improved; also nine cases who went to the Cape and Natal, 58 per cent. improved. Both these are omitted for the same reasons as the first (Fox, p. 904).

11. Brehmer reported on 700 cases treated in his sanitarium at Görbersdorf, with 13 per cent. cured in all stages, and 53 per cent. cured in the first stage. As he does not give the total benefited, I could not use this report (Hare's System, vol. i. p. 423).

12. Von Ruck reports 511 cases treated at Asheville and other places; 45 per cent. improved. These are not included, as they could not be put into any one group. I had previously supposed that they were all treated in Asheville at his sanitarium (Hare, p. 429).

It will be seen that if these omitted reports had been added to the tables given, they would have somewhat increased the percentage of improvement in the high over the low altitudes, and would also have obscured the main features of the statistics.

If we extract the percentages of results for all stages of each group and put them in order, progressing from the ocean up to the high altitudes, we see an almost steady rise in the percentage of improvement as we proceed toward the highlands, as illustrated by the chart on page 136.

It will be observed that the rise would be uniform were it not that the results from ocean-voyages run 2 per cent. higher than those from the sea islands, and that the results from the low inland resorts drop back 2 per cent. below the seacoast climates. As has been before remarked, the selection of the cases sent on sea-voyages is usually more carefully made, and their living more out of doors in a purer air than they would do on shore may account for this slightly better result.

The total percentage of the low inland results is lowered by those of Rome and Pau, otherwise they probably would not have shown the slight dropping back of 2 per cent. below the Riviera reports. It is very much to be regretted that the physicians of Southern California give no reports, and also that none can be had from

Thomasville nor from many other good inland places. The few cases reported from Southern California were not on the coast, or they would have been classed with the Riviera group.

ANALYSIS OF 7795 CASES OF PHTHISIS.

	Home	Sea			Lowland			Highland
P.C.	England	Ocean	Island	Coast	Lowland	Sanitari-ums	Desert	Altitude

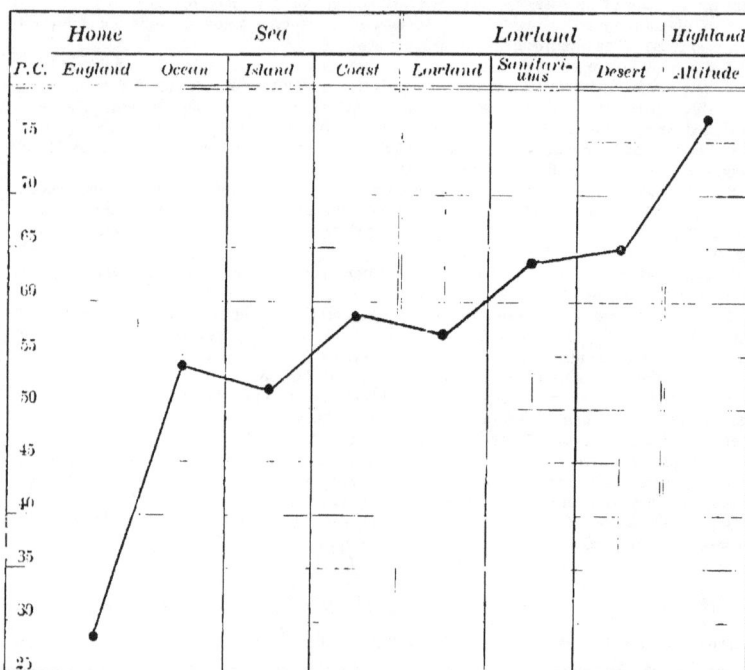

Influence of Sea-air. In making a further analysis of the reports it will be seen that two chief climatic factors exert apparently the greatest influence upon the results, namely, sea-air and mountain-air. If we arrange the reports into three groups, the first consisting of the ocean, the island, and the coast climates (England being omitted, as it does not illustrate climatic change); the second group, composed of the low inland climates, with the sanitariums and the desert climates; and the third, formed of all the high altitude reports, and then compare their total percentages of benefit, we find they range in the following order:

	All stages.	1st stage.	2d and 3d stages.
Sea climates,	57 per ct.	71 per ct.	38 per ct.
Inland climates,	63 "	92 "	57 "
Climates of high altitude,	76 "	89 "	63 "

This shows improvement as the sea-influence diminishes. There is an exception in the first-stage cases in the second group, which is doubtless due to the special selection of the cases before admission to the sanitariums and to the extra care taken of them while there.

These results of the influence of sea-air upon the treatment of phthisis by climate are confirmed by the statistical facts concerning the mortality from phthisis as shown in the tables of mortality in cities, the phthisical death-rate per 1000 diminishing with the distance from the Atlantic and Pacific coast to the Gulf of Mexico.

Dr. R. H. Curtin, in a paper[1] entitled "The Influence of Sea-air on Syphilitic Phthisis," states that he has observed that most cases of phthisis improve in sea-air during the first month after they are brought under its influence, but run down more rapidly afterward. He reports five cases of syphilitic phthisis which were all much better on the ocean than on land. An interesting discussion followed, in which Drs. Bruen, Bowditch, Knight, Shattuck, Musser, Ingalls, and Donaldson took part. The consensus of opinion was that sea-air was usually injurious; that pulmonary tuberculosis sometimes improved on sea islands or on ocean-voyages (either being better than the coast); but that the improved cases were those in which the catarrhal element was prominent.

In a paper which followed this discussion Dr. Boardman Reed said that incipient phthisis usually did well up to a certain point at Atlantic City, but that the second- and third-stage cases did badly. Advanced heart-cases, he thought, also did badly there, because of over-stimulation. On sea-voyages the diet cannot be varied, and sea-sickness is sometimes a most serious complication.

Influence of Altitude. In order to show the influence of high altitudes as compared with that of low altitudes I have combined all the low altitudes, viz., ocean, island, coast, inland sanitariums, and desert, together and contrasted the total percentage with that obtained in high climates. The result is as follows:

	All stages.	1st stage.	2d and 3d stages.
Low climates,	59 per cent.	75 per cent.	47 per cent.
High climates,	76 " "	89 " "	63 " "

This inquiry has clearly demonstrated two things, viz.: that the majority of consumptives do better, other things being equal, the

---

[1] Transactions of the American Climatological Association, 1887.

further they are removed from the sea, and that they do better in high than in low altitudes, wherever situated, the difference in proportionate improvement being here exhibited.

Influence of Sanitarium Treatment on Phthisis. I have been much disappointed in my efforts to gather statistics of the results of treatment in sanitariums. The only reports I received in answer to my requests, which were in such form as to be available for fair comparison in a table, were those of Drs. Trudeau, Von Ruck, and Bowditch. No statistics of Görbersdorf and Falkenstein of recent date and giving the information required could be obtained, nor could I gather any statistics from the many other excellent sanitariums now well scattered over the European continent and the few new ones recently established in America.

In order to make a fair comparison of results obtained in cases treated in open or closed resorts—that is, between patients who resided in sanitariums under supervision, and those who lived in hotels, boarding-houses, or private dwellings without much supervision, where the surroundings were not especially adapted to their peculiar requirements, and their general life and personal conduct were not systematically controlled by a physician—it is necessary to consider briefly the various modifying factors.

The quality of the cases is the first factor of importance. Dr. Hermann Weber, in replying to my request for assistance in obtaining reports from European sanitariums, writes that in his opinion reports of such results are not satisfactory, because sanitariums are largely a last resort for desperate cases, and it is most difficult to get incipient cases to enter them; therefore, a fair comparison with open resorts cannot be instituted. This objection would to a great extent be removed if the reports of results were grouped under the three stages of phthisis. In the cases entering the Adirondack and Sharon Sanitariums the conditions, according to the statements of the physicians in charge, are exactly the reverse, the patients being picked out from a number of applicants as likely to recover.

In the Adirondack Sanitarium, Dr. Trudeau writes me, patients are again examined after three months' residence and only those are retained who have shown an encouraging tendency toward recovery. In his published reports the patients who remained only three months, some of them leaving for pecuniary reasons or causes other than failing health, are classed as a separate group, but the results are not given under each stage. Thus the results at Saranac, in the first stage, as shown in my own tables, are limited to those cases which had passed the second weeding out, and are therefore perhaps abnormally high.

The quality of the cases taken into the Asheville Sanitarium appears to be not so good as at Saranac and Sharon. Dr. Von Ruck recently wrote me concerning this, as follows : "My admissions are all more or less advanced cases, and we do not get more than about a dozen of an early stage in the course of the year. Now, if an institution can discriminate and admit nothing but favorable cases, that institution must necessarily get better percentages."

Again, the conditions of life of the patients admitted to the Adirondack and the Sharon sanitariums are much improved by the change of circumstances, and this may further account for the particularly favorable results obtained. The patients are usually persons of narrow means, formerly engaged in a sedentary city-life, who, having been transported into the country, are compelled to lead an outdoor life and enabled to enjoy comparative luxury under improved hygienic conditions.

Tuberculin was used in a certain proportion of cases at Saranac, but not at Sharon, and the results are given separately (I have combined them in my tables). The percentage of improvement was slightly greater in the cases treated with tuberculin. Dr. Trudeau writes me that he inclines to the opinion that, given the proper conditions, tuberculin is an aid in specially selected cases ; but he believes that it is only of service where the patient could tolerate without fever a slight but gradual increase of dosage, and he used it in only such cases. He goes on to say : "You can see, however, readily, that it may be just because these are insusceptible to tuberculin that their cases turn out well, and not because the tuberculin brings about in them an improvement in their disease or a more or less marked degree of immunity."

Dr. Von Ruck used tuberculin on 36 out of 90 cases treated in his sanitarium at Asheville, and these cases, according to his report, did extraordinarily well, the benefit in all stages being 95 per cent., while in those not treated with tuberculin it was 65 per cent. He, also, does not give it to cases with fever, and in any case administers it only in cautiously increased doses. The greater improvement in his tuberculin cases may therefore be due to the circumstance that the cases upon which it was used were of better material, as suggested by Dr. Trudeau with regard to his own cases.

It is certain that when febrile cases are omitted the most unfavorable element is eliminated; and if those who do not tolerate tuberculin are also removed, we have, in the tuberculin cases, a highly selected class. Acting on this assumption, I have combined the cases treated with and without tuberculin. I am inclined to the opinion that the results were not materially changed by this moderate use of tuberculin, either for better or worse.

The results at Sharon, while not so good as those in the Adirondacks or at Asheville, are probably better than any which could be obtained outside the sanitarium in the same climate. In the Sharon reports it is evidently the lack of elevation and the proximity to the sea that account for the slight deficiency, as my visit to this sanitarium convinced me that the institution and the treatment were quite up to the best standard. In my inspection of both Sharon and Saranac I was much struck with the admirable manner in which the patients were made to live outdoors, and this doubtless helped to account for the good results obtained.

Drs. Trudeau, Von Ruck, and Bowditch all believe that they obtain better results, other things being equal, in the sanitarium than outside, though they have no figures for comparison. My personal experience in sanitarium treatment, while not sufficient to furnish statistics, confirms this opinion; and I believe the great hindrance in all climates to getting better results is due to the mistaken repugnance of most well-to-do patients to enter sanitariums, and the criminal apathy of the State in neglecting to furnish them for the poor, so that their use is extremely limited.

The statements of Drs. Trudeau and Bowditch, then, may be said to show that in sanitariums which are designed for the use of the poorer class of consumptives, and in which the charges are less than the cost of maintenance or merely cover it, the patients are selected with a view to doing the most good by admitting, as far as possible, only curable cases; therefore, the results of treatment are above the average attained under otherwise similar conditions.

The statements of Drs. Von Ruck, Weber, and others show that sanitariums in which the charges are higher and a profit is made are resorted to by the desperate rather than the hopeful cases, and the results are not so good as in the purely benevolent institutions. My observations as to the influence of prudence upon the progress of phthisis, as detailed on page 123, corroborate these opinions of the value of sanitariums.

If we contrast the results of treatment in these sanitariums with that in open resorts in both high and low altitudes, we see that the sanitarium results occupy an intermediate position, as shown in the appended table, except that the results in the first stage exceed those obtained in high altitudes.

	All stages.	1st stage.	2d and 3d stages.
Lowland climates,	58 per cent.	71 per cent.	28 per cent.
Sanitariums,	63 " "	95 " "	58 " "
Highland climates,	76 " "	89 " "	63 " "

The Stage of Phthisis as Influencing Results.

It will be observed, in looking at the first-stage column in the tables, that while the altitude results exceeded all others, except those of the sanitariums, yet the first-stage columns in all the reports were much better in proportion than those of the combined stages or of the second- and third-stage columns. This seems to indicate the correctness of the common opinion that the majority of incipient cases are benefited more or less by *any* climatic change as they usually are by any change of treatment. As to the relative permanency of the benefit, the evidence here produced shows nothing, nor was I able to secure sufficient reports of arrests and cures to make a fair comparison.

In a sense it may be considered as unfortunate that first-stage cases are so easily improved, for these reasons. On dropping their pursuits and changing their environment, by taking rest and fresh air, and by cultivating better habits, their general constitution is very apt to improve and the local signs recede or remain *in statu quo*. As a consequence they, and often their physicians also, look lightly upon the attack, and the patients return prematurely to their former life, believing themselves cured. Accumulated evidence, however, both clinical and post mortem, proves how slow a process is the cure of a tuberculosis, at least of one that has given clear signs, either local or general, to the physician in charge.

There are probably quite a number of persons who pass through a tuberculous infection with so little cough or other disturbance of the general health that they do not even come under the notice of a physician, and no phthisis follows, but local signs may be discovered later during life or after death.

If a patient goes away, gains flesh, feels strong, loses his cough and other symptoms, and his local signs disappear or remain quiescent, it is very hard for the physician, even if he has sufficient wisdom to attempt it, to persuade the patient to change his life or remain longer away, and especially to go to a distant or unknown country. Thus the tide in the affairs of phthisis, which, taken at the flood, leads on to health, is lost, and the next catarrh or other accident starts up the smouldering tuberculosis, and from the lodgement it has already gained it advances with much increased activity.

These second attacks are very frequently the beginning of that sad and tedious progress to the grave which is made by the health-seeker in various climates and under varying treatment. This progress, alternately chequered with hopeful improvement and depressing decline, is beyond the skill of the physician or the benign influence of climate. It is as impossible to arrest the disease as it is to avert its inevitable termination.

There are incipient cases of phthisis for which a comparatively short and slight change of climate is all that is needed ; but these must be carefully selected, and after a return home they should be watched for several years, all their departures from health promptly attended to, and their daily hygiene raised to as high a standard as possible.

In considering the length and character of climatic change for a consumptive, not only the arrest of the disease but also the permanent raising of the standard of the patient's health demands our attention in order that recurrent attacks may be warded off.

Diagnosis.   With reference to the question of the importance of an early recognition of phthisis and of constant watchfulness after the diagnosis has been made, I analyzed the last 100 first examinations I had made just previous to instituting the inquiry,[1] and I found that in 52 per cent. of the cases diagnosis and treatment had been delayed beyond the time when the symptoms were clearly developed, while in 48 per cent. the disease had been promptly recognized and treated. The total average of delay was two years. In those in whom the diagnosis was made and treatment begun in good season the proportion of first-stagers was 48 per cent., while in those in whom there had been delay and neglect the proportion of those in the first stage was only 29 per cent. The difference in the percent-

----

[1] Neglect of the Early Diagnosis and Treatment of Pulmonary Tuberculosis. S. E. Solly, M.D., in Medical News, February 4, 1893.

age of cures between the first stage and the combined second and
third stages, in the high altitudes to which these cases came, is as
follows: first stage, 62 per cent.; second and third stages, 15 per
cent. This negligence has very serious results. In fact, it means
that out of every hundred consumptives who are allowed to drift
into advanced disease only fifteen recover, instead of sixty-two, as
would be the case if there was no neglect.

**Indications and Contraindications in the Climatic Treat-
ment of Phthisis.** If the foregoing pages have been studied, the
general principles by which one should be guided in selecting a
climate for a case of phthisis will be apparent; but as a change to
a high altitude is most desirable for the majority of consumptives
and most dangerous for those who are unsuited for it, it is best to
review briefly the indications and contraindications for the use of
high altitudes in phthisis. After studying these it is easy to deduce
the approximate climate for those unsuited to high ground. They
have been most admirably put by Dr. F. I. Knight, and, as he is
without prejudice in the matter, and has also had a remarkably
successful and large experience in the use of high altitudes for
phthisis, I believe I cannot do better than give an epitome of his
views:[1]

"1. He limits the age of those resorting to altitudes to fifty years.
In temperament he prefers the phlegmatic to the nervous with an
irritable heart, frequent pulse, and inability to resist cold; and with
the latter, he says, we must be careful not to include those who show
nervous irritability from disease, not temperament, as they are generally benefited in high places. As regards disease, he first considers cases of early apical affection with little constitutional disturbance, and, although these generally do well under most conditions,
yet considerable experience assures him that more recover in high
altitudes than elsewhere.

"2. Patients with more advanced disease, showing some consolidation, but no excavation nor any serious disturbance. When both
the apices or much of one lung is involved, and the pulse and temperature are both commonly over 100, it is best to begin with a low
altitude.

"3. Hemorrhagic cases, early cases with hæmoptysis and without much fever or much disease, are benefited by high altitudes.

[1] Transactions of the American Climatological Association, 1888.

"4. Patients with advanced disease, those with cavities or severe hectic symptoms, should not be sent to high altitudes. A small, quiet cavity is not a contraindication; hectic symptoms are contra-indications.

"5. Patients in an acute condition should not be sent.

"6. Cases of fibroid phthisis are not suitable.

"7. Convalescents from pneumonia or pleurisy are usually well suited to elevated regions.

"8. Advanced cases of tubercular laryngitis, if good local treatment and freedom from dust can be obtained, may do no worse than elsewhere.

"9. In cases complicated by other diseases much care is needed. Cardiac dilatation precludes high altitudes; so also does hypertrophy for the most part, though with exceptions. A cardiac murmur resulting from a long-past attack of endocarditis with no signs of enlargement or deranged circulation should not prevent. Nervous derangements of the heart are usually contraindications. In renal disease and in chronic hepatitis the local physicians claim that benefit is often obtained. Intestinal ulceration does not bar out, but benefit is doubtful. Heredity to phthisis is no bar to high altitudes, but diabetes renders them objectionable. Syphilis is no contraindication, though in phthisis the combination always makes the prognosis bad."

In the paper read by me before the American Climatological Association, 1889, entitled "Invalids Suited for Colorado Springs," I expressed very similar views. In closing this chapter upon phthisis I will repeat some remarks I made in addressing the Arapahoe County Medical Association, December, 1892:

It must especially be remembered that all sorts and conditions of men may become tuberculous, and you must study the individual and his circumstances as well as the type and stage of the disease before you can plan a rational and systematic scheme of treatment for him. All this enters particularly into the question of change of climate. The well-to-do, whose domestic ties permit it, are most safely advised to seek at once the climatic changes best suited to their case; and where such luxuries as have become necessities to them can be obtained, providing that they are or can be brought into a condition to stand the change. There are many, however, who, on account of their circumstances, cannot leave home, and should not be forced to do so unless home-treatment is not succeed-

ing. Hereditary cases should *always* be sent away. In cases arising
where the conditions of life, except the climatic, are good, the necessity for change is clearly indicated. Cases resulting from pneumonia,
pleurisy, or bronchial catarrh, which is slow in clearing up, should
also have change. In the hereditary the change must be permanent;
in the second group it must often be so; there are a certain number
of cases that have arisen under peculiarly unfavorable conditions of
life, in which removal would be a hardship, and which, if they can
be placed under good hygienic conditions at home, may often be
safely left, at all events until the physician has gained a clear knowledge of the tendencies of the case. Those also in whom the tuberculosis is not advanced or active, and whose depression of health
appears to be due to digestion or other causes which can be equally
well treated at home, may remain if they are carefully watched for
a time.

Often a change to a different house or soil or different surroundings, physical or social, will start a patient on the road to recovery,
if conjoined with other wise measures. The choice of climate must
be made, not on the usual happy-go-lucky methods, nor from the
results in a single case, but by studying the general principles of
climatology and then obtaining reliable information concerning
the particular resort that is likely to be suitable. Climatology (in
which the profession as a whole is little learned) is not the pure
empiricism that many think, but is a science founded on natural
laws and strengthened by rational experience.

CHAPTER IX.

FORMS OF DISEASE OTHER THAN PHTHISIS AS INFLUENCED BY CLIMATE.

Tubercular Laryngitis.

THE only reports which I could find of the influence of climate upon tubercular laryngitis are those of Dr. Robert Levy, of Denver, and my own, which deal with the effect of the Colorado climate upon this disease. Various opinions have been expressed on the subject, and most of them advocate a mild, equable climate and especially condemn high altitudes. Dr. Clinton Wagner, however, after some experience in Colorado Springs, wrote as follows: "I think that cases of laryngeal phthisis may safely be permitted to remain at high altitude resorts, provided improvement in the pulmonary trouble and general conditions has already taken place." [1]

Dr. Levy, from an extensive experience of several years as a laryngologist in Denver, reports upon seventy-two cases of tubercular laryngitis; thirty cases were reported to the Colorado State Medical Society[2] and forty-two cases to the Pueblo County Medical Society, May, 1895.[3] Both reports are extremely interesting and valuable. They did not admit of tabulation with my own, but the results show a close resemblance. His conclusions as here quoted are also very similar. While agreeing with Heryng that climatic change cannot supplant local treatment, and advocating surgical procedure when possible and suitable, he says: "Still, judging from these seventy-two cases and many others treated in hospital and dispensary of which records have not been kept, I cannot but believe that under proper treatment laryngeal improvement will go *pari passu* with that of the lungs."

Feeling that the relative effect of the treatment of tubercular laryngitis could only be estimated when the quality of the cases and their relation to lung-tuberculosis, which almost invariably accompanies this sort of laryngeal disorder, are taken into consideration, I

[1] Transactions of the New York Academy of Medicine, October 20, 1887.
[2] Transactions, 1891.   [3] New York Medical Journal, July 20, 1895.

reported before the Pan-American Congress, 1893,[1] upon the cases of laryngeal tuberculosis occurring among 250 cases of phthisis treated by me in Colorado. Of these, 45 exhibited unmistakable tubercular laryngitis. 25 of the laryngeal cases showed clear signs of tubercular infiltration, which had not, however, proceeded to the stage of ulceration at the time of the first examination, though some of them did subsequently, while 20 cases had ulceration as well as infiltration when first seen.

Average Duration. The average duration of the non-ulcerated cases, from the date of their first symptoms up to the present time or until death, was six years; while of 17 of these cases which improved and are living the average duration is thirteen years, and of the 8 which are worse or have died it was but three years and ten months.

The total average duration of the 20 cases with ulceration was three years and two months. Of 5 cases which improved and are living the average duration was eight years and five months; while of the fatal and deteriorated cases it was two years. Of the deteriorated cases, with and without ulceration, the average duration was two years and seven months. This is somewhat longer than the two-years' limit given by Bosworth.

Non-laryngeal. 78 cured, 38 greatly improved, 30 improved, 59 worse.

Laryngeal. Non-ulcerated: 6 cured, 7 greatly improved, 4 improved, 3 worse, 5 died.

Ulcerated: 2 cured, 2 greatly improved, 1 improved, 15 died.

Total laryngeal: 8 cured, 9 greatly improved, 5 improved, 3 worse, 20 died.

Grouping together the cured, the greatly improved, and the improved, under the head of "improved," we find that of the 205 non-laryngeal cases 72 per cent. improved; but of the 45 laryngeal cases only 49 per cent. improved. The non-ulcerated cases, however, showed 68 per cent. and the ulcerated only 25 per cent. of improvement.

Taking the condition of the throat, without regard to the ultimate fate of the patient, the results were much better, there being local permanent arrest of the disease in 64 per cent., beside 5 cases which healed temporarily. Among the non-ulcerated alone 68 per

1 Therapeutic Gazette, November 15, 1893.

cent. showed a return to a normal appearance in the larynx, while among the ulcerated cases 50 per cent. healed permanently and 3 additional cases temporarily.

With regard to the position of the ulceration, the results were: commissure, 33.3 per cent. improved; true chords, 30 per cent.; epiglottis, 17 per cent.; while of the arytenoids and false chords none improved.

To recapitulate, it may be said that of the whole number of cases, viz., 250, rather more than two out of three improved.

Of the 45 which had laryngeal disease, one out of two improved.

Of the 25 cases in which there was laryngeal tubercular infiltration without ulceration, rather more than two out of three improved; while of the twenty in whom there was laryngeal tubercular ulceration, only one in four improved; but of the 205 cases without laryngeal disease there was improvement in nearly three out of every four cases, the exact reverse of the laryngeal ulcerated cases.

This shows, as was to be expected, that the laryngeal complication reduces the chance of improvement, and, when it has proceeded to the stage of ulceration, does so to the extent of three to one. But even so, according to the opinions expressed by laryngologists practising in low altitudes, these are far better results than have usually been obtained, and show, I believe, that similar effects, beneficial, retarding, and often curative, which have been demonstrated in pulmonary tuberculosis treated in Colorado and other high altitudes, are exhibited in laryngeal tuberculosis.

It is, of course, to be expected that the results are not actually so good as in cases of simple pulmonary tuberculosis, though relatively they are so, because in all these cases—and I have never seen a laryngeal tuberculosis without an accompanying pulmonary tuberculosis—there was the double disease and therefore the double burden to bear.

Moreover, it is hardly to be doubted that these laryngeal complications almost always indicate a tendency to a free dissemination of tubercle and generally an absence of any self-limiting features. There are undoubtedly some cases in which the laryngeal tuberculosis is derived from an inoculation caused by the lodging of the sputum upon the abraded laryngeal membrane; but clinical observation leads me to believe that in the great majority of cases the infection starts from within and not from without.

Taking the results upon the laryngeal disease alone, irrespective of the ultimate recovery or deterioration of the patient on account of the accompanying lung-disease, we find that in 64.2 per cent. there was arrest; and if we consider also the five in whom there was temporary healing, which broke down again under the strain of the last weeks of fatal pulmonary suppuration, we see that the percentage of improvement in the local laryngeal symptoms is not very far short of that in the simple pulmonary cases.

While I believe that, generally speaking, high altitudes are really beneficial to tubercular laryngitis (contrary to an impression once commonly received that, as regards that disease, they were positively injurious), it is nevertheless true that no such results as I report could be reached, in the majority of cases, without careful local treatment. This remark especially applies to those cases which show ulceration. As Bosworth truly writes, after advocating topical measures, "in no ulcerative process, probably, are we able to detect in a less degree any reparatory effort on the part of nature than in tubercular ulceration, and yet instances of spontaneous cicatrization have been reported by Bouveret, Virchow, Jarvis, and others."

In conclusion, it may be said that the foregoing facts indicate that while tubercular laryngitis is always a grave complication, at an altitude as elsewhere, and that when advanced it is almost invariably fatal, yet in the earlier and medium cases high altitudes, with appropriate treatment, afford relatively, though not actually, as good a chance for arrest or delay in laryngeal as in pulmonary tuberculosis. With respect to the influence of low altitudes, I believe that where most relief is afforded to the accompanying pulmonary tuberculosis the lesions are most likely to heal under appropriate treatment.

Tuberculoses other than those of the Respiratory Tract.

My experience leads me to believe that the question of climatic change for all other forms of tuberculous disease must be answered upon the same grounds as that of change for pulmonary tuberculosis. There are, however, certain modifications to this statement which will be discussed under the head of Scrofulosis.

Scrofulosis is a diathesis which renders its victims peculiarly liable to tuberculosis and induces chronicity of inflammatory pro-

cesses and inflammation of the adjacent glands.   It tends to produce
hyperplasia and caseation of the lymphatic glands.   The object in
the treatment of a scrofulous person, whether child or adult, is to
increase the vital resistance, especially to bacillary invasion and consequent tuberculosis.   The open lymph-spaces and feeble vitality
allow the bacilli to spread readily through the body, and where
hyperplasia or caseation has occurred a peculiarly congenial soil is
presented for the development of tuberculosis.

Some years ago, during my visits to the Infirmary for Scrofula,
at Margate, England, a resort possessing a beach with a good sand
and gravel soil and endowed with a mild, sunny climate, I was
much struck by the remarkable improvement in scrofulous children,
particularly in those cases where the disease had attacked the bones.
Later experience has confirmed me in the opinion that a warm
seaside-resort, where much outdoor life and sea-bathing can be obtained, is the best for counteracting the scrofulous habit and early
tuberculosis; but when the tuberculosis is advancing an elevated
climate is generally more beneficial.

Sea-bathing.   With regard to its benefits for the scrofulous, Dr.
D'Espine[1] writes that of 308 children, from four and one-half to
fifteen years old, who were sent to the Mediterranean coast of France
because of various scrofulous affections, 48 cases were cured, 215
improved, 44 were unimproved, and 1 died.   They each had about
forty-five sea-baths during a stay of six weeks.   Jacobi and Casse
and many French authorities are favorable to sea-bathing.   Treves
thinks the effects are chiefly mental.   Dr. Walter Chrystie[2] writes
as follows : " The writer is convinced that salt-baths stimulate
nutrition, and is in the habit of directing scrofulous persons to add
a small amount of sea-salt to the bath once or twice a week."

Sea- and Mountain-air.   With reference to climate, Dr. Chrystie
goes on to say : " The efficacy of sea-air in eradicating the scrofulous diathesis and its manifestations is admitted by all writers, and
is especially dwelt upon by Deligny, who made a large number of
observations upon scrofulous children treated in L'Hôpital de Berck
on the coast of France.   Frederick Treves also writes that the
records of the Margate Infirmary for Scrofula support the fact
that sea-air possesses a curative influence upon scrofula by showing a large number of cures and a still larger number of cases of

[1] Revue méd. de la Suisse Romande, September, 1888.
[2] Hall's System of Therapeutics, vol. i. p. 930.

marked improvement. But he adds that ' the greatest advantage is observed in instances of acquired struma, in cases where the disease has developed in the purlieus of a great town, and in those patients, in fact, to whom sea-breezes and outdoor exercises offer the most striking possible contrast to their previous surroundings.' "

Cases will occasionally be met with which do not improve in a coast climate or which improve exceedingly slowly, notwithstanding the utmost care in the selection of diet. Such cases will often improve rapidly if removed to a dry, moderately elevated, mountainous region.

It is extremely difficult and often impossible to determine which climate will be the most beneficial in a given case. Treves thinks that the cases which do not improve at the seaside are usually those with a "phthisical tendency," also many cases of eczema, some cases of strumous ophthalmia, and a few cases of lupus. The author's experience has been that cases in which there is marked anæmia, muscular debility, or exhaustion from the presence of complications, such as glandular suppuration, bone-disease, etc.—in other words, cases with well-marked tubercular disease, excluding pulmonary phthisis—usually improve more rapidly in sea-air, and that cases with the scrofulous diathesis alone are better in mountain-air.

Many cases gain in weight and appearance in a coast climate for a few months, and then come to a standstill or even retrograde. Under such circumstances a removal inland is usually followed by renewed improvement. Indeed, in some cases frequent change of climatic surroundings is necessary in order that the improvement may be continuous, and, in all cases, a week or two at the seashore can do no harm if it be preparatory to a more prolonged sojourn in a mountain climate.

Affections of the Respiratory Tract other than Phthisis.

Croupous Pneumonia. In this affection the use of climate has only to be considered during the stage of convalescence. Resolution of the pneumonia and a return to health are frequently assured, and generally hastened by change of air.

The climates most convenient to the place in which the sickness has occurred and which afford the most opportunities for the invalid to rest in the open air without danger of catching cold, are usually the best. A warm seashore, or a dry, sunny, inland spot with a sandy soil, is generally the most suitable choice for the early days of convalescence, and when strength has been sufficiently regained to allow of taking exercise it is wise to seek the more bracing air of the uplands to complete the cure. When, in spite of these moderate changes, resolution is delayed or tuberculosis supervenes, the high altitudes should be resorted to, as there the anæmia, which is the most common cause of delay in recovery, is most quickly and surely combated.[1]   On the other hand, if the disease has occurred on high ground, a change to a lower climate is indicated.

In deciding the question of sending a patient with an unresolved pneumonia to elevated ground the condition of the heart and breathing should be carefully considered, the heart being examined not only during rest, but after exercise, and especially after climbing stairs.   A moderate degree of weakness can be taken care of in those under thirty-five years of age, proper precautions concerning exercise being taken.   For those who are older it is wisest to select elevations under 3000 feet until experience shows that the patient may safely be advanced to higher ground.   In all cases the amount of unused lung of course enters into the problem, obstruction in the left lung being more detrimental than the same amount in the right. If the lungs are seriously involved or the heart manifestly weak, it is always best for the patient to go to bed immediately on arriving and to send for his physician before taking any exercise.   The late Dr. A. L. Loomis dwelt upon the injurious effects to the heart in certain cases sent to the Adirondacks at an elevation of 1600 feet, and in Colorado Springs (6000 feet) I have seen such serious injury result to patients whose hearts had been considered normal, from their walking, on first arrival, a few hundred yards up a slight incline from the station, as to make it necessary to send them back in a short time to sea-level.   However, when the cases are wisely selected and properly warned as to exertion there is little risk, and the results, in cases of retarded pneumonia, whether simple or tuberculous, are generally much more brilliant in high altitudes than elsewhere.

Catarrhal Pneumonia.   Convalescents from this affection are often given moderate change of air with benefit at an earlier stage than the croupous cases.   In these, as a rule, warmer and more

---

[1] In selecting a high climate, it must be remembered that the drier atmosphere is generally the better, and, if the invalid has progressed to the point of taking exercise, and evinces fair powers of reaction, the climate chosen should also be moderately cool on account of the greater tonic effect.

equable climates of moderate elevation are best. Those cases in which the tendency to fever and irritability of the membranes is slight are better suited to higher and drier ground, even though the climate is neither so warm nor so equable as that lower down ; while for those with the contrary tendencies the sea or a slight elevation, such as is to be found on an upland or hillside, or a moorland, is better than such an altitude as that of a mountain or high plateau. When the patient is also tuberculous higher ground can more often be sought with benefit than in cases of simple catarrhal pneumonia, the other considerations referred to being allowed due weight.

Pleurisy. When the convalescent stage is reached, the fluid having been absorbed or removed by tapping, it sometimes happens that recovery is unduly delayed owing to anæmia or to a bound-down lung, or both, and that a moderate change of air is not sufficient to complete the cure. In such cases a visit to the higher altitudes is nearly always the most effective and speedy remedy; but here, again, the condition of the heart, especially if that organ is displaced, must be as carefully considered as in a case of pneumonia. If the patient goes to an altitude, the direction of his exercise must at once be intrusted to a physician.

Empyema. Cases of delayed recovery do remarkably well on high ground, as they do also on a warm, sheltered seashore. The climate to which the patient is sent should be, as far as is possible, a contrast to the one in which the disease arose ; and each case, whether tuberculous or simple, must be decided with a proper consideration of the same principles referred to as being applicable to pneumonia.

Bronchitis, unless tuberculous—and a majority of the cases of chronic bronchitis become so—is usually soonest cured in a dry, warm, inland climate of very moderate elevation where there is little wind.

Asthma,

which is generally either bronchial or nasal (hay-fever), and at times both, may be defined as a spasm of the bronchial tubes, which has a predisposing, a determining, and an exciting cause. The predisposing cause is a peculiar neurotic habit. The determining cause is some disorder or disease of the respiratory tract which can generally be classed as bronchial or nasal. The exciting causes

are very numerous, and are either intrinsic or extrinsic. The intrinsic cause may be an ordinary catarrh or inflammation of some
portion of the respiratory tract, or indeed of any other portion of
the body, or organic disease, such as that of the heart, independently of true cardiac dyspnœa; or it may be simply a derangement
of function, as in dyspepsia, or a nerve-storm, or other disturbance,
physical or mental. The extrinsic causes are also various, being
generally, for bronchial asthma, changes in atmospheric conditions,
and, for nasal, the inhalation of the pollen of a particular plant, or
of dust, or of some animal exhalation.

Bronchial Asthma. In selecting a climate for an asthmatic it must
be determined which of the causes it is most important to remove, as
one climate excels another in its influence over one or more of these
causes. It is impossible, in actual practice, to separate these causes
and to treat each in its appropriate climate; but it will perhaps be
well to try to explain why so many different climates and conditions
are good for asthmatics, and to assist in making the best possible
compromise when a climate is finally chosen, thus causing the
selection to be less empirical and experimental than it usually is.
We must presume, in the first place, that the physician or surgeon
has done all that lies in his power to remove or modify one or more
of the three causes. Let us suppose that the neurosis is the cause
to be especially remedied; the climate must then be chosen for its
quality of sedation or stimulation. If the hyperæsthesia is apparently innate in the individual and customary to him, and is not
brought about, or largely aggravated, by anæmia or some other
depressing cause, a sedative climate is generally most beneficial,
because it modifies the neurotic habit. Further, as the catarrh or
chronic inflammation of the respiratory tract is, in what is termed
a born neurotic, usually of the irritable or inflammatory type, a
climate of a sedative character, in which the air is warm, humid,
and equable, would be the most beneficial; but the climate selected
should also possess tonic qualities in proportion as the respiratory
condition permits a lessening of this warmth and humidity and
consequently of climatic equability. If there is chronic anæmia, a
stimulating climate, cooler, drier, and more variable, would be preferable.

The one meteorological element, however, which has the most
marked effect upon the neurosis of asthma is barometric pressure,
either markedly increased or decidedly decreased. Both conditions

are of the greatest benefit, as is shown by laboratory experiments and also by the actual experience of asthmatics in depressed parts of the earth's surface, such as the valley of the Dead Sea and the sink of California; and at great elevations, such as the high plateaus of the Alps, the Rockies, and the Andes. With regard to the use of increased barometric pressure, while its effects at the time are striking, they are seldom lasting, because, if it be used in the laboratory, its application is necessarily intermittent, and a residence in depressed climates is almost unendurable from heat during the summer and is monotonous and depressing at all times. On the other hand, life at an altitude is pleasant and can be continued without interruption. It is possible to vary the locality, and in many places the pursuit of business or amusement may be carried on profitably and agreeably. The height, latitude, and exposure of the elevated region can be selected in accordance with the temperature, humidity, and variability best suited to the particular condition of the respiratory tract, and a locality which is free from the pollen, dust, or animal exhalation peculiarly obnoxious to the individual asthmatic may be more readily found. I believe it may be said that, other things being equal, an altitude is best for the cure of the neurosis; but if, for special reasons, the diminished barometric pressure is unsuitable, the next best climate for the neurotic habit is in a sink.

In deciding upon a climate with a special view to the cure of the determining cause—that is, the disorder of the respiratory tract—the choice must be made according to the principles dwelt upon in discussing pneumonia, bronchitis, etc., and a sedative or stimulating climate selected accordingly. If, however, the exciting cause appears to be the most important, that must be especially considered. For instance, in a cardiac case the elevation of the resort is the prime factor; but in the case of a dyspeptic the ability to obtain appropriate food is an absolute essential. If, again, the exciting cause be pollen, dust, or animal exhalations, pure air must be sought, and this will be found, first, on the ocean, and next on the mountain-side or desert-plain.

Cases of chronic bronchitis without marked emphysema and with a heart in which permanent dilatation or other lesion is not apparent, are usually most benefited by a change to high ground. Before sending to a high altitude any cases in which the condition of the heart is questionable, it is well to try the effect of a moderate elevation, such as is found between 1000 and 2500 feet, where the air and soil are fairly dry and the skies sunny.

As is well known, an adult asthmatic is rarely permanently cured ; but, to modify Sydney Smith's remark about a Scotchman, " Much can be done with an asthmatic if he is caught young." It is in youthful asthmatics that I have seen nearly all of the comparatively few permanent cures, and among these the use of high altitudes had been the most successful element. This is probably due to the breaking up of the neurotic habit and to the cure of the anæmia generally present in young asthmatics.

Dr. Denison, of Denver, reported[1] 52 cases of asthma treated in Colorado, of which 72.5 per cent. were more or less improved. Of those that got worse or remained stationary all but one had emphysema. This agrees with the experience of my colleagues and myself.

Hay-fever (*Nasal Asthma*). With respect to this form of asthma, I may say, without entering into a full discussion of its causes, that my observations in Colorado have convinced me that vegetation, in some form or other, is the most frequent exciting cause, although we know that in a few cases certain animal exhalations and inorganic dusts play the most important part. The particular kind of vegetable growth, the pollen or dust of which excites an attack of hay-fever, is different for different individuals. While it is probably true that a larger number of cases of hay-fever are benefited by a change to a high altitude, such as the Alps or Colorado, than to lower climates, yet many cases at altitudes have to shift their ground at certain flowering-seasons, and the relief thus afforded is probably due to the amelioration of the neurosis by diminished pressure, to the general purity of the air, and to the comparative sparseness of vegetation.

Geographical Distribution. The following remarks concerning the geographical distribution of hay-asthma are well worthy of attention :

"One thing that markedly distinguishes hay-fever from other catarrhal maladies of similar nature is its geographical relations. It does not exist over the whole of the United States or of Great Britain, yet it would be a difficult task to define its exact limits. Numerous portions of England are immune, especially the highland and seacoast, and all or nearly all of Wales and Scotland.

[1] Transactions of the American Climatological Association, 1890.

In America it obtains to the north of Lake Ontario in a limited degree, but not on the upper side of the St. Lawrence; scarcely at all in the province of Ontario north of the Welland Canal until the Detroit River is reached, and it is wholly unknown to regions above the outlet of Lake Huron. In Michigan, however, it follows Lake Huron to above Saginaw Bay, finding victims even at Alpena, though residents of Buffalo, Cleveland, Detroit, and Cincinnati are here usually immune. On Lake Michigan its effects are lost above Ludington, while over the Mississippi, in Wisconsin, it is felt as far north as the junction of the Chippewa, and in some seasons extends in a mild form to St. Paul, Minn. To the south it extends to the latitude of Memphis in the west, Knoxville in the central area, and Cape Henry on the Atlantic. In all this area there are immune districts at high altitudes, such as the Green, White, Adirondack, Allegheny, and Catskill Mountains, and the southern New York region. Isolated spots where the malady prevails are found about Galveston, Texas; St. Augustine, Florida; Montgomery, Alabama; and Milledgeville, Georgia. Beyond the Mississippi evidence and data are almost wholly lacking; but several persons have suffered at Denver, Colorado Springs, and Golden City, though denizens of cis-Mississippi regions here find relief."[1]

Where the attacks are chiefly dependent on vegetation the surest remedy[2] is naturally a sea-voyage; but if the noxious plant be known, the flora of the proposed resort should be studied before changing a land climate.[3] Where the chronic catarrhal condition is prominent as a cause and cannot be removed by treatment, the climate, with respect to its humidity and temperature, must be selected as the experience of the individual or a general inference dictates.

Nasal and Pharyngeal Catarrhs.

The surgeon, on the one hand, and the physician, on the other, have so narrowed the field of treatment by climate of these affections that there is very little to be said specifically. Change of air is often of benefit, especially as a supplement to surgical or medical treatment; but its selection must depend upon the general or local condition of the patient, one being usually of greater importance than the other.

---

[1] Medical Age, Detroit, 1896. [2] Ibid., September 10, 1896.
[3] It has been noticed in Colorado Springs and Denver that, when the weeds are allowed to grow freely, cases of hay-fever are less often relieved, and perhaps the greater amount of watering done in cities has an influence.

For instance, a stimulating air may be best for the general health, while the mucous membrane may be more benefited by a sedative climate. Nasal and pharyngeal catarrhs are common in all climates where there is much variability, whether they be warm or cool, moist or dry, high or low, and the greatest benefit is usually derived from the climate which is in sharpest contrast to that in which the catarrh was contracted.

Chronic Laryngitis.

When this affection is not entirely amenable to suitable treatment at home it is usually benefited by a change of air opposed in character to that in which the condition first arose, and the question has to be considered on the same principles as those referred to concerning nasal and pharyngeal catarrhs. Tubercular laryngitis is discussed in the chapter on Phthisis.

Diseases of the Heart and Great Vessels.

We can all remember the time when a patient who was pronounced to have heart-disease was warned to avoid all exercise, and was hardly considered to be fulfilling his natural destiny unless he died suddenly. Fortunately, the study and treatment of the underlying causes of heart-lesions or irregularities have done much to prolong the life of patients and have led to more frequent cures.

There are two points of view from which change of climate for a heart-case must chiefly be looked at : the first is the general condition and tendencies of the patient; the second, the mechanical condition of the heart or pump. In the heart the strength and resiliency of the cardiac muscle are of the greatest importance, and the quality of the climate and the amount of exercise allowed should be determined by these conditions rather than by the nature of the valvular lesion. The diagnosis can often be aided by experimenting cautiously with different exercises before leaving home.

Cardiac cases are often slightly affected by a moderate change of air even when this is unaccompanied by an increase in elevation, and they are generally very much affected when climatic change and increase in altitude are united. It is therefore of the utmost importance that, after the resort has been selected with due care and the journey properly planned, the patient should be advised, from the first hour of his arrival, by a local physician especially as to his exercise, habits, and diet.

When a heart-case is to be sent to an elevated resort the temperament of the patient enters into the question; for, if he be heedless and likely to neglect the advice of his physician, he is certainly safer on low ground. Again, if the patient be of the erethic type, he is apt to have an irritable heart, in which case a sedative climate is more suitable.

The age of the patient has much to do in the selection of a climate. In the young, moderate dilatation, a frequent accompaniment of phthisis, does not prevent a patient's being sent to an altitude if it is otherwise expedient; for, when great care is exercised or absolute rest prescribed for the first month, such cases generally recover more quickly and completely than at sea-level. On the other hand, when the dilatation is well marked or if the invalid is approaching middle-life, a high altitude is dangerous. Hypertrophy generally, though not necessarily, contraindicates a change; but this depends upon its extent and duration and the attendant circumstances. Any disease or irregularity of the heart or great vessels in elderly persons, particularly if there are indications of atheroma, should be treated at sea-level and in a sedative, warm, equable climate.

Cases of aneurism in the aged or in prematurely old persons should be treated in the same way; but I have observed in Colorado that most patients suffering from aneurism enjoy a greater sense of well-being and apparently improve more than when living at sea-level. Several cases now residing in Colorado have remained in a stationary condition far longer than I expected. The observations of Weber and other clinicians, and particularly those of Archibald Smith in the Andes, agree with this opinion.[1]

In choosing a climate for a case of cardiac or vascular disease it has first to be determined whether it is reasonable to expect cure or arrest, or only palliation. In the first case the more tonic and stimulating climates may be cautiously tried, beginning with the bracing air of a seashore-resort and advancing, first, to a dry, desert climate like Egypt, then to a slight altitude, and then to a high one through the intermediate climates, and choosing from these according to the degree of stimulation thought to be safe or found by experience to be so.

On the other hand, when the disease is of such a character as to make it unwise to attempt to level up the patient's health and to

---

[1] The physiological causes for this are dwelt upon on page 114.

put extra work upon the peccant organ, levelling down must be carried out and the ship of life made to sail on an even keel.

Again, in choosing a climate for cardiac and vascular cases, not only does the condition of the kidneys, skin, and lungs demand the gravest consideration, but also the effect of the proposed change upon them. If there be one organ which, more than another, requires the most careful scrutiny before an invalid is sent from home to any positive climate whatever, it is the heart, and this scrutiny must be something more than auscultation while the patient is calmly sitting in a chair. Some attempt must be made to reproduce temporarily the climatic conditions to which it is proposed to subject him. The examination of the kidneys is next in importance to that of the heart, and is a necessary adjunct of it. Both of these inquiries are too frequently neglected, or are at most passed over in a perfunctory manner, often to the injury of the patient and the discomfiture of the health-resort physician.

Again, in choosing a resort for a heart-case the effect of the climate upon the liver requires some thought, for it is well known that through the portal circulation the hepatic functions influence the cardiac very markedly in some cases.

As a change to a high altitude is the most extreme that can be made for these cases, I will quote from the opinions expressed by two of the leading authorities—Dr. Hermann Weber, of London, and Dr. Frederick I. Knight, of Boston.

Dr. Weber[1] writes: "The diseases which are unfavorably influenced include most cases of organic disease of the heart and vessels, though a heart moderately enlarged, with a weak muscle, even with a bruit, is often improved, and generally much more so than on the coast. Even cases of aneurism are often relieved. Atheroma and senile affections generally contraindicate a high climate."

Dr. Knight, after prefacing his statements by saying that he refers particularly to altitudes from about four to six thousand feet, writes thus:

"1. Cases of valvular disease with sufficient cardiac enlargement or derangement of the circulation to make the diagnosis certain. While a great difference in risk in such cases must be admitted, dependent upon the compensation, the age of the patient, etc., it is safer to forbid the change to such patients, for we know that in

---

[1] Ziemssen's Handbook of Therapeutics.

any case, even at sea-level, compensation, for some unknown and unexpected reason, may suddenly cease, and of course this may be precipitated by such a change of pressure as would be experienced by a change in altitude of five thousand feet. I say this, knowing that Dr. Solly and others have had young cardiac patients with good compensation who have been not only uninjured, but invigorated and improved by residence in Colorado.

" 2. Cases of chronic myocarditis or fatty degeneration. These should be rigidly excluded, and these are the cases about which more care should be exercised. They are the cases among which there is such sudden fatality, caused apparently by slight variations from a dull routine of life. A sudden change of conditions under which the heart is laboring stops it altogether. This change may be one of atmospheric pressure, a change of nerve-influence, a sudden excitement of joy or grief, a fall, a shock of any kind, mental or physical. There may or may not have been symptoms calling attention to the heart, or the symptoms may have been wrongly interpreted—as, *e.g.*, calling an attack of angina pectoris 'gastralgia.' Many cases have been declared sound on superficial examination chiefly because no cardiac murmur was discovered, when a more careful examination would have resulted in the probable diagnosis of one of the above-mentioned conditions. A careful examination should be made of the area of cardiac dulness. It will often be found enlarged in cases of chronic myocarditis. The character of the first sound at the apex should be carefully studied. A very valvular first sound, an almost entire loss of the booming or muscular quality, with a weak and irregular pulse in a man no longer young, especially in connection with any subjective symptoms, points to myocarditis or fatty degeneration. Breathlessness on slight exertion and a feeble, irregular pulse are strong confirmatory signs. It is much more important to keep this kind of a case out of high altitudes than those of valvular disease. A person over fifty years of age with marked cardiac symptoms or any of the signs mentioned above must not be allowed to make the change.

" 3. Cases which present a murmur anywhere in the cardiac area, but who have never had any symptoms, and who on physical examination show no further evidence of disease. The murmur is fast losing the undue importance which was attached to it for many years, and is falling into its proper place as only one link in a chain of evidence. Even life-insurance companies accept some applicants with

11

cardiac murmurs. Patients with systolic murmurs which are known to have existed for many years without any enlargement of the heart or any alterations of its normal sounds may be allowed to go into high altitudes. Patients with murmurs in diastole must be advised much more cautiously, as these murmurs are more surely indicative of serious organic disease.

"4. Cases of nervous palpitation. Patients with functional palpitation cannot be considered in one class. In many such palpitation is quite temporary, due to errors in diet or mode of life, which can be easily set right. When these errors have been corrected, of course there can be no objection to the patient's going to a high altitude. Affections of this kind, due directly to some morbid condition of the nervous centres, may be divided, as by Eskridge, into two classes—that of patients who have inherent nervous temperaments and that of those who are nervous from malnutrition. The latter, as is pretty generally conceded, are likely to be improved, and consequently the trial of a high altitude may be recommended to them; but the former, those of inherent nervous temperament, are usually made worse by it, and consequently should be forbidden the high altitude."

I will close this subject with some remarks of my own.[1] The decided effect of a high altitude in exciting and disturbing the action of the heart in sensitive persons, even without disease, is shown by many who, after living comfortably for several years at Colorado Springs (elevation, 6000 feet), ascend to the summit of Pike's Peak, 8000 feet higher. These persons, while there, except for a feeling of light-headedness, are all right so long as they do not attempt to walk, but even a few steps will sometimes bring on rapid and often irregular beating of the heart, generally slight headache, often nausea and diarrhœa, and in some slight precordial pain; the headache and pain over the heart often persist more or less for twenty-four hours after the descent to lower ground.

Nervous Disorders.

Organic disease, when progressive, requires a sedative climate at or near sea-level, with a humid air and a moderate rainfall, where warmth and equability are present. When the disease is stationary, mildly stimulating climates may be cautiously used, and in some exceptional cases even the high altitudes. I have, for instance, seen

---

[1] Article upon Climate in Hare's System of Therapeutics.

improvement in some stationary cases of progressive locomotor ataxia, brought into Colorado, and also in certain cases of other chronic paralyses. It is always risky to send cases with well-marked organic lesions to high ground, particularly where the patient is of full habit or inclined to inflammatory or irritable nervous conditions. It is only when the destructive processes are arrested that an accompanying anæmia can sometimes be best treated at an altitude; as a rule, however, such cases do better during summer months at elevations of from 1000 to 2500 feet.

Meningitis is especially unsuited to treatment in high climates, and even a seacoast air is apt to be too stimulating.

I find in medical literature very few references to the influence of climate upon nervous diseases, and these are of the most general character. Dr. J. T. Eskridge, of Denver, from the standpoint of a neurologist enjoying the opportunities of a large practice of several years in Colorado, wrote a valuable and conservative paper[1] which throws much light upon the influence of a high altitude in nervous disorders, and indirectly explains by inference the effects of lower climates. I will therefore give the gist of some of his remarks:

According to Eskridge, numerous careful observations which he has taken of the temperature of the surface of the body, and especially of the head, show that the surface-temperature averages half a degree higher than at sea-level.

Insomnia. A larger percentage of overworked people, Eskridge states, sleep better in mountain climates than near the sea. The cases which do best at high altitudes are those who sleep well from the time of their arrival, and they are the majority. But there is truth in the popular belief that the influence of the Colorado climate ceases to be effective in this respect after a few years of continued residence, especially if the individual be either too little occupied or overworked. A change to sea-level for a short time results in renewed improvement on returning.

Eskridge also states that stimulants usually produce their effects more readily, and with this I agree; but I believe he should have added that when the vitality is depressed, and alcohol is suitable, larger amounts are needed and can be safely taken.

Patients with active hyperæmia of the brain, with insomnia or sleeplessness from organic brain-changes, do badly, whereas those with passive hyperæmia are benefited.

[1] Transactions of the American Climatological Association, 1891.

Inherent nervous irritability is increased, especially in women. Hysteria is apparently as frequent and severe as at sea-level.

Chorea, Eskridge believes, yields as readily to treatment in Colorado as elsewhere. The movements on first arriving are perhaps exaggerated; he is uncertain about their frequency. My own belief is that they are less frequent.

Neuralgia, he states, is, on the whole, less frequent because of the absence of malaria. I think, however, that the pain is more severe in proportion to the conditions.

Migraine. The first effect of a high altitude upon this disorder is beneficial; but it frequently returns, in which case a change to lower ground is advisable.

Epilepsy. Eskridge refers to 21 cases, 5 of which originated in Colorado ; 16 were improved. Treatment seemed to have about the same effect as at sea-level. In all he noticed some improvement, on first coming, if they kept quiet. The writer has seen several epileptics very much improved in Colorado, and a few apparently recovered who were doing badly before they came; others have become worse. My impression is that where there is no local condition beyond the reach of climatic treatment, and the case is one for which an elevated climate is otherwise suited, the strong tonic and alterative qualities of altitude are often of the greatest service. Possibly one reason of improvement is the fact that epileptics are better able to take bromides with less depression.

Insanity. He says that the opportunities for observation have not been sufficient to form an opinion as to whether Colorado is generally beneficial or otherwise. Acute mania is apparently aggravated by the climate, and he has known such cases benefited by being sent to sea-level. Insanity originating in Colorado is more apt to be of the depressive than of the expansive type. In a recent conversation he said that he found from his own experience and that of other practitioners in Colorado that the proportion of melancholia to mania is about three to one, while Spitzka and other writers state that elsewhere it is about equal.

Inflammatory Lesions of the Brain and Cord. So far he could not say whether there was any difference or not. Infantile paralysis is rare. Meningitis, tubercular and non-tubercular, is also rare, the latter in proportion to the number of tuberculous persons.

Chronic degenerative conditions of the cord are perhaps indirectly improved.

Chronic degeneration of the brain is quite as frequent in old-time settlers as it is in Wall Street men, and is in both cases probably due to the exciting lives led by them, to which, in the case of the Colorado men, was added much physical hardship. This explanation is reasonable, as their wives, who are under the same climatic influence, rarely develop such diseases. Autopsies reveal arteritis as the usual cause, and this was several times found associated with slight chronic meningitis.

Apoplexy, while apparently not frequent, is certainly more likely to occur when the vessels have been previously weak.

Sunstroke, Eskridge states, is practically absent. This agrees with my own observation, and I found the same thing true in Egypt, at least south of the Delta. I conclude that this immunity is due to the dryness of the atmosphere in both climates, which favors evaporation and gives relief to vascular congestion.

Leaving these interesting observations concerning altitude and nervous diseases, we will take up the subject of neurasthenia at greater length, as this nervous disorder oftenest demands climatic change and is most frequently benefited by it.

Neurasthenia. The definition and etiology of this affection, so clearly and concisely given by Osler, are as follows:

"*Definition.* A condition of weakness or exhaustion of the nervous system. The term, invented by Beard, covers an ill-defined, motley group of symptoms, which may be either general and the expression of derangement of the entire system, or local—that is, limited to certain organs. Hence the terms cerebral, spinal, cardiac, and gastric neurasthenia. In certain respects it is the physical counterpart of insanit As the essential feature in the latter condition is the abnormal response to stimuli, from within or without, upon the higher centres presiding over the mind, so neurasthenia appears to be the expression of a morbid, unhealthy reaction to stimuli acting on the nervous centres which preside over the functions of organic life. No hard-and-fast line can be drawn between neurasthenia and certain mental states, particularly hysteria and hypochondria.

"*Etiology.* Although the causes are apparently varied, they may be classified as hereditary and acquired.

"(*a*) *Hereditary neurasthenia.* We do not all start in life with the same amount of nerve-capital. Parents who have been the subjects of nervous complaints or of mental troubles transmit to their children an organ tion which is defective in what, for want of a better term, we must call ' nerve-force.' Such individuals start handicapped and furnish a considerable proportion of our neurasthenic patients. So long as they are content to transact a moderate business with their capital, all may go well; but there is no reserve, and in the emergencies which constantly arise in the exigencies of modern life these small capitalists go under and come to us bankrupt.

"(*b*) *Acquired neurasthenia.* The functions, though perverted most readily in persons who have inherited a feeble organization, may also be damaged by exercise which is excessive in proportion to the strength —*i.e.*, by strain. The cares and anxieties attendant upon the gaining of a livelihood may be borne without distress, but in many persons the strain becomes excessive and is first manifested as *worry.* The individual loses the distinction between essentials and non-essentials, trifles cause annoyance, and the entire organism reacts with unnecessary readiness to slight stimuli, and is in a state which the older writers call irritable weakness. If such a condition be taken early and the patient given rest, the balance is quickly restored. In this group may be placed a large proportion of the neurasthenics whom we see in this country, particularly among business men. Other causes more subtle, yet potent and less easily dealt with, are the worries attendant upon love affairs, religious doubts, and the sexual passion."[1]

After the due regulation of rest and exercise for a given case of neurasthenia, the essentials of climatic treatment are the degree of stimulation or sedation needed, the removal of the patient from the rut into which he has fallen, and keeping him in pure, open air.

One of the most common and distressing symptoms of neurasthenia is insomnia, and it may be stated, as a general rule, that where the neurasthenic sleeps the most he does the best—at least till that symptom is overcome and he is thoroughly rested.[2]

In deciding where to send such cases a close inquiry must be made into their past experience in changing air, whether they have recuperated best at the seashore or on the hillside, in the woods or among the lakes, on the warm, dry desert-plains, or at high altitudes, warm or cool.

Rest for both body and mind is the first essential until the anæmia

[1] Osler: Practice of Medicine.
[2] There are exceptions to this, as it is not infrequent to find neurasthenics living in malarial districts who sleep too heavily, and who, while they improve in an elevated climate, yet sleep more lightly because the malarial poison is being eliminated.

is fairly overcome. The ideal rest to both mind and body for those with whom the sea agrees is obtained on a well-found sailing-ship in a warm latitude or on placid waters.[1] This necessitates a good digestion for monotonous and simple food, a non-susceptibility to sea-sickness, a pleasant tolerance of the general boredom of ship-board-life, and a philosophy sufficient to overcome the difficulties sometimes encountered in a forced companionship with uncongenial spirits. Possessing these qualities, the neurasthenic can vegetate in great security from the irritation of contiguity to the busy life or worrying circumstances among which his malady began, and he is also free from the temptation to overexertion.

Again, with those to whom the sea-air is beneficial, but who prefer a shore-life and can endure more exercise, a quiet island in the ocean, where steamers rarely stop and which is free from the requirements of fashion, allows a pleasant alternative, with more varied interest and opportunities for social intercourse and exercise.

If a more stimulating sea-air is needed, it is next best obtained on a pleasant seacoast, because there is less equability of temperature and more mobility of atmosphere, owing to the change from sea to land breezes and to the difference in radiation from the surfaces of earth and ocean. Here, again, both exercise and amusement can be more diversified than on the island or the ship.

To those with whom sea-air does not agree, and who chiefly require rest, a camp-life in the woods or among the lakes is most desirable, with hunting, fishing, and boating or pursuits connected with the study of natural history, according to the individual's strength.

Again, in such cases as are unsuited to the sea, but who cannot, owing to expense or inconvenience, or because of some physical contraindication, visit high altitudes, the anæmia will be more speedily removed by a change to moderately high ground than by one involving no change in altitude. Inland places of medium elevation are best suited to those for whom the seashore or the higher altitudes are too stimulating.

Where, however, anæmia is the prominent symptom or the underlying cause, and where the nervous susceptibility is acquired, not inherent, and there is no cardiac disorder, an altitude between 4000 and 8000 feet affords the surest, speediest, and most permanent cure.

---

[1] See extract from Robert Louis Stevenson, Section III.

A few years ago I reported[1] certain cases of neurasthenia, in some of which the trouble had been continuous, while in others it had taken the form of recurrent attacks. All of these had previously tried low altitudes or the ocean, and in all the improvement was much more rapid and lasting in Colorado than elsewhere.

When we consider the blood-changes induced by the diminished pressure, the greater dryness, and the increased brilliancy of the sunlight, this is not surprising. All writers on the effects of altitude have noted this; therefore, where the other physical conditions are favorable, and opportunity offers, a neurasthenic should be sent to a high climate.

The cooler climates are the more bracing, the warmer more sedative.

Patients seeking climatic change must be especially warned against over-exercising and neglecting to obtain at once the advice and guidance of a local physician.

In the case of those who cross the sea with a view to finding diversion in foreign travel, much harm is done by an excess of sight-seeing, especially indoors in galleries and chilly churches.

In no malady more than neurasthenia is the advice of Dr. Julius Braun pertinent, that "you must consider not only the individual sickness, but the sick individual."[2]

Kidney Diseases.

Prevalence of kidney diseases. This subject has been studied very carefully by several writers, especially by Hirsch, Davidson, Hjaltelin, Martin, Chambers, Morehead, and Dickinson. Dickinson, availing himself of the British Government reports, both civil and military, and of the researches of other physicians, has given, in his valuable work upon *Albuminuria*,[3] a digest of the literature upon the prevalence of kidney diseases. He shows very clearly, and in this he has the support of other authorities, that renal diseases are most prevalent in temperate climates, least so in cold climates, and comparatively rare in warm climates, if lardaceous cases be eliminated.

There are several factors other than the climatic which tend to

---

[1] Influence of Colorado upon Nervous Diseases. S. E. Solly. Transactions of the Colorado State Medical Association, 1870.

[2] See chapter on Personal Equation.

A Treatise on Albuminuria. By W. H. Dickinson.

cause this greater prevalence in temperate climates, such, for instance, as the greater use of alcohol, the fact that the inhabitants lead more sedentary and more anxious business-lives, and that a larger proportion of them are syphilitic and gouty; but, nevertheless, a thorough inquiry into the facts clearly shows that the influence of climate is marked. As an explanation of this, Dickinson states the axiom that the liability of an organ to disease, particularly of an inflammatory character, bears a proportion to its functional activity. The kidneys have to do most work in a temperate climate, the skin, liver, and bowels in hot countries, and the lungs in a cold climate.

Statistics of eight years for the British army show that the most prevalent diseases are tubercular and kidney troubles in temperate climates; in the tropics, malarial affections and diseases of the bowels and liver; and in the arctic regions, catarrhal disorders of the respiratory tract.

Dickinson further demonstrates, by a table[1] and by other evidence, that where the temperature of two climates is the same, the least renal disease will be found in the one which has the lowest humidity.

Influence of the Several Climatic Factors. Weber writes on this subject as follows: "The separation of water by the lungs and skin is diminished in great dampness, and more work devolves upon the kidneys, while their activity is less called upon when the air is dry and warm, and we must always pay great attention to this in affections of the kidneys." Elsewhere he says: "As cold air contains less vapor than warm, therefore there is more loss of water by the lungs when the air is cold than warm." [2]

In warm air there is more secretion from the skin and less from the lungs, and in cold air the reverse holds true; and the drier the air is, whether cold or warm, the more secretion there is from one or the other of these organs. As cold air is always proportionately drier than warm, the secretion from the lungs in a cold air would be relatively greater than that from the skin in warm air; but how the greatest possible amount of water excreted through the lungs compares with the greatest possible amount excreted through the skin has not been determined.

Again, as the action of cold in contracting the cutaneous vessels increases the action of the kidneys and the amount of urine passed,[3] it may be assumed, theoretically at least, that the kidneys have

[1] Albuminuria, p. 281. [2] Ziemssen's Climatotherapy.
[3] Landois and Sterling: Physiology, p. 426.

somewhat more relief in their work in a warm, dry, than in a cold, dry air.

Climatic Treatment. From the clinical knowledge at command, it would seem that while the difference in value between the two varieties of climates is not great, yet if a warm, dry climate is otherwise suitable, it should be preferred, as a general rule, for the relief of renal disease. Cases may occur, however, in which the bracing effect of dry cold is so desirable for the cure of an accompanying anæmia that, in weighing the advantages, a dry, cold climate is preferable.

In both warm and cold climates the question of wind is of the utmost importance, it being the one meteorological element most harmful to a kidney-case.

While any sudden change of temperature is attended with risk, those expected and customary differences between sunlight and shadow, and day and night, which are always present in dry climates, hot or cold, according to their degree of humidity, are of less consequence, as they can be prepared for; indeed, they are often beneficial through their general tonic effect. Changes of temperature in warm, damp climates, however, even when much less extreme, are more dangerous, and, being more subtle, they are not so easily guarded against; and such changes in cold, damp climates are the most dangerous of all.

With respect to the influence of elevated climates upon kidney diseases, there has been very little recorded. Dickinson does not refer to them at all; and most physicians, when mentioning altitudes in this connection, condemn them utterly, as did Dr. J. C. Wilson in his otherwise excellent paper upon " Climate and Bright's Disease."

Analyzing briefly the physical elements of elevated climates likely to influence the development and progress of kidney diseases, we begin with diminished atmospheric pressure. While it cannot be positively asserted that this element decreases the flow of urine, it may reasonably be inferred, for we know that there is a general tendency of the blood to flow to the skin and lungs and thus lessen the blood-pressure upon the kidneys; and that lowering of blood-pressure reduces the urinary secretion. Further, it is stated by Jaminet, Foley, Lange, Pol, Pravazy, and others that increased barometric pressure has the opposite effect.[1] Also the general

---

[1] A. H. Smith: Compressed Air, p. 40.

equalizing and lowering of the blood-pressure caused by diminished barometric pressure, in addition to the effects before mentioned, warrants the belief that diminished barometric pressure directly lessens the work of the kidneys.

Again, the improved quality of the blood and the increased tissue-change brought about by the same climatic factor should be favorable to combating the dyscrasia accompanying renal disease. The increase of sunlight should also be of benefit to the general ill-health. The dryness of the air, which is usually a marked feature of such climates, should, as shown before, be of especial value, and might to a great extent obviate the objections which could otherwise be urged against high climates if cold, or their depressing effects when warm. It must be admitted that wind is generally a feature of elevated climates, except at special seasons or in certain favored resorts (such as Davos) situated in valleys. As most valleys, however, are theoretically and practically unsuited to renal cases, it follows that the benefits ascribed to altitudes are usually offset by the presence of wind, even if it be dry and therefore as unobjectionable as possible. It may be inferred that, if an elevated climate is determined upon, one should be selected which is in all respects medium—that is, neither very cold nor windy. The altitude should not be very great, the resort should be situated on open, sloping ground, where the soil is dry, and the storm-shelters must not lie so near as to cut off the sunlight nor to surround the locality on all sides. These theoretical conclusions agree with my clinical experience in Colorado and other elevated climates. Colorado Springs is often too cold and windy in the winter to be free from danger to these cases, and the ideal elevated climate for them is to be found further south and somewhat lower down ; but my experience, even in Colorado Springs, has convinced me of the benefit of an elevated climate in certain renal cases. Some years ago I reported a few cases which had resided both in Colorado Springs and at sea-level, and the results were by no means adverse to the altitude.[1]

Later experience has confirmed the conclusions I then formed, which were as follows :

First. *Acute nephritis*, like all acute inflammations, is not infrequent in Colorado, especially in the mountains, where the inhabitants, not being natives, are careless of climatic extremes. When

---

[1] Bright's Disease of the Kidneys as Influenced by the Climate of Colorado. S. E. Solly. Transactions Colorado State Medical Society, 1881.

nephritis occurs it is, moreover, like all inflammations at an altitude, more than usually acute. Acute nephritis is not especially induced by the climate, but once established, it is, at the beginning, apparently aggravated by the altitude.

Second. The direct tendency of altitude in *chronic nephritis,* as in most chronic diseases, is toward its cure. This beneficial influence is doubtless mainly exerted through the increased action of the lungs, and, to a smaller extent, of the skin, which lessens the work of the kidneys and diminishes the renal congestion by the general stimulating and equalizing of the circulation.

The late Sir Andrew Clark stated to me that he found that kidney affections frequently developed among consumptives residing in the high Alps.

Lardaceous Kidney. I have been unable to ascertain the average percentage of cases of lardaceous kidney among consumptives generally, therefore I cannot say what is the comparative influence of climate upon the disease. In 150 cases of phthisis which I treated in Colorado and traced up, there were 4 with this complication—2 who were affected before coming to Colorado and 2 in whom the trouble developed after their arrival. In all of the patients the urine was frequently tested. It is quite common in Colorado to find in fatal cases, toward the end, more or less œdema, but the test-tube and microscope generally fail to show any renal affection; doubtless the œdema is usually caused by the weakening heart. Where no test is made this œdema is often attributed to the wrong cause, and I believe this to be the explanation of the somewhat prevalent opinion that lardaceous kidney is frequent in Colorado. In the scrofulous cases, and cases with bone tuberculosis, lardaceous disease is perhaps at least as frequent here as elsewhere.

With respect to *nephritis, acute or chronic,* among the 150 cases referred to, there were three cases; two began in Colorado—one acutely, the other slowly. The former, who showed unmistakable evidence of a tuberculous kidney, is still in Colorado and much improved, though his urine contains a little albumin and a few casts. The latter case died of his pulmonary disease and before his death the renal symptoms were more marked.

In the present state of our knowledge it is undoubtedly safest to try first a warm climate which may have a high humidity, but must have a moderate rainfall. The age of the patient as well as the stage of the disease should greatly influence the choice between a warm and a cool climate; and, again, it is essential to know whether one may anticipate a cure or only an alleviation of the malady before deciding the matter.

With regard to the effect of the Colorado climate upon *healthy kidneys*, persons during the first few days after their arrival have frequently called my attention to the "thickness" of the urine, which, upon examination, proved to contain an excess of solids, especially the urates. These persons have usually thought also that they perhaps passed less water, but certainly more frequently. This phase usually disappears in a few days, and the urine returns to a normal condition. This would suggest that at first the excretion of water by the lungs and skin was much increased at the expense of the kidneys, and the condition of the mucous membranes and the parched skin bear out this opinion.

My friend, Dr. S. A. Fisk, of Denver, informed me that from quantitative and qualitative analyses made by him upon specimens of urine taken from healthy persons who were permanent residents he found the urine usually normal, except that the specific gravity was slightly higher. This would indicate that in healthy persons the balance between the excreting organs is in time regained.

"Certain life-insurance reports gave to kidney diseases in all deaths occurring on the Rocky Mountain plateau and Pacific slope a percentage of but 1.6 per cent. against nearly 9 per cent. of all deaths occurring in New England or the same in New York." Thus writes Dr. Francis Atkins, of Las Vegas, New Mexico. He believes that early albuminurics do well on the Rocky Mountain plateau. He finds that the specific gravity of urine ranges from 1024 to 1030 in the life-insurance examinations he made in Las Vegas.[1] Dr. Fisk[2] says that he was told by the vice-president of the Santa Fé Road that at the stock-yards in Chicago it was rare to find diseased livers or kidneys in cattle coming from the Rocky Mountain region.

Scarlatinal Nephritis. "I have frequently examined the urine during convalescence from scarlatina, but only remember three cases of Bright's disease as a sequel of the fever. In one case there had been exposure and neglect, and there was much dropsy. All the cases, however, made a good recovery while remaining in this climate. In practice in London previously I found albuminuria

[1] Transactions of the American Climatological Association, vol. vii. pp. 250, 251.
[2] Ibid., p. 253.

a common sequel of scarlatina. In our county society the opinion of my colleagues was that scarlatinal nephritis is decidedly rare in this district."[1]

Hepatic Complications. Sir Henry Thompson, some years ago, drew attention to the importance of taking the condition of the hepatic functions into practical consideration in the proper treatment of chronic renal diseases. This, I believe, is an especially important factor in the choice of climate.

Diabetes. My views with regard to this disease remain what they were when I expressed them in Hare's *System of Therapeutics.*[2] "Diabetes, both mellitus and insipidus, would appear not to be diseases *per se*, but symptoms resulting from disturbance of the functions, or change in the structure of some portion of the nervous system or of one or more of the digestive organs, notably the liver. Change of climate, though not usually of the first importance in the treatment of this disease, is often of use in aiding other therapeutic measures."

The choice must then be determined by the primary causes of the affection; thus when, in a full-blooded person, the disease appears to have resulted from high living, a low, unstimulating climate at a spa, such as Carlsbad, where the waters are rich in the salts provocative of increased metabolism, is indicated.

When, again, digestive or excretory deficiencies are at the root of the matter, and the patient is of the anæmic type, a mountain climate is more beneficial.

Where the nervous system is most involved, and when an actual lesion is suspected, a warm seashore climate is generally indicated; while in cases in which the nervous symptoms are functional, and more or less anæmia is present, an altitude will probably do more to retard or cure the disease.

In the statistics compiled by Purdy[3] the rate of mortality from diabetes was very much higher in cold, moist, than in mild, dry climates, and it also increased with the altitude.

I have watched with interest the progress of several cases of diabetes during their residence in Colorado, but have been unable so far to come to any positive conclusions as to the influence of altitude upon the disease.

I have seen some cases of diabetes which had not advanced far,

[1] Author's article upon Climate in Hare's System of Therapeutics.   [2] See page 448.
[3] Purdy on Diabetes, Philadelphia, 1890.

and with whose hepatic and nervous systems the climate seemed to agree, do well in Colorado ; but a diabetic should be carefully examined before being sent to a high altitude, as the lack of equability in the climate increases the risk of intercurrent attacks, and any nervous lesion may be aggravated.

Conclusions. While my experience leads me to prefer warm, low, inland climates, such as those of Egypt and Arizona, for the majority of cases of chronic renal disease, yet I have found that in some a warm seashore, and in others dry, elevated regions, both cool and warm, are better suited, and I believe this is often because the climate insures a more perfect working of the liver. The instructions of the physician at the health-resort must be very definite and should be strictly followed, for in no disease is personal prudence of greater importance.

Diseases of the Liver.

Organic Disease. In serious cases any change must be made with caution and extreme climates should be avoided, except when the patient is in a particularly bad climate, such as that of a tropical country or malarial district, in which case the needed change may be to a climate which offers the sharpest possible contrast.

In functional disease, on the other hand, a radical and extreme change is usually of the greatest benefit, as, for instance, one from mountain to seashore, from land to ocean, from a cold climate to a warm one. As pertinent to the subject I will make use of an address which I read before the Colorado State Medical Society :

In estimating and recording the effects of Colorado upon the various organs it is necessary to take into consideration the race and temperament of the individual, and it is probable that these will generally explain apparent contradictions. Take the liver, for example. It is frequently asserted, on the one hand, that the climate of Colorado causes *biliousness*, and, on the other, that it relieves it. I have formed the impression that what is called a bilious person, one who at home frequently exhibits signs of functional hepatic irregularities, is, while resident in this climate, less subject to them, and his skin is apt to lose its yellowish tinge ; but persons of a sanguine or nervous *temperament*, in whom such attacks had previously been rare, are, in Colorado, not uncommonly subject to such disorders. The explanation for this assumed fact appears to me to be briefly as follows : in the individual of bilious temperament the cause of the derangement was mainly from lack of sufficient stimulus to the liver through its circulatory or nervous supply, or both, and the acknowledged activity of these two systems, induced by the climate, was the cause of the better working of the organ. In the individuals of sanguine or nervous temperament the circulatory and nervous supply of their livers was already sufficiently active, and the climatic stimulus would be, for them, excessive and induce a condition in which slight exciting causes would readily give rise to congestion of the liver and subsequent derangement.[1]

In a dry climate, such as Colorado, the rapid and constant loss of moisture from the tissues, while beneficial and stimulating to the action of the liver in one class of individuals, would probably be irritating and hampering to it in another class, unless especial care were taken to compensate for this by drinking water freely. Again, while a dry, warm air induces free perspiration, yet the evaporation of the water of the sweat, as soon as it reaches the surface of the skin, is so rapid that the sweat is seldom noticeable ; but it is probable that, unless the skin is kept in an active condition by more than usually frequent washing, rubbing, and general exercise, the salts of the perspiration clog the pores, and elimination through the skin becomes lessened instead of increased. Consequently waste-products which should be excreted in this way may be returned by the circulation to the liver and so interfere with its action.

Dr. Weber and other writers on climatology have spoken of the favorable influence of elevated climates upon sluggish circulation in the abdomen generally, and it has been my observation that passive congestions in anæmic persons were usually benefited at an altitude which failed to agree with the full-blooded or those in whom there were active congestions or marked nervous irritability.

The direct influence of climate is perhaps more marked upon the functions of the liver than on those of any other organ. Speaking broadly, persons may be divided into those who become bilious when they go to the seashore and those upon whom the mountain-air has a similar effect. It will be found, as a rule, that with fair and full-blooded persons the liver acts best at the seashore, while with the dark-skinned and anæmic the hepatic functions are most regular in mountain-air.

---

[1] Presidential address. S. E. Solly, M.D. Transactions Colorado State Medical Society, 1887.

Attacks of Gallstone. To remove the conditions upon which the formation of gallstones is dependent certain drugs are usually required, notably sulphate or phosphate of sodium; and clinical experience indicates that these are most effective when taken in mineral water. A suitable spa should therefore be sought; but to get the best effects one should be selected which possesses the climatic conditions most suited to the patient upon the general principles already laid down. The success of treatment at Carlsbad largely depends upon a proper consideration of these general principles, and its failure is often directly traceable to an ignorance of, or lack of regard for, them.

As balneotherapy does not lie within the range of this treatise, I cannot pursue this subject much further; but I will add that I have been gratified by the beneficial results in the treatment of gallstones from the use of the Shoshone Spring at Manitou, Colorado, and this in cases where the patients had previously been treated at Carlsbad without much benefit. The spring at Manitou is much poorer in the requisite salts than the one at Carlsbad, and often requires an artificial addition; therefore I cannot but believe that a mild mineral water in a suitable climate is sometimes more beneficial than a stronger spring in a place of which the climate is not so appropriate for the case.

The question is, Does the nervous or the circulatory system call for a sedative or a stimulating quality to provoke or regulate the necessary metabolism in order that health may be restored?

Affections of the Stomach.

Dyspepsia is usually amenable to treatment at home. When chiefly due to the condition of the mucous membrane the regulation of the diet and habits of the patient is of the most importance, aided as may be required by drugs or other therapeutic measures, or by change of climate for the purpose of taking the waters at some special spa.

Catarrh, dilatation, and ulceration of the stomach do not usually of themselves call for climatic change; but when a change is made it is necessary to consider particularly the food and cooking to be found in the resort recommended. The change needed may be of a purely negative character, perhaps to induce the patient to drop a too engrossing occupation, habits of irregular and rapid eating or of over-indulgence, and to take him away from the temptations of luxury to a plain table and simple living.

Prolonged indoor occupations particularly foster dyspepsia, and therefore the place where much pleasant outdoor life and exercise is to be obtained is most desirable, especially one possessing a climate suited to the action of the patient's liver.

If anæmia be present, the climate must be selected with a view to curing it.   This disorder demands a bracing air, and the seashore, the moorland, or the mountain should be chosen, in accordance with the strength and idiosyncrasy of the patient.

While most gastric cases require good and varied diet, plain food and rough, open-air life are best even for many delicate dyspeptics.

An iron spring which contains also a small quantity of alkaline salts and is well charged with carbonic acid is often of the greatest service to the anæmic dyspeptic.   It must not be situated in a hot, low valley, but on the moor or mountain.   San Catarina, at San Moritz, and the Iron Ute, at Manitou, are notable examples of the best quality of iron springs in a bracing climate.

If, on the contrary, the dyspeptic tends more to plethora and obesity, low, inland climates possessing springs strong in alkaline and purgative salts without iron, which induce active metabolism, are the most suitable; the resort should also possess resources for the carrying out of intelligent hydrotherapy.   Saratoga and Glenwood on this continent and Carlsbad in Europe are excellent examples of such spas.

When the dyspepsia is largely nervous in character the climate must be selected with a view to its probable effects upon the nervous system, and all the attendant circumstances must be carefully studied.

Diseases of the Bowels.

Chronic diarrhœa is sometimes cured by a radical change of climate without other therapeutic remedies.   The proper change would be from a damp to a dry, from a warm to a cool, from a high to a low, climate or *vice versa*.   Heat is generally more provocative of diarrhœa than cold, and dampness than dryness; but these are not universal rules.   In warm, humid climates or seasons and sometimes also in a dry atmosphere, sudden changes of temperature will cause diarrhœa or even *dysentery*.   Such variations are less frequent in cold climates, especially if the air be dry; but if it be damp, they occur oftener, and are more to be dreaded.

Chronic constipation is often benefited by change of climate, usually because the patient lives more out of doors, takes more exercise, and is very apt to drop temporarily any unhygienic habits. In all affections of the bowels the influence of climatic change upon the hepatic functions must be especially considered.

Tuberculous disease of the bowels or of any portion of the abdomen is benefited or increased by climatic change according to the same principles as those set out in the chapter upon Phthisis.

Gout and Rheumatism.

Both of these diseases are frequently benefited by climatic change. Gout is more especially influenced by diet. Next to this in importance come certain drugs and mineral waters, and then exercise and bathing.

The geographical distribution of gout is chiefly due to the diet, drinking, and the social habits of the residents in countries where it is most prevalent, as in England, for example.

Climate, however, has some influence in its causation and cure. The appropriate climate for aiding the treatment of a particular case of gout depends mainly upon its effect on the general health of the patient; but in deciding the question certain points dwelt upon by Dr. Haig should be considered.[1]

He says: " I have pointed out that the excretion of uric acid is greater in summer than in winter, and Sydenham speaks of gout as a winter disease. Now, there is no doubt that a laboring-man has, so to speak, summer all the year round. His exertions keep his skin constantly active. He gets rid of a large amount of acid in this way, hence the acidity of his urine rules low, and the alkalinity of his blood is well maintained; he therefore excretes uric acid freely and retains but little in his body; and so, as observed by Cullen, he but rarely suffers from gout, and this is so almost without regard to his diet, for he excretes all the uric acid he introduces as well as all that he forms. As to the acidity of the sweat, see also Heuss, *Monatsch. für prakt. Dermatol.*, Band xiv. Nos. 9, 10, and 12; and *Lancet*, 1892, vol. ii. page 118. A sedentary man has not only higher acidity and retention of uric acid, but his circu-

[1] Uric Acid, third edition, London, p. 179.

lation, especially in peripheral parts, like the hands and feet, is less well maintained, and as a result the alkalinity of the fibrous tissues in such parts is also less well maintained.

" Though, as I have said, I used to have headaches when my life was not sedentary, I have no doubt that I could indulge in meat and beer with comparative impunity if I lived the life of a laborer.

"After what I have said about the effects of summer, it is not surprising to find Sir A. Garrod saying (page 235), 'Gout is undoubtedly much less prevalent in hot than in temperate climates,' though no doubt, as he remarks further on, food and habits have also something to do with its absence.

" The reverse effects of cold need hardly be gone into; but Sir A. Garrod says (page 247), ' When cold acts as an exciting cause the effect is due, at least in part, to its arresting the secretion of the skin and checking the escape of acid from the surface,' and my experimental experience is in complete accord with this statement."

Rheumatism. Articular rheumatism is more common in a cold climate than in a warm one, and in a damp than in a dry atmosphere; but it occasionally occurs in a dry climate, especially if it be also hot, like that of Lower Arizona.

Muscular and nervous rheumatism are perhaps nearly as frequent in dry as in damp climates. Altitude does not prevent, and sometimes even aggravates these affections.

I believe, as has so often been said in regard to other diseases, that the climatic effect upon the functions of the liver has especially to be considered. While the war still rages round the question of the identity or kinship of gout and rheumatism and the part played by uric acid in their causation, it is difficult to speak definitely as to the *rôle* taken by the liver in these diseases; but, as has been said in discussing liver disease, I believe that both humidity and dryness have a great and direct influence upon the secretion and nervous action of this organ, and consequently upon the excretion of urea or its accumulation in the tissues. Again, climate undoubtedly has an indirect influence upon the action of the liver through its effect upon the functional activity of the skin Variability in a damp climate, or the exaggerations of meteorological variations resulting from excessive dryness, often cause a chilling of the surface of the body; thus the pores are closed, the excretory functions of the skin temporarily interfered with, and the work performed by the liver increased.

It may be stated as suggestive that it is a common saying in Colorado that persons who have rheumatism when they go there lose it, and those coming without it acquire it. This, although an exaggeration, has a basis of truth.

In rheumatism it is particularly important, before sending a patient to a resort, to inform one's self as to the humidity tables, the range of temperature, the nature of the soil, and the presence or absence of fogs.

SECTION III.

INTRODUCTION.

THE physician who has studied the principles outlined in the first section of this book can, on consulting the facts contained in this portion, form for himself a very fair opinion of the climate of the resorts mentioned, and be independent of the statements of ignorant or prejudiced persons. The main facts necessary to the forming of an opinion are the elevation, latitude, distance from the ocean, proximity of large bodies of water or mountains, aspect, configuration (whether flat, irregular, or sloping), and the nature of the soil and vegetation. Knowing these, he can estimate approximately its general meteorology, even if he is unable to obtain reliable records; but if these are at hand, let him remember that humidity is more important by far than any other single factor, and their importance as climatic conditions depends upon the place held by it. He must not forget also that *weather* and *climate*, though allied and interdependent, are not the same thing, and that a climate may, in the long run, be beneficial to a patient to whom much of the weather is unpleasant and even adverse if precautions be not taken.

The mode of heating and ventilating houses at resorts is of especial importance, because, unless a proper relation is maintained between the house-climate and that of the open air, catarrhs and other evils are more prevalent than they otherwise would be. The ideal is to keep the surface of the body warm and the air which is breathed cool. In Italy and Mexico the transition from the warm, sunny, open air to the gloomy, vault-like chilliness of the native houses is dangerous, unless one follows the manner of the Mexican, who reverses the usual custom and puts on his overcoat when he comes indoors. Again, in American winter-resorts the steam-baked, overheated air of most hotels unfits the guests for benefiting by the cool, bracing atmosphere outside or combating the trials of that exceptionally stormy, chilly weather which is to be found at times at all health-resorts, not even excepting those of the sunny South.

Physicians very properly inquire into the range of temperature, but they should remember that its good or evil effects can only be estimated fairly when the humidity is studied with it.

Climates may be divided therapeutically into negative and positive. They are negative where it is simply a question with the traveller of avoiding perhaps the damp cold of home and seeking the warm South, hoping to find the poet's land, " where it is always afternoon"; thence he returns, not changed or toned, but satisfied to have escaped, for another season at least, the climatic dangers and disagreeables of his customary abiding-place. Positive climates, on the other hand, are those which produce an alterative or tonic effect. The two extremes of climate, sea and mountain, both do this; but sea-air is, for the most part, sedative in character, while mountain-air is stimulating; and the wise physician chooses on this principle or takes a middle course, and, selecting a medium climate, remembers, " In medio tutissima ibis."

In deciding what order of sequence appeared to be best and most convenient for reference to the climates described in this section, the author selected the geographical, starting from the northwest and going east and then south and returning by the west and north. This plan has been followed so far as was practicable in every quarter of the globe, and it has only been deviated from where the prescribed order would have separated certain resorts from the climatic groups to which they properly belong.

The coast- and island-resorts are generally described before the inland places.

RELIEF MAP OF NORTH AMERICA.

CHAPTER X.

NORTH AMERICA.

It is difficult for those who have not travelled throughout this continent to realize its size and the variations of climate consequent upon its topography and its extent from arctic to tropical regions. As compared with Europe, the stranger would at first be struck with the dryness of the climate and with its great range. Cities in the same latitude as those of Europe are colder by many degrees, a striking example being that furnished by the contrast between Naples and New York. This observation is, however, applicable only to the eastern coast of the continent, the climate of the western coast being much milder and more equable. This is, of course, due to the fact that while the Kurosiwo, or warm Japan current, follows almost the entire western coast, the Gulf-stream, whose warm waters are deflected from the equatorial current near the Caribbean Sea, turns from the eastern coast, after following it for a comparatively short distance, and flows across the Atlantic to warm the coasts of western Europe.

Charles Maclaren, in an interesting article on the American continent, written for the *Encyclopædia Britannica*, gives a very clear explanation of the way in which prevailing winds and mountain-ranges influence a climate, and especially this climate. Between the parallels of 30° north and 30° south latitude are found the so-called "trade-winds," which blow from the east. Beyond these limits, unless local causes suffice to alter the direction, the prevailing winds would be westerly, which is the general direction of winds in the United States. The barriers to the free sweep of wind are: first, the high range near the western coast, beginning in Alaska, known throughout the continent as the Cascade and the Sierra Nevada Mountains, and terminating in the chain which extends through the peninsula of Lower California; secondly, the Rocky Mountain system, the backbone of the continent, which is continued through Mexico under the name of the Sierra Madre; thirdly, the Appalachian system, arising south of the St. Lawrence and terminating in the northern part of Georgia and Alabama.

The general conclusion from this would be that the west coast of Mexico and the eastern portions of the more northerly parts of the continent would be least humid, and, local differences aside, this is, in fact, very much the case. It is true that the western coast of Mexico, as far as 23° north latitude, is fertile and well-watered, because there is no intervening elevation high enough to form an effectual barrier to the moisture-bearing east winds, and, also, it is thought that a branch of the trade-winds is deflected from that which crosses the flat Nicaraguan territory and follows the west coast for some distance; but from 23° to 33° or 34° north latitude the western coast of Mexico and even the narrow peninsula of Lower California are dry, sandy, and practically without vegetation.

A marked change is exhibited on the west coast of the United States. Here the prevailing westerly winds have free sweep over the land until stopped by the Sierra Nevada and the Cascade ranges, and bring not only moisture from the Pacific Ocean, but the tempering warmth imparted by the Kurosiwo. Eastward of these ranges the land is very dry and sandy, and would, according to Maclaren's hypothesis, continue so throughout the Mississippi basin were it not that as the land is level from the Gulf of Mexico northward, part of the trade-wind from that gulf, unable to cross the Mexican Cordilleras, is thereby deflected to the right and carries its moisture to the region drained by the Mississippi and Ohio.

The eastern coast is mainly dependent for its moisture on the east and northeast winds which blow off the Atlantic Ocean and bring to this region most of its storms.

The climatic divisions of the country are, then, three: first, that including the region from the Atlantic Ocean to the neighborhood of the Rocky Mountain system, possessing, especially throughout the Appalachian Mountains, very marked local variations; secondly, the high plateaus and mountainous districts of the Rocky Mountain region, distinguished by a dry, cold climate, lacking, for the most part, in any element of harshness, and showing a very large number of clear days; thirdly, the district extending from the Sierra Nevada Mountains to the shores of the Pacific, and possessing a moderate, equable, humid climate with mild winters. These divisions will be found amply typified in the selected resorts.

Canada.

The superficial extent of Canada is nearly equal to the whole of Europe. It runs from about 45° north latitude to the Arctic Ocean, and has the two great oceans of the world—the Atlantic and the Pacific—for its eastern and western boundaries.

The climate of British Columbia partakes of the general character of that of the Pacific coast, and is consequently much milder than is the climate of the Atlantic provinces.

The climate of the Atlantic provinces resembles that of Sweden and Norway. The winters are colder and the summers hotter than is the case in corresponding latitudes in Europe. January and February are the coldest months in the year. Snow finally disappears in Quebec about the middle of April, and in Ontario almost a month earlier. While the summers are hot the nights are usually cool. The dangerous times for invalids in Canada are during the melting of the snows and in the late autumn, when snow and sleet are apt to come with alternating thaws. When the winter snows finally lie upon the ground the air, although cold, is dry and bracing.

In northeastern Canada, especially, the change from winter to summer is very sudden.

Extensive forests cover the greater part of Canada. The peninsula formed by Lakes Erie, Huron, and Ontario has a fertile soil which yields large crops, especially of wheat.

It is extremely difficult to speak with definiteness and justice of the claims of Canada as to its health-resorts, because, even in regard to general climatic characteristics, the area to be covered is so great and accurate information is so scanty and difficult to obtain. The local variations, too, are many, and would require close investigation before their exact merits and demerits from this point of view could be decided upon.

Dr. P. H. Bryce, Secretary of the Ontario Board of Health, to whom the author is indebted for much of his statistical information on this subject, divides the country into three districts: first, that which includes the provinces of Ontario and Quebec, and is characterized by the climatic conditions peculiar to low-level districts; secondly, that including the prairie-land extending from Calgary to the foot of the Rockies; and, thirdly, the Rock Mountain region itself, possessing the characteristics of " high-level " climates.

As to the first region, a portion of it, " the central plateau of

Ontario," has a greater exposure to winds, its forests having been cleared away. Its soil is a clayey or gravelly loam overlying limestone rock. The remaining portion, the Muskoka district, has wooded areas, and its surface is divided by rocky ridges with valleys lying between. The forest-growths include balsam, hemlock, and spruce trees.

The second or prairie region, having an elevation of from 400 to 5000 feet, is characterized, as are all such regions, by absence of humidity, great amount of sunshine, rapid radiation of heat during the night, and high winds. Variability from day to day, more than between seasons, is the main feature of such climates.

The third or mountain region, marked by valleys through which flow streams of greater or less volume, presents, necessarily, the greatest variety of climatic conditions. One valley differs from another in aspect, in elevation, and in exposure to the sun and wind. The sides of the foot-hills are often almost barren, and yet, as we rise higher, we find that the climate is damp enough to support a considerable forest-growth far up the slopes of those loftier peaks whose summits are always snow-covered.

Dr. Bryce, as an instance of local variability, mentions two climates between Vancouver and Kamloops, one of which has an annual rainfall of 35 inches, while the other records but 11 inches.

Kamloops, situated in British Columbia, in a broad, open valley and on the line of the Canadian Pacific Railroad, has an altitude of 1100 feet. The neighborhood offers many inducements to sportsmen and many incentives to an active outdoor life, as the fishing and shooting are good and the natural surroundings attractive. The climate is dry, the annual rainfall being 11.05 inches, and the number of rainy days per year only 75. The mean annual temperature is 46.3° F., nearly 10° higher than that of Calgary, and the average annual range is less, being but 22.8°. The mean daily range is as follows:

August.	September.	October.	November.	December.
29.2°	27.9°	17.9°	11.0°	11.9°

The soil is light and gravelly and dries quickly, and the water-supply is pure and abundant. It is stated that the hotels are good and the charges reasonable.

Banff, lying on the line of the Canadian Pacific Railroad and surrounded by the most beautiful mountain-scenery, possesses many natural advantages. There is also a handsome hotel with spacious accommodations and a good *cuisine*. Many pleasant excursions may be taken through the neighborhood. Banff, being situated in the Canadian National Park, has a government museum which is well worth visiting; although shooting is forbidden within the limitations of the reservation itself, the vicinity affords many attractions to the sportsman, and there is good fishing close at hand. From this point many of the wonderful parks and glaciers of the Selkirks can be visited. Banff also possesses hot sulphur springs, which are recommended in cases of rheumatism, etc.

Calgary, situated in the province of Alberta, at an elevation of 3500 feet, has a dry, sunshiny climate. The annual rainfall is but 11.54 inches, and the number of rainy days during the year is 90. The mean annual temperature is 36.9° F., and the average annual range, from the records for four years, is 26.1°. In a pamphlet upon Calgary, Dr. George Macdonald gives the following monthly means:

January.	February.	March.	April.	May.	June.
6.3°	11.8°	14.9°	37.5°	48.8°	55°

July.	August.	September.	October.	November.	December.
59.6°	58.1°	50.9°	40.5°	24.7°	14.7°

During the winter months snow frequently lies long upon the ground, affording good sleighing, and, as the thermometer may remain below zero for several weeks together, sometimes dropping to —25° F., there is excellent skating. There are no available wind-records, but it is stated that the " chinook " sometimes blows here. A visitor to Calgary reports that there is a light daily breeze from the northwest. The mean daily range in temperature is:

August.	September.	October.	November.	December.
28.9°	30.4°	27.1°	23.7°	21.6°

There is no record of humidity, but all the evidence points to the conclusion that the climate is dry. There are very few cloudy days, particularly during the winter season. The climate being, therefore, sunshiny, dry, and bracing, is well suited to tuberculous patients who are sufficiently robust to exercise.

The city is lighted by electricity and is furnished with waterworks and a system of sewerage. There are comfortable, but simple, accommodations, and the expense of living is moderate.

St. Lawrence River Resorts. The pleasures of a trip down the St. Lawrence to Quebec and from there up the Saginaw are well known.   There are a number of charming resorts from Kingston to Brockville, on the shores of the St. Lawrence opposite the Thousand Islands, all along which are numerous fine hotels and villas.   This district, as a summer residence, with a cool, equable, and medium-moist climate, is hard to surpass, and one may live as luxuriously or as simply as the individual inclination prompts.

Gravenhurst, in the province of Ontario, is the distributing-centre for the resorts of the Muskoka Lake district.   It is situated some miles from Montreal, and has an elevation of 500 feet.   The snowfall here is heavy.   The annual rainfall is 36.77 inches, and the number of rainy days throughout the year is 143.   The mean annual temperature is 41.8° F., and the average annual range is 21.3°.

This country is much resorted to for summer camping, boating, and fishing, and enjoys a reputation similar to that of the Adirondack region for healthfulness and pleasantness.

Caledonia Springs, in the province of Ontario, have been resorted to for many years by sufferers from rheumatism, disorderd liver, kidney-troubles, etc.   It is expected that the Montreal and Ottawa Railroad will be completed to Caledonia within a year; it has not heretofore been reached directly by rail.   The accommodations are good, and many amusements are provided for visitors. There are four springs, a strong saline well and one not so powerful, a carburetted hydrogen gas spring, and one known as the white sulphur spring.   The waters are used for drinking and bathing.   The summer climate is said to be pleasantly cool, and great heat is rare.

St. Agathe, in the province of Quebec, is situated about seventy miles from Montreal, upon a branch of the Canadian Pacific Railroad, among the Laurentian Mountains.   Its altitude is about 1200 feet.   The hotels are only moderately good, but cottages can be rented, and here many of the residents of Montreal have their summer-homes.   The village is situated on a lake studded with islands; the adjacent country is covered with pine-forests and the scenery is beautiful.   It is much used as a health-resort, and is considered to rank, climatically, with the Adirondacks.   The country offers many attractions to sportsmen and supplies good fishing and shooting.

St. Leon Springs, located between Montreal and Quebec and lying about five miles from the St. Lawrence River, is situated in a charming district of country.   The hotel-accommodations are good and all the usual amusements of a summer-resort are provided for guests. The waters are used internally and externally, and are said to possess aperient and tonic properties and to be efficacious in disorders of the digestive tract and in renal troubles.

Dalhousie (New Brunswick), on the Baie des Chaleurs, at the mouth of the Restigouche River, has a mild summer climate and affords excellent bathing; there is a first-class hotel, the " Incharron." Here is the starting-point for those who are going, in search of sport, up the widely celebrated salmon rivers.

The Region of the Great Lakes.

The district lying on the frontier between Canada and the United States, with a climate cool, moist, and equable, abounds in pleasant, quiet, inexpensive resorts which afford all the sports and diversions that make a seashore-life so attractive. The climate, too, possesses some characteristics in common with that of the coast, chief among which are the increased ozone, the purity of the air, and its comparative freedom from dust-particles and bacteria. A trip on these inland seas, starting, for instance, from Duluth, crossing the lakes and following the St. Lawrence River down to Quebec, with an occasional stop-over at some of the pleasant resorts *en route*, is a most agreeable and healthful experience. The steamers are fine boats, and the *cuisine* is good.

Apropos of the voyage between Buffalo and Duluth, the following remarks in a recent number of *Outlook*[1] are worthy of notice :

" Ocean-voyages usually grow monotonous. Not so a trip on a fast steamer on the lakes, for there is a constant succession of novelties to arouse the attention. It may be the historic Mackinac, with its picturesque headlands and its decadent fort, last year finally abandoned by the government. It may be the St. Clair River, with its ' little Venice' of cottages and club-houses, the resort of Detroit's water-loving citizens, or a glimpse into Canada, with a hint of French and Indian in such an inn as the " Ashiganikaning," Léon Bellair, proprietor, or a view of the famous copper country on Lake Superior, with the smoke rising from the ' Calumet' and ' Hecla,' the richest copper-mines in the world; or, here and there, a whaleback, a lumberbarge, or a ' sand-sucker'.

" It may be, at night, the vast dim luminosity on the far horizon that indicates some large city like Cleveland or Duluth, soon to appear to the groups of watchers on deck.

"The tourist who makes the trip of the lakes is sure to come back a wiser, healthier, happier man, with new ideas as to ocean-travel in his own land, so to speak, and with a memory full of pleasant reminiscences to hearten him for his daily work."

The United States.

This portion of the North American Continent, inclusive of Alaska, extends from about 48° north latitude at the northwestern corner of Washington State, near Vancouver, to latitude 25° north at Cape Sable, the extreme southern portion of Florida. British Columbia forms the western half of the northern border, and the Great Lakes and the St. Lawrence River the eastern half. The eastern boundary is the Atlantic Ocean, the entire western coast is washed by the waters of the Pacific, and as Mexico forms only part of the southern boundary, the rest of the south coast bordering on the Gulf of Mexico, the country may be said to possess four water-fronts, three of which touch upon salt water and one upon fresh. All of these are resorted to by health-seekers at different seasons of the year.

The Pacific coast region has the most equable and temperate climate, both in summer and in winter, and this is particularly true below the thirty-fifth parallel, where the climate is pleasant during the entire year, and resembles that of the Riviera, although the temperature is higher in winter and lower in the summer. Above the thirty-fifth parallel the rainfall is much augmented and increases toward the north, and the climate is too damp and the winters too chilly for most invalids, though the summers are usually pleasant and temperate.

The Atlantic seaboard is used in summer as far south as Old Point Comfort, and in winter the resorts from Atlantic City to Palm Beach are much patronized. On this coast the range is greater and the climate less temperate than is the case on the Pacific shore.

The peninsula of Florida is used by invalids only during the cold season ; but, exclusive of Florida, the Gulf coast is frequented both in winter and in summer, the former being the season for Northerners, while the latter brings an influx of visitors from the Southern States. The air of this district is, however, much more humid than that of Southern California or the Riviera, and the rainfall is greater.

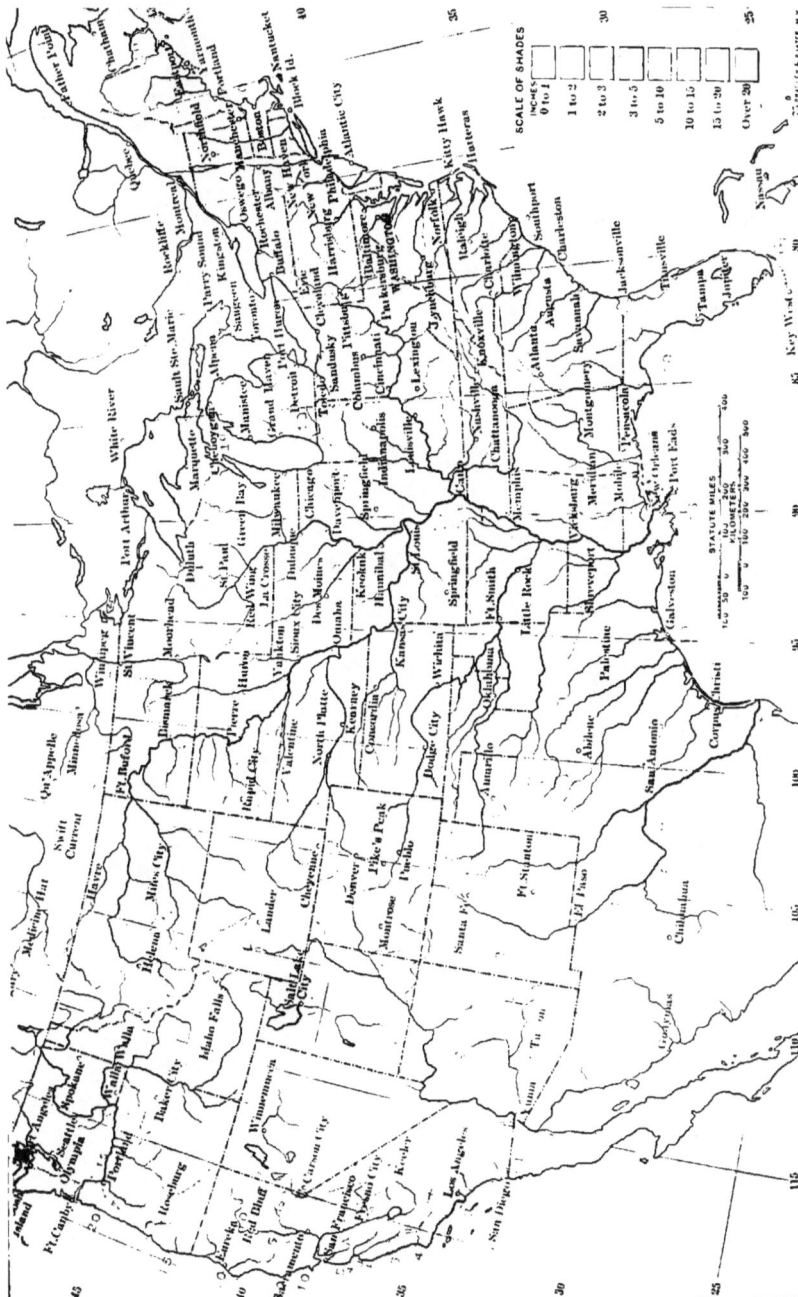

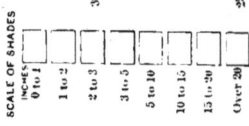

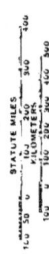

SCALE OF SHADES

INCHES
0 to 1
1 to 2
2 to 3
3 to 5
5 to 10
10 to 15
15 to 20
Over 20

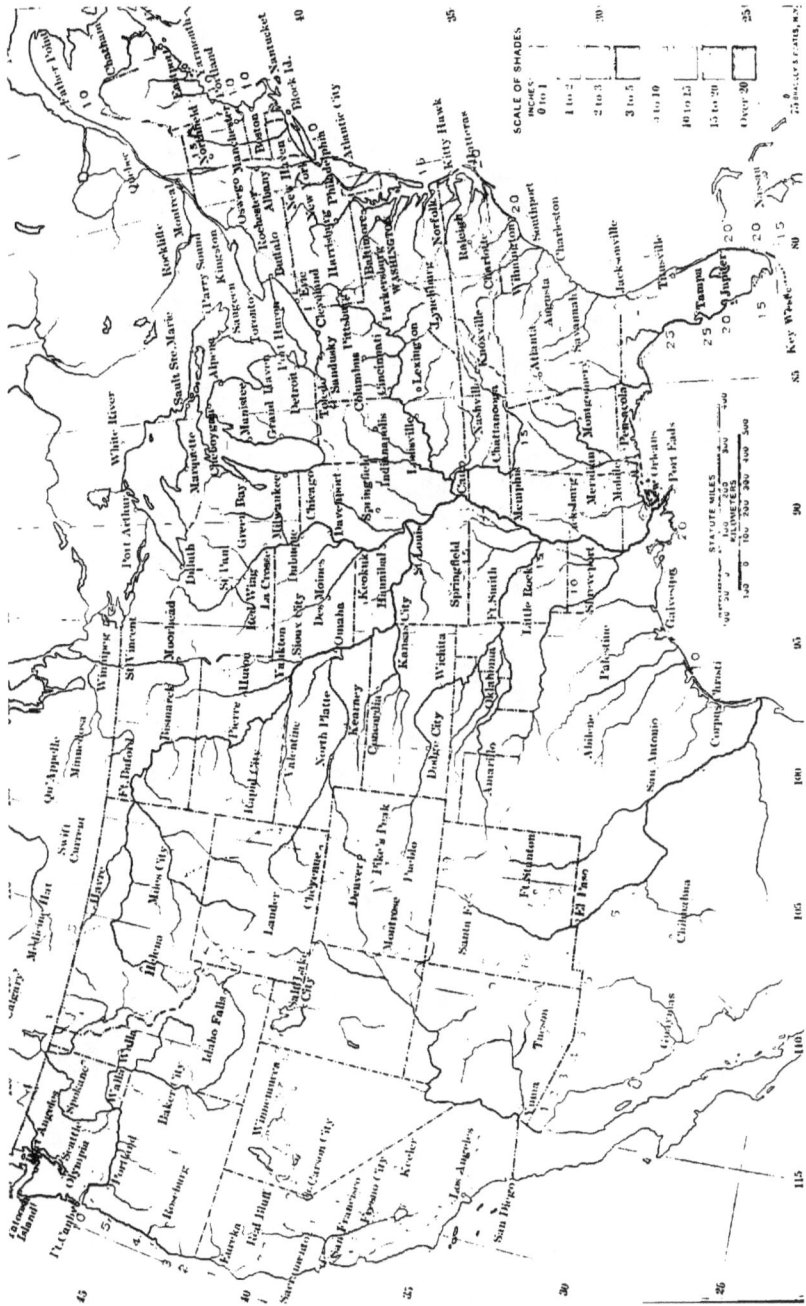

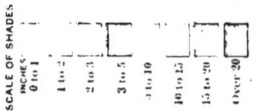

STATUTE MILES

KILOMETERS

120
115
110
105
95
90
85
80
Tatoosh Island
Port Angeles
Seattle
Olympia
Spokane
Calgary
Medicine Hat
Swift Current
Qu'Appelle
Minnedosa
Winnipeg
White River
Port Arthur
Havre
Ft. Buford
St. Vincent
Portland
Walla Walla
Helena
Miles City
Bismarck
Moorhead
Roseburg
Baker City
Duluth
Marquette
Sault Ste. Marie
Idaho Falls
Cheboygan
Alpena
St. Paul
Green Bay
Eureka
Red Bluff
Winnemucca
Lander
Rapid City
Pierre
Huron
Red Wing
La Crosse
Manistee
Grand Haven
Port Huron
Milwaukee
Salt Lake City
Valentine
Yankton
Sioux City
Dubuque
Detroit
Carson City
Cheyenne
North Platte
Des Moines
Chicago
Toledo
Sandusky
San Francisco
Omaha
Davenport
Denver
Kearney
Columbus
Fresno City
Pike's Peak
Concordia
Keokuk
Springfield
Indianapolis
Cincinnati
Montrose
Hannibal
Keeler
Pueblo
Kansas City
St. Louis
Louisville
Lexington
Dodge City
Wichita
Los Angeles
Springfield
Cairo
Santa Fe
Nashville
Knoxville
Amarillo
Oklahoma
Ft. Smith
Chattanooga
San Diego
Memphis
Little Rock
Ft. Stanton
Yuma
Tucson
Atlanta
Augusta
Vicksburg
El Paso
Abilene
Shreveport
Meridian
Montgomery
Savannah
Palestine
Mobile
Pensacola
San Antonio
New Orleans
Galveston
Port Eads

SCALE OF SHADES
INCHES
0 to 1
1 to 2
2 to 3
3 to 5
5 to 10
10 to 15
15 to 20
Over 20
STATUTE MILES
KILOMETERS
Seattle
Olympia
Spokane
Portland
Walla Walla
Baker City
Roseburg
Eureka
Red Bluff
Sacramento
San Francisco
Fresno City
Keeler
Los Angeles
San Diego
Carson City
Winnemucca
Idaho Falls
Helena
Miles City
Havre
Swift Current
Medicine Hat
Calgary
Qu'Appelle
Salt Lake City
Lander
Cheyenne
Denver
Pike's Peak
Pueblo
Montrose
Santa Fe
Yuma
Tucson
El Paso
Chihuahua
Ft. Buford
Bismarck
Moorhead
St Vincent
Winnipeg
Port Arthur
White River
Rapid City
Pierre
Huron
Valentine
North Platte
Kearney
Concordia
Dodge City
Wichita
Amarillo
Oklahoma
Abilene
San Antonio
Corpus Christi
Palestine
Galveston
Shreveport
Little Rock
Ft. Smith
Springfield
Kansas City
Omaha
Yankton
Sioux City
Des Moines
Duluth
St Paul
Red Wing
La Crosse
Dubuque
Keokuk
Hannibal
St. Louis
Davenport
Chicago
Milwaukee
Green Bay
Marquette
Cheboygan
Alpena
Manistee
Grand Haven
Detroit
Port Huron
Toledo
Sandusky
Cleveland
Erie
Buffalo
Rochester
Oswego
Albany
Boston
Northfield
Eastport
Portland
Yarmouth
Nantucket
Block Id.
New Haven
New York
Philadelphia
Atlantic City
Harrisburg
Pittsburg
Columbus
Cincinnati
Indianapolis
Springfield
Louisville
Lexington
Cairo
Memphis
Nashville
Chattanooga
Knoxville
Lynchburg
Norfolk
Kitty Hawk
Hatteras
Raleigh
Charlotte
Wilmington
Southport
Charleston
Augusta
Atlanta
Savannah
Montgomery
Vicksburg
Meridian
Mobile
Pensacola
New Orleans
Port Eads
Jacksonville
Titusville
Tampa
Jupiter
Key West
Nassau
Baltimore
Parkersburg
WASHINGTON
Toronto
Saugeen
Kingston
Parry Sound
Montreal
Rockliffe
Quebec
Chatham
Father Point
Sault Ste. Marie

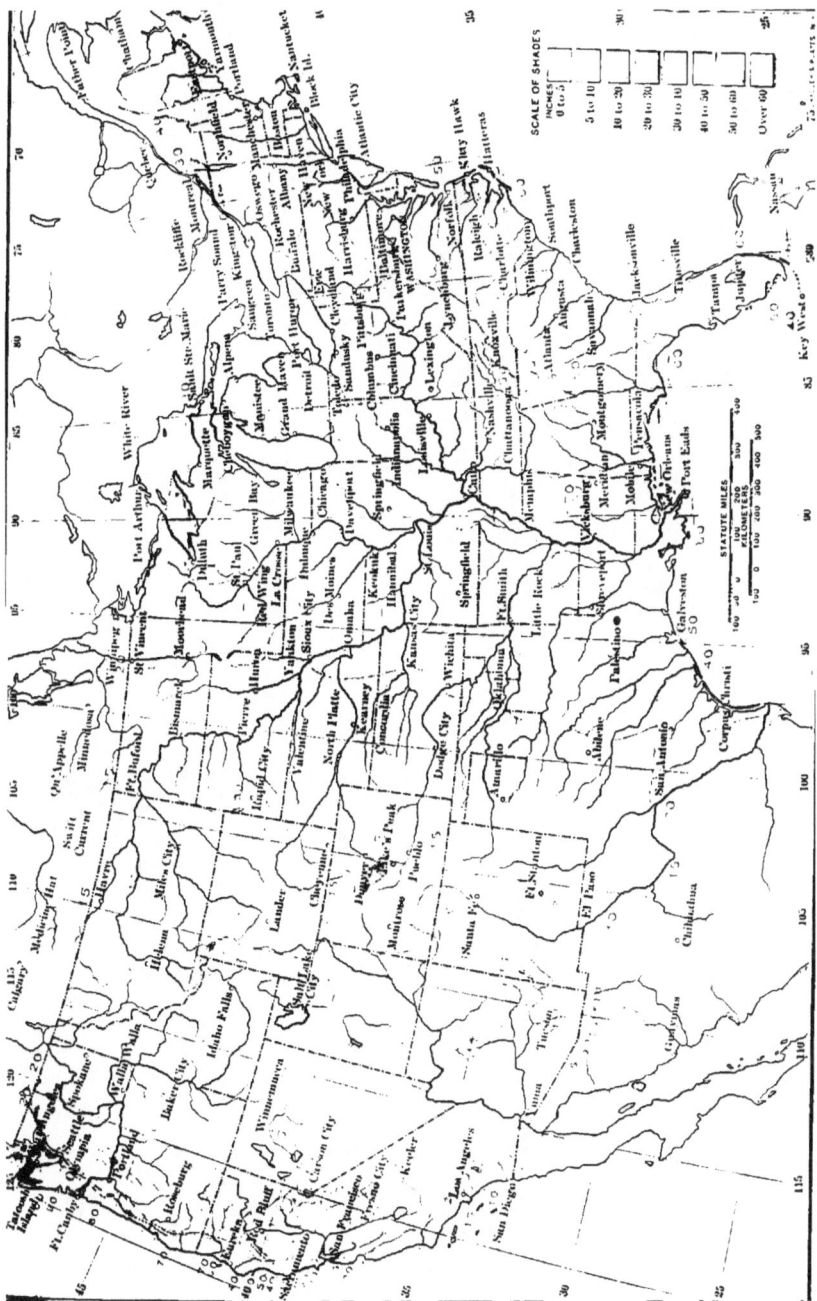

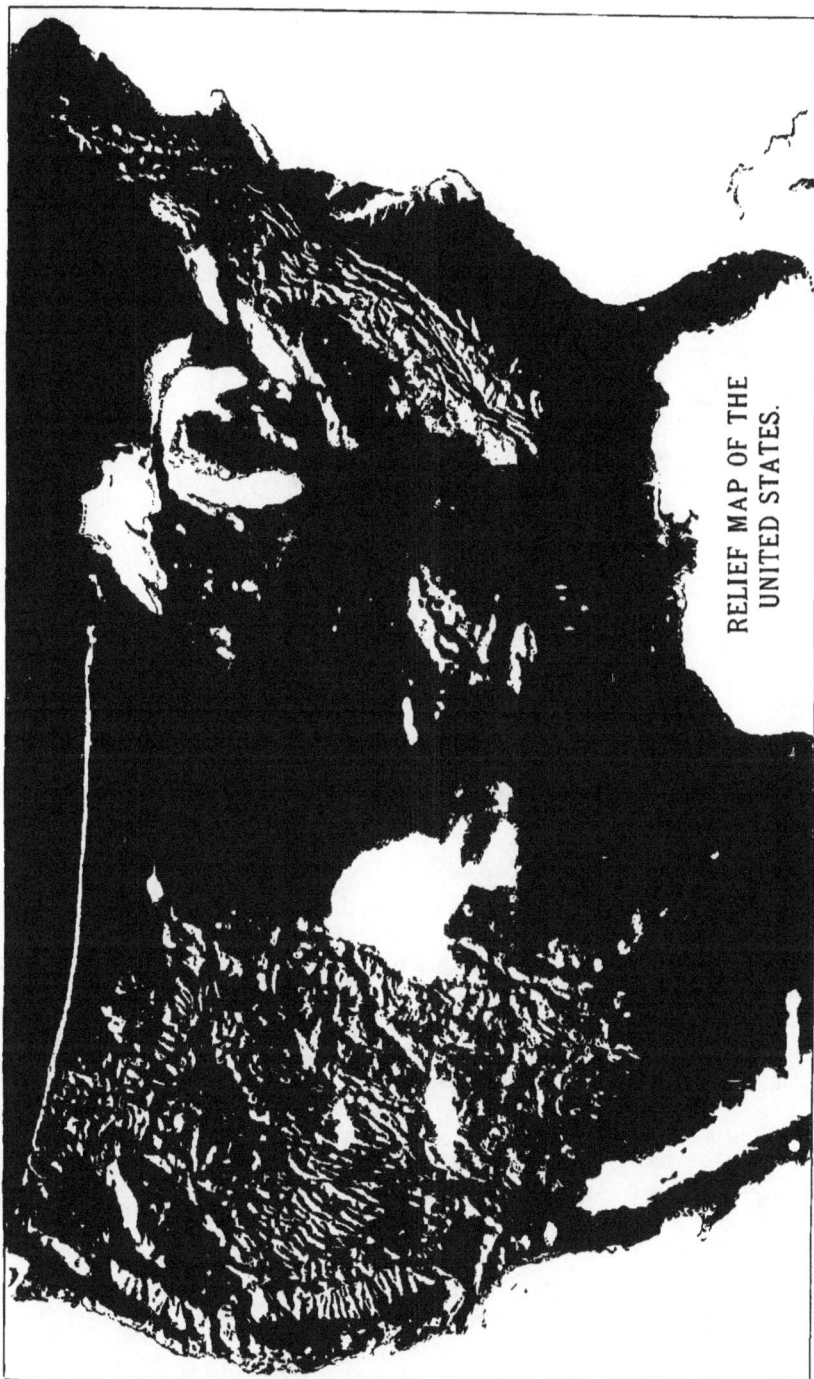

RELIEF MAP OF THE
UNITED STATES.

These climatic features and the temperature are also somewhat increased on the Florida peninsula, although there is more equability than on the Gulf coast.

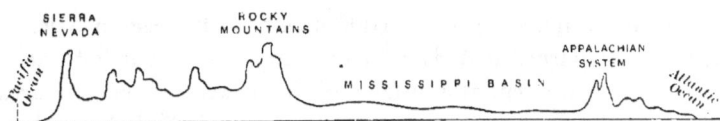

As regards the interior of the continent, if reference is made to the relief-map opposite, and to the profile shown here, it will be seen that there is high ground near the eastern coast—the Appalachian system. Through these uplands are situated numbers of valuable summer-resorts, with cool, though somewhat moist climates, at elevations ranging from a few hundred to three thousand feet. A few of these resorts, such as Saranac, in the Adirondacks, and Asheville, in North Carolina, are used also in winter.

In the South, running east and west, are the Ozark Mountains, of a more moderate elevation and possessing a somewhat drier climate, but not at present much resorted to by invalids.

In the western half of the continent is the Cordilleran system, in which are situated the high-climate stations of Colorado, New Mexico, and Arizona. In these resorts, which are situated at altitudes ranging from 3000 to 8000 feet, the precipitation and humidity are less, the temperature usually higher, the hours of sunlight longer, and the wind-movement and the dust greater than in the Alpine stations of Europe. Still further to the west are the Sierra Nevada and Coast Ranges of California, which are beginning to be much used in summer, but are too cold and snowy for pleasant winter-residence except at very moderate elevations.

The lowlands of Arizona and Southeastern California resemble in climate Egypt, Syria, and other desert-countries of Northern Africa, though they are inferior to those countries in interest and accommodations.

In comparing the resources of American and European resorts it may fairly be said that while the former rival and even excel those of Europe in places which are well established and most frequented, this is not the case with the smaller, less-used, and more recent resorts, though there are notable exceptions and a general improvement is noticeable.

CHAPTER XI.

EASTERN CLIMATES.

ATLANTIC COAST RESORTS.

THE direct influence of the Gulf-stream on the eastern Atlantic coast is less marked than is commonly supposed. The hot waters flowing north are kept at a distance of from twenty to one hundred miles from the coast by the counter-current from Baffin's Bay, which skirts the shores of North America. The inner limit of the Gulf-stream is well defined by a bank or wall, where the waters of the opposing currents meet in passing. One of the mildest and driest of the sea-climates north of St. Augustine is found on the south-eastern shore of Nantucket. The Gulf-stream approaches nearer to the coast at this point, and it is warmer in winter and cooler in summer than any other point on the northern Atlantic coast.

TABLE I.—ATLANTIC COAST CLIMATES IN THE UNITED STATES.[1]

Locality.	Mean monthly temperature.			Annual rainfall.	No. years of record.	
	January.	July.	Annual.		Temp.	Rainfall.
Eastport .	20°	60°	41°	48.4 in.	18	18
Portland (Me.)	23	69	46	42.7 "	18	21
Nantucket .	33	67	48	42.0 "	5	16
Wood's Holl	31	68	49	44.9 "	9	13
Newport .	30	...	50	50.9 "	3	5
Block Island	31	68	49	44.4 "	11	12
Philadelphia	33	76	54	40.9 "	18	22
Baltimore .	34	78	55	44.8 "	19	21
Atlantic City	32	72	52	42.8 "	18	19
Cape May .	34	74	53	46.7 "	13	10
Norfolk . .	41	79	59	52.7 "	19	22
Charleston .	51	82	66	56.3 "	18	21
St. Augustine	57	81	69	49.2 "	20	17
Pensacola .	53	81	68	59.9 "	12	13

Cool summer marine climates are found along the New England seacoast, from the Islands of Campobello and Mt. Desert and the Isles of Shoals, to Cape Ann and the Manchester and Beverly shore. South of this the temperature is higher and the air more

[1] For temperature, rainfall, etc., of other ocean coast-stations—Boston, New York, Jupiter, Key West, San Francisco, Santa Barbara, San Diego—see Table V.

humid, and of the seashore-resorts on the south coast of Massachusetts Dr. V. Y. Bowditch says "that they are not often beneficial to people with pulmonary disease, the prevalence of wet fogs being a serious drawback to the climate in that region. The quality of the climate is distinctly relaxing, and is often beneficial to patients suffering from nervousness and insomnia."

Eastport, lying very near the Canadian border at the southeastern extremity of Maine, is first described because it is the most northern resort mentioned and also because it has a weather-station ; its report is inserted in Table I. Its meteorological conditions are more or less typical of this part of the coast. It is a pretty place, with a population of about 4000, and is located on a small island on the western shore of Passamaquoddy Bay. It has many natural advantages, among them being, of course, facilities for yachting and fishing. Summer-life there is, however, subject to the disadvantages contingent upon residence in a town.

Mount Desert, an island far-famed as a summer watering-place, of which Bar Harbor is the most important and fashionable resort, lies off the coast of Maine and rises from sea-level to a height of over 1500 feet. On the west the slope is gentle and gradual, but on the east the hills rise precipitously from the sea. The northern coast is very near the mainland, to which it is joined by a bridge. The island is reached over the Maine Central Railroad, which connects with the Bar Harbor ferry, or by steamer from New York, Boston, Eastport, and Bangor, or via Rockland.

Mt. Desert is formed of granite-rock, and the soil is notably dry and has great power of absorption. The climate during the summer is cool, refreshing, and very equable, the mean temperature being 70° F. for the days and 64° for the nights.[1] Fogs are rather frequent. No records of humidity are obtainable, but the observations furnished by Dr. Longstreth would indicate that it is by no means excessive for a place so situated. He writes that the prevailing wind is from the west, but that the "high" winds blow from the southwest and bring with them the greatest degree of humidity, and that, under these conditions, the mountains are often capped with fog-clouds. During many dense fogs the humidity, as shown by instruments, is 25 or 30 points less than it is during a southwest wind with general sunshine. It is only fair to say that Dr. Long-

[1] New York Medical Record, June 13, 1896.

streth's remarks refer exclusively to Bar Harbor and its immediate vicinity, and perhaps would not, he says, be wholly applicable to all parts of the island because of difference in relative position with regard to the mountain elevations, the ocean, and the wind. The water is usually too cold for bathing, but the boating and fishing are of the best. Accommodations are good all through the island, and at Bar Harbor they are as luxurious as could be desired. The roads and foot-paths in this vicinity are very tolerable. Bar Harbor affords a delightful social life and every kind of outdoor and indoor diversion; but those who wish to pass a rather monotonous, purely restful summer will find more suitable quarters in other parts of the island.

Portland, situated on the southwestern coast of Casco Bay, is the commercial centre of Maine, and is also the centre of a group of attractive seaside-resorts. The weather-report is to be found in Table I. It is reached from the south by the Boston and Maine Railroad and from the north by the Maine Central Railroad. Its summer climate is very pleasant because of the sea-breeze, which blows steadily. There are some fine buildings and churches in Portland, and the accommodations are good. The drives through the suburbs and environs are most enjoyable.

Scarborough Beach, Old Orchard Beach, Kennebunkport, Wells Beach, and York are coast-resorts to the south. The hard, gradually sloping beach at Old Orchard affords particularly good and safe bathing.

Portsmouth, New Hampshire's only seaport, is situated on a peninsula near the mouth of the Piscataqua River, and is almost surrounded by water. It is, historically, one of the most interesting towns of the coast. The streets are broad and beautifully shaded, and the residences are large and comfortable. The hotel-accommodations are very good.

The Isles of Shoals, nine rocky islands lying from six to nine miles from the mainland, are reached from Portsmouth by steamers which leave and arrive several times daily. They are much resorted to because of the pure sea-air, mild and even temperature, and their freedom from that pest of summer seaside-life—the mosquito. There are two large hotels on these islands. The Isles of Shoals have been rendered famous by the writings of Lowell and Celia Thaxter.

New Castle, Rye Beach, and Hampton Beach, lying to the south of Portsmouth, are well-known resorts. All of them afford good hotel-accommodations, boating, fishing, bathing, and driving, and Rye Beach is, perhaps, the most fashionable seaside-resort on the New England coast.

Cape Ann, projecting from the east coast of Massachusetts, and Gloucester, on the same peninsula, are resorts of the quieter order. They have been much visited by artists and authors on account of their quaintness and picturesque interest. There are two hotels.

East Gloucester, reached from the station by an electric tramway, is much resorted to by invalids and by persons who desire a quiet summer-life. The accommodations, though of a simple order, are good.

Plymouth is a charming old town which lies on the coast forty-six miles south of Boston. It was the landing-place of the Pilgrim Fathers. It is reached by the Old Colony Railroad. Besides its claims to historic interest it possesses all the attractions of a watering-place, and is, moreover, situated in a charming country which gives opportunity for many delightful drives and excursions. The accommodations are good.

Cape Cod, which extends eastward from the coast of Massachusetts for miles into the Atlantic Ocean, is, for the most part, a flat, sandy expanse, devoid of rocks and trees. It is traversed throughout its entire length by the Old Colony Railroad, and is much resorted to during the summer. Its lovers asseverate that there is an especial charm about the resorts on "the Cape" not to be found elsewhere. Sea-fishing, yachting, and surf and stillwater bathing are enjoyed here in their perfection. Good accommodations of all sorts may be had at various points on Cape Cod.

Wood's Holl, a small maritime village situated at the southeastern extremity of Buzzard's Bay, is reached by the Old Colony Railroad from Boston, and from New York *via* Fall River. It is a station of the United States Fish Commission, has a marine biological laboratory, and is an attractive resort of the quieter order. It has one hotel and good accommodations may be obtained. For weather-report of Wood's Holl, the reader is referred to Table I.

Martha's Vineyard. Off the southern coast of Massachusetts lies Martha's Vineyard, an island twenty-three miles long and, at its widest, ten miles across. Its inhabitants, like those of Nantucket, were formerly engaged in the whale-fisheries; but it has long owed its importance to its advantages as a summer-resort. Most of the summer visitors go to Cottage City, on the northeastern shore of the island. A narrow-gauge railway runs southward along the east coast to Edgartown and Katama. There is every facility for fishing, sailing, and stillwater bathing, and the roads are excellent, so that both driving and wheeling are popular amusements. At Martha's Vineyard are the great camp-meeting grounds, where from twenty to thirty thousand Methodists gather every August.

Nantucket. The island of Nantucket, sandy and treeless like Cape Cod, lies about fifteen miles east of Martha's Vineyard, far out in the Atlantic. Steamers ply daily between Nantucket and Cottage City and New Bedford, and there is a weekly steamer to Portland and New York.

Dr. Harold Williams says: "The soil is chiefly sand—very dry and porous. For the summer of 1894 the highest temperature was 85° F., and the lowest 51° F. The greatest diurnal range was 19°. The mean relative humidity was 84 per cent. for July, August, and September, 1894." The yearly mean of relative humidity for five years is 81 per cent. The mean monthly temperature for summer is 64°; relative humidity 83 ; seasonal rainfall 8.5 inches. There is about the same amount of rainfall as at Atlantic City. The air is unusually dry for sea-air and more stimulating than that of the adjacent coast. Dr. Williams says, further, that there are frequent fogs, and that the wind is constant and often high—blowing, of course, off the ocean, no matter from what quarter it sets. In 1894, beginning with July 1st, there were 92 consecutive pleasant days.

There are excellent hotels and boarding-houses. One may also rent or buy houses, and the rates of living are low. The settlements are quaint and picturesque, and the island bears a great variety of wild flowers. The amusements comprise bathing, rowing, sailing, fishing, shooting, tennis, golf, riding, and driving; the blue-fishing is especially fine. A narrow-gauge railway runs from the village of Nantucket to Surf Side and Siasconset. Dr. Williams says: "Nantucket claims to be especially desirable as a summer health-resort because of the purity of its air, its coolness, the smallness of its diurnal range of temperature, and the particles of sea-salt contained in its air."

Newport, one of the capitals of Rhode Island and "Queen of American seaside-resorts," is situated on a low plateau in the southwestern part of the State. The town is an old settlement of much historic interest, but it is chiefly known as the most fashionable summer-resort in America. The scenery and surroundings are

beautiful, and the climate is equable and balmy, but humid and
relaxing. Newport has fine buildings and churches, a casino, and
a library of 40,000 volumes. The hotel-accommodations are good.

Narragansett Pier, situated on the west shore of Narragansett
Bay, is also a noted and fashionable resort. It has a very fine
bathing-beach, many magnificent hotels, and a large casino.

Watch Hill is a favorite resort on the extreme southwestern
corner of Rhode Island. Here are to be found good bathing and
fishing, beautiful scenery, and hotel-accommodations of much excellence.

Block Island. To the south of the Rhode Island coast, at a
distance of ten miles from the mainland, lies Block Island. It is
reached by steamer from Stonington, from New York direct, or
from Providence and Newport. The mean temperature for the
summer months is 73° F. On the north shore are good beaches
for surf-bathing.

The Connecticut shore and both the north and south shores
of Long Island are dotted with attractive little summer-resorts
affording all the usual seashore diversions and, in addition, charming drives.

The coast of New Jersey is lined with resorts which are frequented by seekers after health and by those who desire to escape
from the disagreeables which form an accompaniment to summer
residence in the flat, inland country of New Jersey, Pennsylvania,
and Delaware. There are to be found at most of these resorts good
hotel-accommodations and all the diversions which such places
usually afford—boating, surf-bathing, and fishing. The bathing-
beaches of the New Jersey coast must be especially commended.
The great disadvantage of summer-life upon this coast is found in
the presence of swarms of mosquitoes; but this is mitigated whenever the breeze blows from the sea. Of the New Jersey resorts
may be mentioned the following:

Monmouth Beach is chiefly a collection of private cottages, with
a club-house and a casino.

Long Branch, with which we may include Elberon, Hollywood,
and West End, is one of the most popular watering-places on the
continent, and one of the most expensive. The number of summer-
guests rises as high as 50,000. The original settlement is a small
village situated on a bluff about thirty feet above the beach. At
Elberon are located most of the fine villas. The hotel-accommoda-

tions are, as might be expected, comfortable and luxurious in accordance with the requirements of one of the most fashionable resorts in
the country. The Hollywood Hotel, surrounded by trees, is especially excellent, and is open all the year round.

Asbury Park and **Ocean Grove** are neighboring resorts, equal
in natural advantages, the former being frequented by those who,
liking the locality, object, nevertheless, to the exclusively religious
management of Ocean Grove. Each settlement has a plank-walk
about a mile long bordering the beach, and the accommodations
are good in both places.

Other attractive resorts are **Sea Girt, Squam, Barnegat** (notable for the good shooting which it affords), **Beach Haven,** and
Brigantine Beach.

Atlantic City is situated on a long and narrow island on the
New Jersey coast, where the coast-line bears sufficiently to the west
to afford a southeasterly exposure to the sea. This resort has a
permanent population of 12,000, which is increased during the
summer to about 60,000. It is ninety-five miles, or about four
hours by rail, from New York City, and fifty-six miles, or one and
one-quarter hours, from Philadelphia.

The soil is sandy. Snow seldom remains on the ground for any
length of time. The water-supply is considered good. Besides
house-cisterns for storing the rain-water there are town waterworks, which bring spring-water from the mainland, seven miles
distant. There is a town-system of sewerage.

The mean monthly temperature and total rainfall for Atlantic
City, by seasons, are as follows:

	Mean temperature (18 years).	Total rainfall (mean of 19 years).
Winter	. . 34°	11.1 inches.
Spring	. . 47	10.0 "
Summer	. . 70	11.4 "
Autumn	. . 55	10.4 "

There is a yearly average of 6 days above 90° temperature and
127 days below 32°. Number of cloudy days, 110. The mean
annual relative humidity (three years, 1891–'93) was 81 per cent.;
for winter, 81 per cent.; for summer, 83 per cent.

The sea-breeze usually begins to blow about 11 A.M. and continues until nightfall. The mean annual hourly wind-movement
for the three years 1891–'93 was 11.9 miles.

In 1885 a report of the mean yearly wind-movement for five years for Atlantic City and Cape May, published by Mr. B. A. Blundon, observer at Atlantic City, was as follows :

Atlantic City	82,630 miles.[1]
Cape May	130,055 "

The total wind-movement for Atlantic City for the year 1891 was 108,624 miles; for 1892, 105,120 miles; for 1893, 92,492 miles. It was in each case higher than the above mean for five years.

In the pamphlet just referred to, two reasons were advanced on which the claim of "dryness" could be made for the climate of Atlantic City. One was that the rainfall was less than at any other place on the coast, and the other that the records of the hygrometers were not significant, as the instruments at that time (1885) were but thirteen feet above the sea and "affected by the spray, during the strong winds off the water, and by occasional morning mists which do not extend back into town." For several years past (1895) the instruments used by the Weather Bureau have been located sixty-eight feet above sea-level. The percentage of relative humidity does not, however, appear to read any lower than in the former records.

Atlantic City has a number of advantages as a resort. It is easily accessible from the large cities of the East. The town is well built, possessing markets and shops, miles of streets and suburbs, street-cars, and churches, and hundreds of hotels and boarding-houses of every grade. There is a good beach and in summer excellent sea-bathing.

In a pamphlet entitled *Atlantic City as a Winter Health-resort,* Dr. Boardman Reed says : "There is no body of fresh water nearer than the Delaware River—distant about sixty miles—and the salt-water bays to the landward side are nearly always open, ice seldom forming, except for a short time, occasionally in the severest winters." Dr. Reed also notes that the land-winds pass for long distances over dry and porous, sandy soil before reaching Atlantic City.

The rainfall is surprisingly regular during the four seasons. It averages 3.5 inches for each month in the year.

The climate of Atlantic City is mild for its latitude. The winds

---

[1] Atlantic City as a Winter-resort. B. A. Blundon, Sergeant Signal Service, U. S. A., 1885.

are bracing, and in winter the winds from the sea are warm winds. The soil is dry, and an absence of malaria is reported.

During the cold season, when the temperature is low and the winds come from the west and northwest, the amount of humidity in the air is much smaller than that at some southern and warmer stations.

During the summer, with the mean temperature about 70° F. and the relative humidity in the neighborhood of 80 per cent., the amount of actual humidity is high, as may be expected in all marine climates.

Atlantic City has a plank-walk bordering the beach for four miles. Among the attractions are fishing and wild-fowl shooting, and there is a very pretty casino with reading-, smoking-, and dancing-rooms and an enclosed piazza overlooking the boardwalk and the ocean.

Cape May. This well-known summer-resort is situated at the southern end of the State of New Jersey, opposite the entrance to Delaware Bay. The soil is gravelly, with sand under the gravel, and below another layer of gravel.

The water for domestic use is obtained from wells and distributed after the Holly system. It is soft and pleasant to the taste.

There is an absence of extreme temperatures at Cape May. In winter there are rarely any readings down to zero. Dr. Huntington Richards, in the article on this resort in Buck's *Reference Handbook*, says that the equability of the temperature at Cape May is at all seasons more marked than at Atlantic City.

From the same authority the following table is obtained, based on observations taken from 1871 to 1883:

METEOROLOGICAL TABLE FOR CAPE MAY—RECORD FOR 13 YEARS.

Observations of temperature taken at 7 A.M., 3 P.M., 11 P.M.

	Mean monthly temperature.	Mean relative humidity. Per cent.	Total rainfall. Inches.	Wind. Miles per hour.
Winter .	. 36°	77	11.9	15.4
Spring .	49	75	11.1	14.5
Summer .	72	80	12.8	10.2
Autumn .	58	75	11.7	13.7
Year .	. 53	77	47.6	13.4

Average number of cloudy days in a year, 117.

It is very windy—in fact, Cape May is one of the windiest stations in the country.

The distance from Philadelphia is about eighty miles, a ride of two hours by rail.

There are good hotels and boarding-houses. During the summer the sea-bathing is one of the greatest attractions. Cape May may be said to be rather more fashionable than Atlantic City. It has a magnificent hard bathing-beach five miles in length.

Norfolk, mentioned because its weather-report (Table I.) may be taken as representing this part of the coast, is the second city of **Virginia,** and is surpassed among the Atlantic ports south of the Chesapeake only by Savannah. There are three good hotels. Eighteen miles from Norfolk is **Virginia Beach,** a seaside-resort surrounded by pine-woods. **Currituck Sound,** thirty miles to the south, is much resorted to because of the wild-fowl shooting it affords.

Old Point Comfort, situated on a peninsula north of Hampton Roads, is the site of the great fortification known as Fortress Monroe. There are two hotels, comfortable, but rather expensive, of one thousand beds each, and some cottages. The winter-temperature is rarely below 40° F., nor does the summer-temperature often exceed 80°. Good bathing, boating, and "crabbing" are among the amusements. The social life of the place is, partly because of the presence of the garrison, very gay. Old Point Comfort may be reached by railroad or by steamer. The hotels have sun-galleries protected by glass. During the winter and spring the place is resorted to by visitors from the North who wish to escape the inclement home-season, but during the summer it is frequented chiefly by health- and pleasure-seekers from the Southern States. The climate is considered beneficial to patients recovering from bronchitis and for sufferers from nervous troubles.

St. Augustine, latitude 29° 53' north, has a resident population of 5000, but during the winter the population of this popular resort is increased to 10,000. It is thirty-eight miles from Jacksonville, two hundred and forty-four miles north of Jupiter, and rather more than two hundred and fifty miles from Lake Worth. St. Augustine is the oldest town in America, and is situated on the Atlantic coast of Florida, on a peninsula opposite Anastasia Island. The harbor is small and shallow, but, with its miles of connecting rivers, is well adapted for small boating. The surrounding country is flat and sandy and is overgrown with scrub-palmetto. The average elevation of the town above tidewater is twelve feet. The older portions of the town have narrow streets and quaint old houses built of "coquina" or shell-limestone. In the modern town are some of the finest hotels in America. There are also beautiful parks and semi-tropical gardens. North Beach is a favorite driving-resort. St. Augustine is a United States military-post, and has guard-mount daily and frequent parades with the military band.

The climate is mild, equable, and humid. A record of the mean temperature of St. Augustine for twenty years (1824–'53) was quoted by Dr. J. P. Wall, of Tampa, in a paper read before the American Climatological Association in 1891. The following is a summary : monthly mean, winter, 58° F.; spring, 68°; summer, 80°; autumn, 71°; annual, 69°; January, 59°; July, 81°.

A record of rainfall in St. Augustine for seventeen years shows an annual mean of 49.2 inches. The greatest yearly precipitation was 67.4 inches, in 1880, and the smallest 33.9 inches, in 1851.

The winter climate of St. Augustine is partly shown in the following report, which is adapted from a paper prepared by Dr. Frank F. Smith in 1887.

Mean of records for ten seasons from 1877–'78 to 1886–'87 :

Temperature, November, 64°; December, 58°; January, 55°; February, 59°; March, 62°; April, 68°; mean, six months, 61°.

During these six months the average temperature for ten seasons was at 7 A.M., 56°; at 2 P.M., 67°; at 9 P.M., 59°.

Mean temperature for winter (December, January, and February) 57°.[1]

There was an average of 152 days in each season above 60°, of which 90 days were over 70°. The wind blew from the east about half the time.

Mean rainfall for six months (November to April, inclusive) was 33.5 days, of which rain fell at night on 19.1 days, and during the daytime on 14.5 days out of 181 days in each season.

INLAND RESORTS.

Maine.

The Rangeley Lakes. Among the inland summer-resorts of New England should be mentioned these famous trout-lakes, which afford the attractions of camp-life, beautiful scenery, and clear air. There are half a dozen lakes connected by waterways. The eleva-

[1] The formula is $\frac{7 + 2 + 9 + 9}{4}$.

tion of the highest lake is 1511 feet above the sea. Black-flies and mosquitoes are troublesome until after July.

The climate is cool during the summer. There are numerous hotels and camps around the lakes, which are reached by a narrow-gauge railroad from Farmington—forty-seven miles distant—in about four hours.

The little village of Rangeley affords good hotel-accommodations.

The Maine Woods. The climate of the forest country of Upper Maine resembles that of the Adirondacks, and has been visited for many years, both in summer and in winter, by those who wished to live a rough, hearty, outdoor life. The principal gateway of this region is Greenville, at the southerly end of Moosehead Lake, which can be reached by railway.

Moosehead Lake (elevation, 1023 feet) is about thirty-five miles long, with an average width of ten miles. It varies, however, from one to fifteen miles in width. Half-way up the lake is the Mt. Kineo Hotel, which has accommodation for 500 guests. The lake and its tributary streams afford good fishing.

From the north end of the lake there is a two-mile carry to the west branch of the Penobscot River. By means of canoes long trips can be made around Mt. Katahdin (5385 feet) and further into the pine-forest, or down the stream to Bangor.

Black-flies and mosquitoes are very troublesome throughout this region in summer.

There are no detailed weather-records for the forest country. It is a land of severe winter cold and heavy snows. The trees are pine, spruce, hemlock, and fir, with some hardwood growth in the highlands. There are good hotels to be found in the village of Greenfield. Good hunting can still be found at a distance from the settlements.

In the **Aroostook** farming country, northeast of Moosehead Lake, a broken record of temperature for two years can be given for Houlton, a town of 4000 inhabitants, situated near the New Brunswick line about one hundred miles from Moosehead:

	Winter.	Spring.	Summer.	Autumn.	Year.	Max.	Min.
Mean for two years, } 1892–'93.	11°	35°	64°	43′	38°	97°	—28°

January, 1892, and June, 1893, missing.

Annual rainfall about 30 inches.

Poland Springs, a favorite inland watering-place, lies five miles from Danville, and is reached from there by means of six-horse coaches. It has an elevation of 800 feet, and the views are very fine. There are two hotels. The chief attraction of the place lies in its mineral springs.

Vermont.

Breadloaf Inn, in the Green Mountains, is twelve miles from Middlebury. It is very well kept and the prices are extremely moderate. The hotel stands at an elevation of 1600 feet, and the air is dry and bracing.

New Hampshire.

Bethlehem (elevation, 1459 feet). Among the summer-resorts in the White Mountains Bethlehem ranks high as possessing a cool, pure, and clear atmosphere; a supply of pure, soft water, free from all danger of contamination; adequate drainage; freedom from malarial and typhoid fevers; good accommodations; and beautiful scenery. The place is remarkably free from insect-pests.

Dr. W. H. Geddings, in 1891, prepared a paper for the American Climatological Association, descriptive of Bethlehem and of Maplewood (one mile distant), from which much of the following information is taken:

The plateau upon which Bethlehem and Maplewood are located, formerly known as Bethlehem Street, is protected by a range of high hills from warm south winds. There is a fine view over the beautiful valley of the Ammonoosuck, which is over 200 feet lower, and insures good drainage. In the early morning the valley is often filled with fog and mist, while the plateau above is bathed in sunshine. On the east the town is protected by Mount Washington (elevation, 6923 feet), twenty miles away, and by the peaks of the Presidential Range. On the west the country is rolling, a succession of hills and valleys, with a distant view of the Green Mountains. Water is supplied from a reservoir fed by springs behind the town. A drain-pipe runs through the streets and into the valley below. The soil is rich but rocky, and is usually covered with boulders.

There are a number of hotels and boarding-houses in Bethlehem and a large hotel and cottages in Maplewood.

The population of Bethlehem is 1000. The summer visitors are said to number 10,000 or 15,000.

The season is from the 1st of July to about the 1st of September.

Few people remain throughout the fine weather of September and October.

The mean temperature from three daily observations (7 A.M., 2 and 9 P.M.) for nine years' records was: July, 66° F.; August, 65°; the first half of September, 63°. The average temperature for the season was 65°. The temperature once reached 90°, but it rarely rises above 87°. The mean daily range for July and August for two years was 10°. The mean relative humidity for five years was: July, 64 per cent.; August, 65 per cent.; first half of September, 68 per cent. This record is lower than that of any other summer-resort known to the author east of the Rocky Mountains. (See Table V.)

The normal rainfall is at the rate of 3½ inches for each month of the season, the amount being usually greater in August. The prevailing wind comes from the southwest. There is no record of the wind-velocity.

Mosquitoes are troublesome in June, but are rarely seen after the 1st of July. Bethlehem has for years been especially noted for affording exemption from hay-fever. Its climate is considered cool and tonic, with a tendency to sudden changes.

Jefferson (elevation, 1440 feet), situated in the Franconia region, is to be classed with Bethlehem as to accommodations, and is equally popular. It is a station on the Concord and Montreal Railroad, and is two miles from the Maine Central Line. The view from Jefferson is considered by many to be the finest general view of the White Mountains obtainable. The drives and walks in this vicinity are very good.

The villages near Lake Winnipeseogee are much used as summer-resorts. The lake is traversed by small steamers.

Gilmanton, situated on a high tableland, is a favorite resort. There are no good hotels, but there is one first-class boarding-house.

Dublin is a beautiful village, finely situated on the northwestern base of Mount Monadnock, at an altitude of 1500 feet. It may be reached by stage from Peterboro', or *via* Harrisville over the Boston and Maine Railroad. There is one first-class hotel and one thoroughly good boarding-house, and Dublin possesses also Episcopal and Unitarian churches and a fine public library. The roads are good and the surrounding scenery particularly beautiful. A body of water, called Monadnock Lake or Dublin Pond, lying within easy reach, affords boating, swimming, fishing, and bathing.

The social life of the place is easy and pleasant, and includes dances and concerts as well as the usual outdoor diversions of a summer-resort. There are frequent strong west winds, but fogs are rare. It is said that the climate is characterized by the purity of its air and its tonic qualities. Below is a table giving meteorological data (humidity not recorded):

1893.	Max.	Min.				
January	45°	—14°	Clear days 13 ;	part cloudy		8
February	45	—10	" " 13 ;	"	"	5
March	53	1	" " 11 ;	"	"	9
April	61	17	" " 12 ;	"	"	5
May	85	32	" " 14 ;	"	"	5
June	86	44	" " 10 ;	"	"	7
July	84	46	" " 18 ;	"	"	9
August	86	47	" " 16 ;	"	"	9
September	72	37	" " 12 ;	"	"	6
October	70	23	" " 10 ;	"	"	3
November	58	11	" " 10 ;	"	"	5
December	51	— 8	" " 12 ;	"	"	7

Franconia Village, while it does not in itself offer attractions to the visitor, is the point for reaching several pleasant summer-resorts. Of these, one—Forest Hills Hotel—looks down from an elevation of 1100 feet upon the village. The hotel faces south, and the rooms and table are excellent and reasonable in price. The views are varied and beautiful. The soil is dry and sandy. Dr. Francis Bacon, of New Haven, considers it one of the most desirable places in the White Mountains, being especially dry and bracing for that region. There are golf-links, tennis-courts, etc., and pleasant walks and drives. Across the valley on Sunset Hill are good hotels and many private cottages.

Peterboro', Jaffrey, and **Rindge** are resorts lying in the same district. Good farmhouse-accommodations may be obtained throughout this vicinity.

Massachusetts.

Princeton, Worcester County, is sixty miles from the sea, and has an elevation of about 1000 feet. The situation is open, with a good, free exposure, and stands well up above the surrounding country, which, while hilly, in this immediate neighborhood slopes rapidly to the plain, except in the direction of Mt. Wachusett. The vicinity abounds in beautiful views and pleasant drives and walks. The air is bracing and markedly drier than that of the seacoast. There is a good country hotel, and excellent plain boarding can be obtained.

Mt. Wachusett, in the immediate neighborhood, is somewhat higher than Princeton. It has one or two good hotels situated above the town, and therefore affords to those who desire to benefit by it, the more rarefied air of a comparatively mountainous district.

Petersham, at a little distance from Princeton, possesses the same general climatic conditions, and, like Princeton, is an excellent resort during the summer and early autumn for people suffering from pulmonary trouble.

Sharon is a small town eighteen miles from Boston, on the Providence Railroad, with an elevation of 300 to 400 feet. The soil is gravelly and the water-supply pure, and there is an additional advantage in the presence of pine-woods. The east winds of the coast are much tempered, and the air has a medium bracing quality not noticed at lower points. It has a local reputation for healthfulness.

Dr. V. Y. Bowditch, of Boston, selected this region as the best accessible spot for a sanitarium, which is referred to more particularly in the therapeutical portion of this book. It is an admirably conducted and pleasant home for consumptives of limited means, and is situated on sloping ground at the edge of pine-woods, about a mile from the village.

New York.

Adirondacks. This elevated plateau lies in the northeastern portion of the State of New York. Roughly outlined, it extends from the Mohawk Valley, on the southern boundary, northward one hundred and fifty miles to the St. Lawrence valley and the Canadian line, and is bounded on the east by Lakes George and Champlain; the high, rolling forest country continues westerly for eighty or one hundred miles, covering an area of 3,588,000 acres. The mountains run in a southwesterly direction from Lake Champlain in five parallel ranges seven or eight miles apart. The greatest width of the mountain-belt is about forty miles. The most westerly of these ranges forms the backbone or divide of the region, separating the watershed of the St. Lawrence River from that of the Hudson River and Lake Champlain. There are a number of peaks rising over 4000 feet, the highest being Mt. Marcy (5345 feet) and Mt.

McIntire (5200 feet).   The mountain-plateau has a general elevation of from 1500 to 2000 feet.

Except on the peaks, the entire wilderness is still well covered with forest, although a vast quantity of lumber has been removed. The proposed reservation, Adirondack Park, will set apart over 2,800,000 acres in the centre of this country, of which over 1,500,000 acres are primeval forest and more than 1,000,000 acres lumber-forest.   Much of this is still under private ownership.

The principal tree-growth is pine, balsam, spruce, and hemlock, and the resinous odor is very strong.

A network of lakes and ponds is one of the attractive features of the Adirondacks, adding diversity to the landscape and affording easy communication throughout the interior country by means of light boats.

The facilities for camping, hunting, and fishing all through this region are well known.   There are nearly two hundred comfortable hotels, boarding-houses, and camps scattered through it.[1]   The principal resorts are from ten to fourteen hours' journey from New York City by rail.   The most important gateways are Plattsburg, Port Kent, and Westport from the east; Saratoga on the south; and several stations on the Adirondack division of the New York Central Railroad from the west.

A meteorological record for Saranac Lake for the year 1894 was kindly furnished to the author by Dr. E. R. Baldwin, voluntary observer, from which the following abstract is taken :

Saranac Lake, Weather Station. Lat. 44° 19′ N. Elevation 1750 ft.	Winter.	Spring.	Summer.	Autumn.	Annual.	
Mean monthly temp. (1894),	18°	43°	63°	44°	42°	{ Mean for Jan.20°, min.—15° " July, 66°, max.91°
Total rainfall,	7 in.	7.2 in.	10 in.	10.4 in.	34.7 in.	Including snowfall, 85¼ in.[2]
Cloudy days.	35 days	35 days	17 days	42 days	129 days	Stormy days, 125

The coldest month was February : mean temperature, 13°; minimum —31°.

One of the best known resorts is **Paul Smith's**, or the St. Regis Lake House (1620 feet), which is situated on the north shore of

---

[1] See hotel-list of the Adirondack Railroad for 1895.
[2] The snowfall is usually measured by melting. An approximate value is one inch of water for each ten inches of snow.

Lower St. Regis Lake, a lovely chain of lakes about five miles in length, with sandy shores and very little rock, well suited for camping. This region is comparatively level, there being only one mountain in the vicinity—St. Regis—about 3000 feet high.

Saranac Lake Village (elevation, 1600 feet) is the largest settlement in the Adirondacks, and the only winter-resort. On a hillside one mile from the village and ten miles from Paul Smith's is the Adirondack Cottage Sanitarium (1750 feet), a cluster of nearly twenty small cottages and other buildings, built up during the past ten years under the management of Dr. E. L. Trudeau. The sanitarium is situated on a shelf-like plateau, seventy-five or one hundred feet above the Saranac River. A hill rises on the northwest to a height of about one hundred feet, and at a considerable distance north and northeast is a range of the Adirondack Mountains.

Saranac Lake can be reached in ten hours from New York by the Adirondack division of the New York Central Railroad, which joins the main line at Utica.

The climate of this region shows the temperature to be low and steady during the entire winter. Much snow falls, and there are many windless, cold, snowy days. The normal annual rainfall was stated by Dr. A. L. Loomis, several years ago, to be as high as 55 inches.[1] Rain- and snow-storms are frequent, although they are said to be less severe than on the coast. Snow lasts from the middle of November to the middle of March or April, varying in different seasons. The soil is porous and dries quickly.

There is a preponderance of cloudy weather at all seasons, especially during the winter.

There are no records of the relative humidity nor of the wind-velocity, but the movement of the wind is usually gentle, as might be expected in a forest-region, while the percentage of humidity is high.

Ampersand (elevation, 1600 feet) is a large and excellently kept hotel on the slopes of Lower Saranac Lake, one mile from the Saranac Lake Village. It has attached to it a little village of tents. No meteorological data for this resort were obtainable.

Saranac Inn is to be found at the head of the Upper Saranac Lake. It is reached by the New York Central Adirondack Railway. It is a comfortable hostelry, situated upon one of the most picturesque and largest of the Adirondack lakes.

[1] The Adirondack Region. A. L. Loomis, M D. Transactions of the American Climatological Association, 1879.

Blue Mountain Lake (elevation, 1800 feet) is thirty miles by turnpike from North Creek (*via* Saratoga), or it can be reached from the Adirondack division of the New York Central and Hudson River Railroad, *via* Fulton Chain, by lakes and carries. The lake is two miles in diameter, and lies in a basin formed by Blue Mountain on the north and a ridge of mountains on the south side, and on its shores are two large hotels. The pine-forests are said to be very extensive in this region.

A table of seasonal temperature from records for three years[1] is as follows:

	Winter.	Spring.	Summer.	Autumn.	Annual.
Mean monthly temperature,	20°	37°	61°	41°	40°

Mean for January, 18°; for July, 61°; maximum, 86°; minimum, —26°.

The prevailing winds are from the northwest. The winds are occasionally high.

The fly-nuisance is usually over by July. While visiting this country in June the author was agreeably surprised to find it not noticeable.

Lake Placid (elevation, 1860 feet), four miles long by two miles wide, with its close companion, Mirror Lake, and with numerous hotels and cottages and fine mountain-scenery, is a deservedly popular resort. The beautiful Keene Valley (1015 feet) contains a village and hotels. In this valley lie the Ausable Lakes, with St. Hubert's Inn, a most comfortable hotel; Raquette Lake (1775 feet) and Adirondack Lodge (2160 feet), in a secluded corner of the forest, must also be mentioned. In the Fulton Chain regions are the domains of the Adirondack League Club and other clubs which attract the sportsman, but are not readily accessible to invalids. All these and dozens of other equally attractive resorts must be dismissed for lack of space. They will be found fully described in guide-books.

The climate of the Adirondacks can only be described in the most general way, as there are no complete meteorological reports for a series of years covering any portion of this well-known region. Such records as exist are far from adequate, and are for limited periods. The stations sometimes used as a basis for estimating the values of the Adirondack climates—Plattsburg, on Lake Cham-

---

[1] The Southern Adirondacks. E. T. Bruen, M.D., 1886.

plain, and Malone and Potsdam, in the St. Lawrence valley—are beyond the limits of the true Adirondack region. A record for Saranac Lake for one year has been given above. This weather-station was established November, 1893. From the report of the Director of the State Meteorological Bureau of New York for 1893 the following data were obtained, taken from the valuable article on the "Climate of the State of New York," prepared by Mr. E. T. Turner, Meteorologist to the New York Weather Bureau :

Northern Plateau.	Winter.	Spring.	Summer.	Autumn.	Annual.	
Three stations.[1] Mean elevation, 1578 feet. Mean monthly temperature	18°	37°	62°	43°	40°	January, 16° July, 64° Records of temperature, average for 3 years.
Eight stations.[2] Mean elevation, 973 feet. Total rainfall,	8.79 in.	9.17 in.	10.87 in.	10.14 in	38.97 in.	Records of rainfall, average for 15 years.

The records of extreme temperatures are incomplete, but the maximum reported temperature is 89°F. for Constableville (in 1889–'90–'91), and 88° for Fenton's Hotel, or "Number 4" (in 1889 and 1892). For Saranac Lake the maximum is 91° (1894) and the minimum —34° (1889).

The above mean annual rainfall of 39 inches includes melted snow. The amount of seasonal rainfall is indicated by the following record: Constableville, mean depth, average of three years, 137 inches ; Number 4, mean depth, three years, 108 inches ; Saranac Lake, mean depth, two years, 75 inches. The snow lies less long in comparatively cleared districts, such as Saranac Lake, than in the more wooded regions.

In summer thunderstorms from the eastern Adirondacks often pass down the St. Lawrence valley. The interior of the Adirondack region, with its high mountains and numerous streams and lakes, appears favorable for the development of thunderstorms.

Summer-rains are frequent in the Adirondack highlands. Although there are local variations, a general increase of rainfall is

[1] The three weather-stations were Lyon Mountain (1917 feet), about thirty-five miles north in an air-line from Saranac Lake ; No. 4, Fenton's Hotel (1571 feet), about seventy miles south-west in an air-line from Saranac Lake ; and Constableville (1246 feet), some thirty miles south-west from No. 4.

[2] The eight stations were Constableville, Lowville, Fairfield, Johnstown, Pottersville, Elizabethtown, Keene Valley, and Dannemora.

found in the mountain region over the Champlain and St. Lawrence valleys.

The number of rainy days averages high during the year, but detailed reports for the forest or mountain-plateau are lacking.

There are no published records of relative humidity for the Adirondacks. Mr. Turner says in his report that the region of least moisture appears to be the Champlain valley. The moisture is much greater in the St. Lawrence valley, and in summer an increase of humidity is noticed, with an increase of altitude in the region of the mountain-plateau.

The humidity appears to be largely determined, however, by local conditions. In the valleys the air is very moist, and there are fogs nearly every morning over the lakes during August and September. Many of the hotels throughout the Adirondacks are located in the valleys and near lakes and rivers. The fact that the humidity is lower in the Champlain than in the St. Lawrence valley would indicate that it is probably lower in the eastern than in the western portion of the Adirondack highlands.

The earliest frosts occur in the eastern highlands. In the vicinity of Keene Valley the first frost of autumn occurs about September 20th and the last frost of spring near the close of May. It is sometimes a week or two earlier in September and late in May or early in June. As this region is almost entirely covered with virgin forest and very sparsely settled, it necessarily possesses great purity of atmosphere, being especially free from dust.

The climate can be characterized as *cold and moist.* It is cold in winter, when the air is dry for most days and the snow lies for months, and cool and moist during the summer.

There are a large number of cloudy days, and the humidity is high. The soil is generally light and sandy.

The region surrounding **Lake George** and **Lake Champlain** is interesting picturesquely and historically. Lake George has been favorably known as a summer-resort for so many years that its attractions need no special mention. Its excellent accommodations and the facilities which it offers for boating, fishing, and camp-life have made it a favorite spot with all classes of summer visitors. It lies at an elevation of 345 feet above sea-level, and is surrounded by well-timbered mountains. The attractions of the Lake Champlain district are of much the same order, but the elevation of this sheet of water above sea-level is only 100 feet. At Bluff Point, beyond

Port Jackson, stands the luxurious Hotel Champlain, splendidly situated so as to command views of the Adirondacks, Green Mountains, and Lake Champlain.

Richfield Springs lies at an elevation of 1700 feet, and is distant only a mile from the head of Candarago Lake. It is reached over a branch of the New York Central and Hudson River Railroad, from Richfield Junction. The scenery of this district is most picturesque; but the chief attraction lies in the sulphur springs, seventeen in number, which are used for both drinking and bathing. There is an excellent bath-house, completed in 1890, which includes a swimming-basin. The accommodations are very good, and boating and fishing, riding and driving are among the amusements. Coaches run to the head of Otsego Lake, making connection with the Cooperstown steamer.

Sharon Springs, lying about sixty miles west of Albany and fourteen miles distant from Richfield Springs, has an elevation of 1350 feet. It is situated in a narrow valley, and the surrounding hills rise to a considerable height, affording beautiful views. The sulphur and chalybeate springs are chiefly used for bathing; the bath-houses are of especial excellence, being spacious and elaborately fitted up for every variety of baths. The hotel-accommodations are very good.

ANALYSIS OF THE WHITE SULPHUR SPRING AT SHARON SPRINGS.

One gallon contains:

Sulphate of magnesium	34.000	grains.
Sulphate of calcium	85.400	"
Bicarbonate of magnesium	24.000	"
Chloride of sodium } Chloride of magnesium }	2.700	"
Sulphurets of calcium and magnesium	3.000	"
Total	149.100	"
Gas: sulphuretted hydrogen	20.50	cub. inches.

Saratoga Springs, in the eastern part of New York, has an elevation of 300 or 400 feet and is situated in a valley which has a dry, sandy soil. Monthly normal temperatures are: January, 21°; February, 22°; March, 30°; April, 44°; May, 58°; June, 68°; July, 72°; August, 70°; September, 62°; October, 50°; November, 37°; December, 25°. The mean daily range is 20°. No humidity-records have been kept, but the climate is said to be comparatively dry. There are few high winds or fogs. Saratoga is, and long has been, one of the most fashionable resorts on the continent. It possesses an electric tramway, schools and churches, a good water-supply, and fine shade-trees. The hotels and boarding-houses are, as might be expected, most excellent, and the place affords all the usual attractions of resorts of this order.

The springs are numerous, some being chalybeate, while others contain iodine or sulphur, but in all of them carbonic acid gas is noticeably strong. Their temperature ranges from 45° to 50° F. They are both tonic and cathartic, and are, for the most part, pleasant to drink.

ANALYSIS OF THE CONGRESS SPRING AT SARATOGA.

One gallon contains :

Bicarbonate of lithium	4.761 grains.
Bicarbonate of sodium	10.775 "
Bicarbonate of magnesium	121.757 "
Bicarbonate of calcium	143.399 "
Bicarbonate of strontium	trace.
Bicarbonate of barium	0.928 grain.
Bicarbonate of iron	0.340 "
Chloride of sodium	400.444 grains.
Chloride of potassium	8.049 "
Sulphate of potassium	0.889 grain.
Phosphate of sodium	0.016 "
Biborate of sodium	trace.
Bromide of sodium	8.559 grains.
Iodide of sodium	0.138 grain.
Fluoride of calcium	trace.
Alumina	trace.
Silica	0.840 grain.
Total	700.895 grains.
Gas : carbonic acid	392.289 cub. inches.

New Jersey.

The Pine Belt. Running through the centre of the State of New Jersey is a sandy strip of land, which was described by Dr. I. H. Platt, of Lakewood,[1] as sixty miles in length and from eight to twenty miles in breadth, reaching from Freehold (Monmouth County) almost to Vineland (Cumberland County). The soil varies from light sandy loam to clear beach-sand. Extensive pine-forests are a prominent feature of this region.

[1] The Pine Belt of New Jersey. Isaac Hull Platt, M.D. Transactions of the American Climatological Association, 1889.

Lakewood lies ten miles from the sea and one mile from the border of the sandy strip under consideration, with which it is identical in soil and climate. It is the only place affording good accommodations, and has several large hotels and a number of cottages. Lakewood is sixty miles, or one and one-half hours by rail, from New York City, and has become well known as a winter-resort. The temperature is usually ten or twelve degrees warmer than in New York. The town is from sixty to one hundred feet above the sea, situated on the southern slope of a low ridge, which forms a shelter from northerly and northwesterly winds. Pleasant walks and drives have been laid out through the woods. The pines are thickest to the north and west.

The town is supplied with water from the Metedeconk River. There is a fine spring of pure water about a mile from the village. The town has a system of drainage.

No meteorological record for Lakewood could be obtained.

Dr. Platt gave a summary of the record for Vineland, which is sixty miles southwest of Lakewood and twenty-five miles inland from the sea. In its physical features, however, it is almost identical with the Pine Belt region, from which it is separated by the strip of damp soil that lies between the town and the Pine Belt.

Vineland's annual record for ten years:

Temperature.			Relative	Rainfall.	
Mean.	Max.	Min.	humidity.	Mean.	Max.
54°	104°	—9°	72 per ct.	47.4 inches.	60 inches.

Average number of cloudy days, 107. Fogs are very rare. Snow seldom remains on the ground more than a few hours.

By seasons the mean monthly record of temperature in Vineland for three years (1891–'93) was as follows: winter, 34°; spring, 49°; summer, 73°; autumn, 55°; annual, 54°. The mean for January was 30° and for July 74°. In 1893 the maximum temperature was 102° and the minimum 9°.

The region west of Lakewood and north of Vineland possesses the natural advantages of a dry, porous soil, an environment of pine-forest, and easy accessibility from the great centres of population. The accommodations are better in Lakewood than elsewhere. There the hotels are large, with glass-screened verandas, and are noted for their comfort and luxury.

Pennsylvania.

The system of mountains known as the Alleghenies, extending in several parallel ranges through Pennsylvania, Maryland, and Virginia and into North and South Carolina and Tennessee, furnishes throughout its entire length innumerable summer- and health-resorts.

The **Delaware Water Gap** lies between Pennsylvania and New Jersey. The scenery of this mountainous district is wild and beautiful, the "Gap" itself being a narrow gorge through which flows the Delaware River. The walls of the gorge attain a height of 1600 feet, and are capped by Mounts Minsi and Tammany. The Water Gap is a famous and favorite resort, and affords many amusements and occupations, among which are driving, shooting, and fishing.

Mauch Chunk, lying in the centre of the Pennsylvania coal-region and surrounded by beautiful scenery, is located on a rocky shelf overhanging the Lehigh River and at the foot of Bear Mountain, which rises to a height of 700 feet above the town. It is reached over the Reading Railroad, and is much resorted to in summer; it abounds in comfortable hotels. The famous Switch-back Railroad, built to bring coal from the mines, but now used only for pleasure, is the chief curiosity of this vicinity.

Glen Summit is thirty miles from Mauch Chunk, lying on the crest of the mountains at an elevation of 2000 feet. It affords the most beautiful views and many charming walks and drives. There are a number of pretty cottages and one good hotel, and the place is much patronized in summer.

Cresson Springs, lying at the summit of the Alleghenies, 3020 feet above the sea, is a popular summer-resort. The scenery in this neighborhood is most beautiful. The summer climate at Cresson is delightful, for the elevation adds dryness to the pure air and insures a low temperature. Coaches run between Cresson and Loretto, which was founded by Prince Demetrius Gallitzin, who for forty years worked as a missionary in this vicinity.

Kane is in the northern part of Pennsylvania, on the watershed which separates the waters flowing into the Ohio from those which empty into the Susquehanna. This region is drier than any other district of Pennsylvania. Kane has an elevation of 2000 feet. The surrounding country is rough and mountainous, and the heights are, for the most part, well timbered to the top. The soil is sandy and absorbent, the climate stimulating, and there are opportunities for pleasant and healthful outdoor life. The population of the town is about 4000, and the chief industry is manufacturing. Kane has an excellent hotel, which is kept open throughout the year.

Eagle's Mere is a summer-resort lying at an elevation of 2060 feet. It has a lake, which furnishes opportunity for boating and bathing, and four hotels; there are also some cottages which may be rented.

Mount Pocono. Three and a half hours by rail from New York and an hour longer from Philadelphia, in Monroe County, is Mount Pocono. This resort has a stony, absorbent soil and a dry, pure air, and the quality of the climate is decidedly tonic. The water-supply is said to be especially pure. The scenery is remarkably beautiful, and there are delightful drives and walks through the surrounding country. Dr. Judd[1] says that the temperature of the plateau is ten or fifteen degrees lower than it is in New York and Philadelphia, and there are no mosquitoes. There are good, plain accommodations.

[1] Transactions of the American Climatological Association, 1895.

CHAPTER XII.

Maryland.

Deer Park. In the extreme southwestern part of Maryland, in the heart of the Alleghenies, lies Deer Park, which is reached by the Baltimore and Ohio Railroad. Its elevation is 2400 feet, and it is situated in a region the beauty of which is unsurpassed. The summer-climate is delightfully cool and fresh, and among the diversions of the place are charming walks and drives and good boating and fishing.

Virginia.

The group of thermal springs known as the Hot, Warm, and Healing Springs lies in a mountainous region at an elevation of over 2000 feet. The waters of the Hot Springs have a temperature ranging from 78° to 180° F. The Warm Springs have a temperature of 98° and the Healing Springs of 84°.

The Hot Springs are provided with the most elaborate modern bathing-accommodations of every kind for invalids or visitors, which compare favorably with those at certain celebrated spas in Europe, such as Aix-les-Bains; the surrounding grounds are beautifully and attractively laid out, and the hotels are acceptable to the most fastidious. The **Warm Springs** and **Healing Springs** are each within a pleasant drive, and the accommodations at these places are comfortable, with a pleasing, old-time air about them. The climate is moderately cool and fairly dry, giving an agreeable change in summer to Northern visitors, for whom this has recently become a fashionable resort.

West Virginia.

West Virginia has several resorts where the chief attraction is the presence of mineral springs. Among them may be mentioned the Jordan Alum Springs, in the Mill Mountains; the **Old Sweet Springs**, in the midst of the Alleghenies, where ample swimming-baths form the chief feature; and the **White Sulphur Springs**, in

Greenbrier County, lying amid beautiful mountain-scenery and well
known for many years as a fashionable resort.

North Carolina.

Asheville (elevation, 2250 feet; population, 10,000). The town
of Asheville is situated on an elevated plateau one hundred and
seventy-five miles long and from ten to fifty miles wide. The
average elevation of this irregular plateau is 2000 feet, rising still
higher in the northern portion. On the south and east are the
Blue Ridge Mountains, while the Great Smoky Mountains form the
northern and western boundary. This plateau is one of the most
beautiful tracts of land in the United States, and has long been a
favorite summer-resort.

It has a few hotels that are open all the year round. The winters are said to be fine, as snow seldom remains long in the valleys.
The soil is mostly red clay—in some places sandy. After rain
there is deep mud, but it dries rapidly. The mountains are covered
with a dense growth of forest. On the mountains the rainfall is
15 to 20 inches, and the relative humidity 10 or 12 per cent. higher
than at Asheville.[1] There are numerous streams of clear water.
Good water is obtained for town-supply from the mountains. The
hotel-accommodations are excellent.

Sulphur and chalybeate springs are found near Asheville.

The average number of fair days is 259. The mean of annual
rainfall is 45.4 inches. Taken from the record for thirteen years,[2]
by seasons the rainfall is: winter, 9.3 inches; spring, 11.2 inches;
summer, 13.7 inches; autumn, 8.2 inches.

The mean monthly temperature for winter is 38°; spring, 53°;
summer, 71°; autumn, 53°. The mean annual relative humidity
is 69 per cent.[3]

The principal seasons are in February and March and July and
August. Seventy thousand persons are said to visit Asheville
annually.

The wind-movement for Asheville is not obtainable. As none of
the peaks are within ten miles of the town there is room for free
air-circulation. If the winds are too keen and penetrating during

[1] Asheville and its Climate. Karl Von Ruck, B.S., M.D., 1891.
[2] Rainfall and Snow of the United States. Prof. M. W. Harrington, 1894.
[3] Records of J. W. Gleitsmann, M.D., for Asheville temperature (thirteen years) and humidity
(four years), as quoted by Buck's Reference Handbook of the Medical Sciences.

the winter and spring, a more sheltered country can be found over the Blue Ridge Mountains, on the southern slopes of the range, in some of the picturesque northern counties of South Carolina. There are no complete meteorological records, but as a guide the following records of temperature for the year 1891 may be compared : Asheville—January, 37°; February, 45°. Spartanburg—January, 42°; February, 48°. Spartanburg is 790 feet above the sea, and has a population of 5500. It is seventy miles, or three and one-quarter hours by rail, from Asheville.

Warm Springs. Thirty-seven miles northwest from Asheville are the Warm Springs of North Carolina. There are two springs flowing at the temperature of 97° and 102°F. They are situated in a small and pretty open park, surrounded by mountains covered with hard and pine timber. For accommodation there are one good hotel, two boarding-houses, and several cottages. The elevation of the Warm Springs is given in Toner's *Dictionary of Elevations* as 1326 feet.

A recent record of the weather for six months—November to April, inclusive—shows 32 days cloudy or rainy, and a mean temperature of 47°F., with a relative humidity of 71 per cent.

Southern Pines. There are a small village and a hotel at Southern Pines Park,[1] sixty-eight miles southwest of Raleigh, and near the line of the Raleigh and Augusta Railroad. The situation is on the summit of an extensive sandy elevation covered with pine-forests. The record of temperature at Manly, on the railroad, seven miles to the northwest of Southern Pines, for one year (December, 1881, to November, 1882) is: for winter, 48°; for spring, 61°; for summer, 79°; for autumn, 62°; mean annual temperature, 62°. For eastern North Carolina the average temperature for a series of years is stated to be : for winter, 46°; for summer, 80°; annual, 69°. The average annual rainfall is 44 inches. Snowfall rare and light. The soil is sandy and porous. Spring-water can be obtained at a depth of forty feet. There is a large hotel at Southern Pines, and at Pine Bluff are a number of cottages which may be rented with or without board.

Pinehurst, situated six miles to the west of Southern Pines, is reached over the Seaboard Air Line, or *via* Aberdeen by the Aberdeen and West End Railroad, and is said to be a comfortable and

---

[1] Southern Pines Park, a New Winter Health-resort. A. N. Bell, M.D. Transactions of the American Climatological Association, 1886.

inexpensive resort. The town is furnished with a sewerage-system and has a plentiful supply of pure water. Pinehurst is connected with Southern Pines by an electric tramway. The average summer temperature is 77° F., and that of the winter 44°, corresponding nearly to those of Southern France and of Genoa and Florence; and snow is rare. The soil is sandy and absorbent, the situation sheltered, and the atmosphere dry. There is a good hotel—the Holly Inn— which has a glass-enclosed sun-room, and there are cottages and apartments to let.

South Carolina.

Camden (population, 3500), situated in the upper pine-belt of South Carolina, has been well spoken of as a winter-residence. It is a gently rolling country, covered with forests of long-leafed pine. The elevation is about 200 feet. There are two hotels. The soil is light and sandy, and the water and drainage are said to be good. The coldest noon-temperature in February, 1890, was 50° F.; in March, 40°; in April, 50°. The warmest noon-temperatures were in February, 83°; in March, 81°; in April, 86°. Mean winter-temperature, 45°; mean spring-temperature, 62°. The mean annual rainfall is 41½ inches. The greatest precipitation is usually in July and August, the least in October and November. The average monthly rainfall for December to May, inclusive, is 3¼ inches. Frosts occur at night only, and snow is exceptional.

Aiken (elevation, 550 feet; population, 2500) is situated in South Carolina near the Georgia State-line, on an elevated plateau between the Savannah and Edisto Rivers, and distant from the ocean a little over one hundred miles. The town lies on sandy soil and in the country of the yellow or long-leafed pine, the balsamic odor of the surrounding forests being very perceptible. Grass grows scantily, but there are beautiful gardens in the town. There is said to be no malaria. Pure water is procured from wells at a depth of from eighty to one hundred and fifty feet.

The mean annual temperature is 61° F., and by seasons: winter, 47°; spring, 59°; summer, 77°; autumn, 61°. The mean monthly temperature of January is 41° from records for the three years 1891–'93. The mean annual rainfall (for twenty-five years) is 48 inches.[1] The greatest annual precipitation was 65.6 inches in 1888, and the least 33.9 inches in 1860. Snow rarely falls. There is

[1] Rainfall and Snow of the United States. Prof. M. W. Harrington, 1894.

little dew. Frosts are light, and usually occur in January and February.

Dr. W. H. Geddings, of Aiken, read a paper before the meeting of the American Climatological Association in 1886, in which he reported the following meteorological data for Aiken:

Monthly means.

	Temp. (11 years).	Rel. humidity (7 years).	Rainfall (11 years).	Cloudy days (11 years).
November,	54°	62 per ct.	3.43 inches.	11 days.
December,	47	59 "	3.28 "	11 "
January,	48	62 "	3.64 "	11 "
February,	50	56 "	3.26 "	9 "
March,	56	52 "	4.86 "	10 "
April,	66	56 "	4.71 "	7 "
Winter: December, January, February,	48	59 "	10.18 "	31 "

The greatest monthly precipitation is in March and April, but the sandy soil of Aiken is porous, and it will be seen that for those months there is no increase of cloudy days.

The mean monthly relative humidity for the year 1877 was 64 per cent.; for the winter it was 65 per cent.[1] The wind-movement in winter averages 3.5 miles per hour. The prevailing winds are from the southwest. There are occasional high winds.

Aiken is about fifteen miles from Augusta, Georgia, which may be described here, as its meteorological record is more complete.

Georgia.

Augusta is an attractive city of 35,000 inhabitants, situated on the Savannah River, about four hundred feet lower than Aiken, although no nearer the sea. It has broad, well-shaded streets.

Its weather-record for two years is as follows:

METEOROLOGICAL DATA, 1892–'93.

	Mean monthly temperature.	Relative humidity.	Absolute humidity.	Total rainfall.	Wind per hour.	Cloudy days.
Winter,	47°	75 per ct.	2.56 gr.	11.80 in.	6 miles.	32
Spring,	63	66 "	3.95 "	7.92 "	6 "	26
Summer,	79	78 "	7.74 "	15.41 "	4 "	22
Autumn,	63	76 "	4.38 "	8.95 "	4.9 "	16
Year,	63	74 "	4.38 "	44.09 "	5.3 "	91

[1] Aiken as a Health-station. W. H. Geddings, M.D. Charleston, 1877.

The annual means for Augusta (for ten years) are as follows: temperature, 64° F.; relative humidity, 74 per cent.; absolute humidity, 4.54 grains ; rainfall, 45½ inches ; number of cloudy days, 99 ; average wind-movement, 3.8 miles per hour.

Summerville. Many visitors and invalids resort in the winter season to the Bon Air Hotel at this point, which is well known for its pleasant features.

Thomasville (elevation, 330 feet ; population, 5500) is situated in the pine-woods in the extreme southern part of Georgia, near the Florida-line. It is about one hundred and sixty miles from the Atlantic Ocean and sixty miles from the Gulf of Mexico. The soil is sandy. Water is obtained from artesian wells, one well having a depth of 1900 feet. There are several good hotels and boarding-houses.

The annual precipitation is 51½ inches. The greatest amount of rainfall is in the spring. There are 97 cloudy days during the year. The mean annual temperature is 68° F. ; relative humidity, 65 per cent.; absolute humidity, 4.56 grains of vapor per cubic foot ; wind-movement, 5.7 miles per hour. For winter Thomasville has 50° of temperature ; 67 per cent. of relative humidity ;[1] 4.6 grains of absolute humidity ; and 11.71 inches of rainfall. The wind-movement in winter is low. The prevailing winds are south and northwest.

Florida.

The peninsula of Florida, bounded on the north by the States of Georgia and Alabama, projects southward with a slight inclination to the east, and separates the Atlantic from the Gulf of Mexico. Its length is about three hundred miles, and its width averages over one hundred.

As set forth in the valuable article by the late Dr. J. M. Keating,[2] Florida may be divided by two methods and into two parts. The first, or geographical, division is into north and south Florida, and is determined by the twenty-ninth parallel of north latitude. The second, or topographical, division is into the low-lands which border the rivers and coasts and the higher tracts found inland.

---

[1] Buck's Reference Handbook of the Medical Sciences quotes temperature- and humidity-records for six years.
[2] Transactions of the American Climatological Association, 1885.

The climate of the peninsula below the twenty-ninth parallel is very different, in both summer and winter, from that of the country immediately north of it. During the summer there is more breeze and the nights are usually cool. In winter, besides the fact that the climate of the southern portion of the State is, of course, milder, there remains the consideration that the cold winds which blow from the northwest reach here only after crossing the warm waters of the Gulf and receiving their tempering influence, and, as a consequence of this, the temperature, which may drop rather suddenly in the northern part of Florida, changes much more gradually in the southern part.

The land is mostly low and flat, though there is a ridge extending about halfway down and ranging from one hundred to three hundred feet in height, upon which pines grow.

Florida is studded with lakes and traversed by numerous streams and salt-water channels.

The soil is very porous, and is for the most part sand over a clay subsoil. Semitropical fruits grow freely.

The climate, as follows from its geographical position, is marine in character, and, though situated in the same latitude as Northern Hindustan, Southern China, and the Desert of Sahara, it is far more equable and temperate.

The mean annual temperature varies from 69° F. at Sanford to 79.8° at Jacksonville; for the winter the variation is from 54.6° at Pensacola to 66.5° at Jupiter. The range, for winter, averages between 14° and 20°. Frost, snow, and ice are very rare.

The mean annual rainfall varies from 53.19 inches at Pensacola to 57.16 inches at Cedar Keys. Of this about one-half usually occurs in the summer.

The mean relative humidity for the year varies from 76 per cent. at Pensacola to 80 per cent. at Cedar Keys. In the winter months it is from 76 per cent. to 87 per cent.

Of the absolute humidity Dr. J. P. Wall, of Tampa, speaks as follows: "The amount of absolute humidity for summer is about twice what it is in the winter, dependent, of course, upon the higher temperature of the summer. According to General Greely's estimate of absolute humidity based on ten years of observation, 1876–1886, the absolute humidity for the month of January is, for the northern part of the State, 3 grains of water to the cubic foot of air; for the base or neck of the peninsula, 4 grains of water to the cubic foot; along the twenty-eighth parallel of latitude from the Atlantic to Tampa Bay on the Gulf, 5 grains of water to the cubic foot; and for the extreme point of the peninsula and Key West, 6 grains of water to the cubic foot. The amount of absolute humidity for the month of July is 9 grains of water to the cubic foot for the entire State.

"Of course, as might naturally be expected of a climate with the temperature and humidity of that of Florida, heavy dews on clear, still nights are always present, and during the winter and spring fogs in the nights and early mornings are not uncommon. Fogs are, however, somewhat worse on the Atlantic coast and on the St. John's River than on the Gulf side of the peninsula."[1]

The average hourly velocity of the wind (for 1887) varied from 7.4 miles at Pensacola to 9.4 miles at Cedar Keys.

The total number of rainy days for the year 1887 ranged from 103.8 at Cedar Keys to 124.1 at Pensacola, and the number of cloudy days for the same year and the same places was 66.8 and 84.5.

Dr. Wall, while admitting the prevalence of malaria, states that there are many places entirely free from it, and that it is generally diminishing.

Dr. Keating expresses the opinion that southern Florida is better adapted for invalids than the northern section of the State. He was not favorably impressed with the resorts upon the St. John's River for health-purposes, and speaks of the interior lake-district around Winter Park and Orlando as best adapted for chest-cases. Dr. J. C. Wilson also writes of this country: "It is safe to predict for this region a useful future in the climate-treatment of diseases of the chest."[2]

Florida was formerly much used as a winter-resort for consumptives, but of late years physicians have recognized the fact that in the enervating air of this beautiful peninsula there is great danger of increasing the anæmia usually found in phthisical invalids; danger also to them through the fact that their depressed vitality renders them peculiarly susceptible to the insidious malarial influences which are present at so many of the Florida resorts at certain seasons of the year.

For elderly people, well-nourished persons with irritable catarrhs,

[1] The Climate of Florida. Dr. J. P. Wall, in the Climatologist for November, 1891.
[2] Transactions of the American Climatological Association, 1885.

and certain valetudinarians and convalescents, Florida is well
suited, and for them it presents peculiar attractions.

It is easily accessible from the Eastern and Southern States,
and has accommodations suited to the millionaire or to the health-
seeker of moderate means.   The weary may rest out of doors in the
temperate and sunny air, and the invalid who is able to live a more
active life may seek diversion in hunting, fishing, and boating, for
which the extensive coast-line and the numerous lakes, rivers, and
forests offer exceptional opportunities.

Jacksonville, on the left side of the St. John's River and
twenty-two miles from its mouth, is a city of nearly 8000 inhabitants.   It is a busy town, well provided with hotels and tram-
ways.   The shell-roads are good, and many charming drives may
be taken through this vicinity.   The town is a much-frequented
resort.   The winter climate is medium moist, mild, and equable.
The mean winter-temperature is 55° F.

Palatka, situated south of Jacksonville, on the St. John's River,
is a very pleasant town of about 3000 inhabitants.   It is a gather-
ing-point for many Northern visitors.   The smaller steamers which
make the trip up the Ocklawaha River start from here.   Palatka
possesses good hotels.

Winter Park, one hundred and twenty miles south of Jacksonville and sixty-five miles north of Tampa, is situated in the narrow
part of the peninsula of Florida, about forty miles from the Atlantic
coast.   It is reached direct by rail from Philadelphia, New York,
and Washington, and is thus easily accessible.   It offers good edu-
cational facilities and possesses a pleasant society.

Dr. Eager,[1] of Winter Park, says that there are good sanitary
arrangements and a pure and abundant water-supply, and that there
is no malaria.   The site and surroundings of the town are the high
pine-lands of the peninsula.   From April 1st to June 1st very little
rain falls, and during these months the temperature from 4 P.M. to
9 A.M. ranges from 65° to 72° F., and from 72° to 85° for the
remaining hours.   The average temperature of the winter-day is
60° to 65° for the twenty-four hours.   Frost is very rare.

A large hotel, the "Seminole," offers first-class accommodations
to visitors.

Orlando, a town of 2500 inhabitants, lies one hundred feet above

---

[1] The Climatologist for July, 1892.

the St. John's River, in an undulating, often hilly country. It is surrounded by pine-forests, and is said to be entirely free from malaria. The soil is very absorbent.

Altamonte, a few miles from Orlando and situated in the pine-forests, has a good hotel with an excellent *cuisine.* The soil is sandy and readily absorbs moisture.

In Orange County, near the Ocklawaha Lakes, are a number of towns—Eustis, Taverse, Leesburg, Mount Dora—situated in a rolling country which is well suited for health-seekers.

Tampa, on Hillsborough Bay, a branch of Tampa Bay, and at the mouth of the Hillsborough River, is a town of over 5000 inhabitants. This part of the country affords good sport, as the waters abound in fish, and inland are deer and other game. Tampa is a favorite resort, and has become more so since the establishment of the Tampa Bay Hotel, which is a magnificent building provided with every possible luxury and convenience. There are also other hotels and fine villas. The neighborhood is interesting, and various pleasant excursions may be taken. At Indian Hill, not far distant, are the singular shell-mounds in which were found human remains.

Los' Pinellas is a small peninsula running southwest between Tampa Bay and the Gulf of Mexico. Its climate has been compared with that of the peninsula of Coronado. It is recommended by Dr. J. C. Wilson, of Philadelphia, for its climate and attractiveness. The climate is colder than that of places in the same latitude on the Indian River. The mean temperature is 72° F.; mean winter-temperature, 62°; mean summer-temperature, 80°. The relative humidity is 85 per cent. The soil is sandy and porous.

The Lake Worth district is much resorted to by visitors who desire to escape the inclemency of Northern winters. The lake is twenty-two miles in length and varies in width from one-half a mile to one mile. Its waters abound in fish, among others the much-talked-of tarpon. The vegetation includes the cocoanut-palm. Many handsome villas have been built at different points along the shore. The hotels are first-class, and include the handsome Lake Worth Hotel and that at Palm Beach.

Gulf Coast from Pensacola to Galveston.

Along the Gulf coast westward from Pensacola to New Orleans and on to Galveston are many shore-resorts where the winters are mild and the summers cooler than the neighboring inland. These are much used by the inhabitants of the Gulf States. Perhaps the most agreeable of such resorts are found where the bay curves from Mobile southwest to New Orleans. The best known and the most frequented in winter by Northern visitors is Pass Christian. No meteorological tables are obtainable, and the information given below was obtained from a local physician and from statements of visiting patients.

Pass Christian lies on the coast of the Gulf of Mexico, in the centre of the curve between Mobile and New Orleans. It is eighty-four miles from Mobile and fifty-eight miles from New Orleans *via* the Louisville and Nashville Railroad. The population is about 3000. It has a large, well-kept hotel, several boarding-houses, and cottages for rent. In the winter-season, from November to May, it is much frequented by invalids and visitors from the North, and in summer by the residents of neighboring cities.

Dr. C. L. Le Roux, of Pass Christian, writes of its situation : "At a distance from shore varying from five to eight miles a belt of islands forms a protective bulwark against the occasional storms that may prevail in the Gulf, making the body of water between the islands and the main shore a sort of inland lake, elliptical in shape and rarely disturbed beyond a ripple. Immediately behind the town . . . lies a pine-forest which extends hundreds of miles in depth." The average mean temperature in summer is given at 85° F., and in winter at 70°. The range is said to be small, and the rains neither frequent nor protracted. An occasional " norther " is experienced. There is an average of six pleasant days a week. The soil is well drained, porous, and dry. There is no malaria, and there are pure artesian-well water and a good sewerage-system. Roses bloom in profusion throughout the winter. There is a shell-drive for five miles along the coast, and good boating, fishing, and shooting are among the attractions.

Tennessee.

Chattanooga (elevation, 700 feet ; population, 30,000), in the southeastern part of Tennessee, on the Tennessee River, is reached over the East Tennessee, Virginia, and Georgia Railroad, connecting with the Norfolk and Western and the Baltimore and Ohio lines. It is surrounded by beautiful and varied scenery, and the atmosphere is pure, stimulating, and bracing. The soil of this region is dry and loamy, and the drainage good. Maximum daily temperature, 101° F.; minimum, 7° below zero. The relative average annual humidity is 71 per cent. The average number of clear days per year is 117, and of fair days 147. Owing to the sheltered situation of the city fogs and winds are seldom experienced, and extreme sudden changes of temperature are rare. The roads in the neighborhood are good, and driving and wheeling are popular amusements. Chattanooga has excellent accommodations as regards both hotels and boarding-houses. Lookout Inn, located on Lookout Mountain, is open all the year. The vicinity is historically very interesting, having been the scene of the famous battles of Lookout Mountain, Orchard Knob, and Missionary Ridge, so ably described by Benjamin Taylor in his *Three November Days*.

Arkansas.

Hot Springs (elevation, 425 feet; population, 10,000) is situated in a narrow valley or ravine in the heart of the Ozark Mountains. It has become one of the most frequented resorts in the United States, the great attraction being the springs, numbering about seventy, the waters of which have a temperature ranging from 76° to 158° F. The land on which they rise is the property of the Government, which has erected a hospital for officers, soldiers, and sailors. These springs discharge about 500,000 gallons of water daily. The amount of solids is very small. The town has good hotels and shops and handsome bath-houses.

The Eureka Springs in Arkansas are also well known.

Texas.

El Paso (elevation, 3700 feet; population, 10,000). The old Mexican town of El Paso is situated on the banks of the Rio Grande, at the foot of the Rocky Mountains, in the extreme western portion of the State of Texas, where it narrows to a point between Mexico and the Territory of New Mexico. Climatically, this district belongs to the latter more than to its own State. It is blessed with cloudless skies and is beyond the "northern" belt of Texas, and consequently less liable to high winds, although, as the valley is not very wide, the full force of the wind is felt as it sweeps through. There is least wind during the winter months, when it averages 5.3 miles per hour. In the three months of spring it averages 6.6 miles, and for the year 5.5 miles per hour. There are, however, occasional severe wind- and dust-storms. The Government record for the year 1892 shows the number of days with gales (wind 40 miles per hour or over) to have been 15 for El Paso, as against 3 at Santa Fé and 3 at Denver.

The principal rainfall is in July and August, about 40 per cent. of the year's supply falling during those two months. Mean of annual precipitation, 9 inches.

Temperature for January, 44° F.; July, 83°; for the year, 64° (means for twelve years). Average number of days above 90°, 94; below 32°, 47. Cloudy days, 39; stormy, 37 (means for six years). The mean minimum temperature for three years (1891–'92–'93) in July was 70°, indicating hot nights.

Fair accommodations can be secured. There is one tolerable hotel, "The Vendome," and comfortable lodgings can be found in the town, while good meals can be obtained at a Chinese restaurant.

The soil is sandy in places, but adobe near the river, and the water is bad unless from artesian wells.

The river bottom-lands should, of course, be avoided. Across the river in Mexico is the town of Juarez, where Mexican customs prevail. There is some dirt and squalor, but a visit is usually considered interesting.

El Paso has limited resources for the entertainment of visitors. It has a fine winter-climate, but after April the midday temperature becomes too high for comfort.

Fort Bliss, a military-post, is about five miles northeast of El Paso. The present Fort Bliss is in a better location than the old post, which was one mile from the town.

San Antonio (elevation, 650 feet; population, 40,000). San Antonio is situated in latitude 29° 27' north, about one hundred and thirty miles inland from the Gulf of Mexico. There are no mountains to obstruct the prevailing winds from the southeast and east, and the town is not, probably, entirely beyond the climatic influence of the Gulf, as the cooling effect of the "sea-breeze" may indicate. As a health-resort it is available only during the months of winter and early spring. It is a picturesque and interesting town, containing many reminders of the early mission-days. Many of the old churches and Spanish-looking public buildings are in an excellent state of preservation, and Spanish names are commonly used. In the suburbs are several picturesque old missions.

San Antonio is on the edge of what was formerly a great cattle-raising country. It can be reached by three lines of railroads.

Two small rivers—the San Antonio and San Pedro—flow through the town. The San Antonio River is a dirty stream about sixty feet wide and spanned by numerous bridges. Formerly a highly objectionable method of draining into these winding rivers was practised. In 1895 an improved system of town-sewerage was put into operation; but a medical writer recently stated that of 8000 houses but 400 were connected with sewers. There are three plazas or public squares, street-cars run by electricity, a social club, six hotels, and a number of boarding-houses and restaurants.

The accommodations for invalids are inferior. For housekeeping a location on the outskirts of the town should be selected.

The town is supplied with pure but hard water from springs at the head of the San Antonio River. Artesian wells have been drilled in the town, which vary greatly in the character of the water they yield. Many of them furnish good, soft water. There are also wells flowing hot sulphur water of marked qualities.

The soil is adobe.

The mean temperature of San Antonio for January is 51° F.; for July, 84°; for the year, 69° (means for thirteen years).

The record of rainfall for twenty-one years gives a yearly mean of 30.6 inches, of which 6.7 inches usually fall during the three winter months, and 7.8 inches in the spring.

The air is moist, the absolute humidity even exceeding that of San Diego.

The annual wind-movement is not high, being a little more than that of Denver—about 7 miles per hour. San Antonio is not exempt from "northers," but they are infrequent and are much modified compared with those felt in the upper portion of the State. They rarely last over forty-eight hours.

The average number of cloudy days during the year is 92; of stormy days, 82. Number of days above 90°, 90; days below 32°, 12 (means for six years).

The winters are mild in San Antonio. Roses usually bloom until Christmas, and sometimes later, and in February they begin again to bloom outdoors.

The summers in San Antonio are hot. In 1893 there were 124 days on which the temperature rose over 90° F. Of these, 85 days were in July, August, and September (out of a total of 92 days), the temperature being over 90° every day in July. The maximum temperature for the year was 103°. The mean minimum, or night, temperature for three years (1891-'92-'93) for July was 73°.

(See Table XI., "Night-temperatures.")

Boerne[1] (elevation, 1670 feet; population, 800). The town of Boerne is situated on the River Cibolo, about thirty miles northwest of San Antonio, on the road to Kerrville (San Antonio and Aransas Pass Railroad; time, one and one-half hours).

The soil is mostly sandy loam, with a gravelly subsoil. The country is hilly. Except for narrow borders of timber along the streams and a few tracts of post-oak and forests of cedar, the principal growth is brush and mesquite.

In a description of the town prepared by a local physician[2] it is stated that the water-supply is chiefly from wells of an average depth of thirty-five feet. This water is considered "fairly good." Water from cisterns is also much used. There are said to be 88 cloudy days during the year. The wind is moderate, coming usually from the south or southeast.

There are two or three medium-sized hotels and a few boarding-houses in Boerne. Boarders are always received at a number of ranch-houses a few miles from town.

The drives are good. There are several waterfalls and other objects of interest in the vicinity.

Three and one-half miles from Boerne, on the Hughes ranch, are the Indian Mineral Springs. The following analysis was made by C. F. Chandler, Ph.D., of New York:

	In one U. S. gallon.
Chlorine in chlorides	0.512 grain.
Equivalent to sodium chloride	0.844 "
Oxides of iron and aluminum	0.093 "
Lime	45.832 grains.
Magnesia	6.435 "
Sulphuric acid (SO_3) in sulphates	67.246 "
Silica	0.355 grain.
Solids by evaporation	138.388 grains.

Dr. I. M. Cline, who has charge of the Texas weather service, has written of the country in which Boerne lies as follows: "Between latitude 29° 45' and 30° and to the 100th meridian on the west the elevation changes rapidly from 1000 to 2000 feet, with considerable irregularity, and is broken with deep ravines and small creeks, and

---

[1] Pronounced Ber'-ney. The town was settled by Germans thirty or forty years ago.
[2] Boerne and Adjacent Country. William Miller, M.D.

along the western border, as the 2000-foot line of elevation is reached, it is much sculptured by erosion. The soil between these elevations is very irregular in its formation, but is to a great extent of the black, stiff soil over the eastern portion, and then blends toward the west with black and red sandy and red loam, with a pebbly soil in some parts; it is also crossed here and there by strips of white, sandy land, with a growth of scrub post-oak."[1]

The monthly mean temperature for eight years, arranged by seasons, is as follows : winter, 50°; spring, 69°; summer, 79°; autumn, 65°. Mean for January, 49°; for July, 81°; for the year, 65°.

The total seasonal rainfall based on nine years' records is: winter, 5.9 inches ; spring, 9.6 inches ; summer, 6.4 inches ; autumn, 4.9 inches ; annual mean, 27 inches.

The mean relative humidity is said to be from 66 to 72 per cent.[2]

Galveston (latitude, 29° 18′ north ; population, 35,000) is situated on the northwesterly end of the island of the same name, four miles from the mainland. The harbor is the finest on the coast of Texas, and the depth of water over the bar has been greatly increased in recent years by means of jetties.

The island is thirty miles long by an average width of three miles. It is level, with sandy soil.

Galveston is an attractive and healthy city, with wide, straight streets, ranging from six and a half to ten feet above ordinary tide-level. There are several parks and squares. Many of the streets are shaded with oleanders.

The climate of Galveston is warm, mild, and humid. Occasionally there are winters when the temperature does not fall below 32°. During the past twenty years there have been thirteen years in which the temperature has not fallen below 24°, and but two years below 20°. The seasonal mean-temperature, from the Government records for eighteen years, is as follows: winter, 55°; spring, 69°: summer, 83°; autumn, 71°. The annual mean is 70°. Monthly mean for January, 53°; for July, 84°. The extreme maximum temperature-record is 98°, and the minimum 20°.

The mean of the annual rainfall for twenty years is 51 inches, distributed as follows: winter, 11.5 inches ; spring, 10.2 inches; summer, 13.3 inches; autumn, 16.6 inches. The greatest precipi-

[1] The Climate of Texas. I. M. Cline, M.A., M.D., Galveston, 1894.

[2] Report of the Committee on Health-resorts. Transactions of the American Climatological Association, 1895.

tation is in September and the least in February and July. The annual mean of cloudy days is 92, of stormy days 108.

The mean annual relative humidity is 77 per cent.; for winter it is 81 per cent. The wind-movement averages for the year 11.1 miles and for the winter 11.7 miles per hour. The prevailing wind is from the south and southeast. The highest winds occur during the winter, and blow from the north, but the average " northers " of Upper Texas are but little felt on the west coast of the Gulf of Mexico.[1]

Galveston is rendered accessible by railroads from the north and west, and by steamers from New York and one or two European ports.

There are good hotels and boarding-houses. On the eastern shore of the island is a fine beach for driving and surf-bathing.

Galveston is supplied with pure, soft water obtained from artesian wells and brought twenty-five miles through thirty-inch pipes.

The natural drainage is aided by a slight incline from the centre of the island toward the bay and the gulf. Sewers empty into both bay and gulf.

[1] Notes on the Climate and Health of Galveston. I. M. Cline, M.A., M.D., Galveston, 1894.

CHAPTER XIII.

ROCKY MOUNTAIN REGION.

MONTHLY RAINFALL OF THE SOUTHWEST.

Before proceeding to the study of the climates of Colorado, Utah, New Mexico, Arizona, and Southern California it is thought best to present a general view of the monthly rainfall over these regions.[1]

January. This is a *very wet*[2] month over all California except the southeastern portion, where it is *wet.* Over Colorado (except the eastern half, where it is *very dry*), New Mexico, and Arizona the amount of precipitation is either about the proportional amount with reference to annual rainfall, or deviates slightly therefrom. The normal rainfall-records for the principal cities are as follows for January : Denver, 0.6 ; Colorado Springs, 0.2; Santa Fé, 0.6 ; El Paso, 0.4; Prescott, 1.4 ; Tucson, 0.8 ; Salt Lake City, 1.6 ; Los Angeles, 3.9 ; San Diego, 1.6 inches.

February. This month is *wet* over California (except the southwestern part of the State, where it is *very wet*). It is *dry* over the Dakotas southward to Western Texas, including Eastern Colorado. Elsewhere the rainfall for February shows but slight deviations from its proportional amount with reference to the yearly range. The normal rainfall-records for the principal cities are as follows for February : Denver, 0.5 ; Colorado Springs, 0.3 ; Santa Fé, 0.9; El Paso, 0.5 ; Prescott, 1.7 ; Tucson, 0.9; Salt Lake City, 1.8 ; Los Angeles, 0.4 ; San Diego, 2.1 inches.

March. A *wet* month for the western part of California. A *dry* month over Western Texas and the eastern part of New Mexico. In

---

[1] "The terms *wet* and *dry*, with reference to months, are something more than relative as used in this report. Here it is defined fully with reference to average rainfall, the same rule being followed as has been employed elsewhere. A *wet* month is one in which 50 per cent. more rain falls than the average, and, in like manner, a *very wet* month is one in which double the usual amount of rain occurs—that is to say, 8.33 per cent. of the annual rainfall is the proportional amount for each month, so that under the definition here given a month with 12.5 of the average yearly rainfall is a *wet* month, and one with 16.7 is a *very wet* month. In like manner a *dry* month is one in which the average rainfall does not exceed 4.2 per cent. of the annual rainfall, and a *very dry* month is one in which 2.1 per cent., or less, of the annual amount occurs."

[2] Adapted from a Report on the Climatology of the Arid Regions, etc. By General A. W. Greely, Washington, 1891.

Eastern Colorado the rainfall is a little less than the proportional share for the month. The normal rainfall-records for the principal cities are as follows for March : Denver, 0.9 ; Colorado Springs, 0.6 ; Santa Fé, 0.7 ; El Paso, 0.3 ; Prescott, 1.6 ; Tucson, 0.9 ; Salt Lake City, 2.1 ; Los Angeles, 2.2 ; San Diego, 0.1 inch.

April. A *dry* month in the southern half of Arizona and New Mexico, with tendencies in localities to be *very dry.* A *wet* month in the interior valleys of Southern California, over Western Colorado, and parts of Eastern Utah and Northern Texas. Usually a *wet* month for Eastern Colorado. The normal rainfall-records for the principal cities are as follows for April : Denver, 2.1 ; Colorado Springs, 1.5; Santa Fé, 0.8; El Paso, 0.1 ; Prescott, 0.9 ; Tucson, 0.1 ; Salt Lake City, 0.2 ; Los Angeles, 1.3; San Diego, 1 inch.

May. A *dry* month over all California, the western half of New Mexico, and the northern half of Arizona, and a *very dry* month in Southern Arizona. A *wet* month over Texas (except in the neighborhood of El Paso) and Northeastern Colorado. In Southeastern Colorado *very wet.* The normal rainfall-records for the principal cities are as follows for May : Denver, 2.8; Colorado Springs, 2.5 ; Santa Fé, 0.8; El Paso, 0.2 ; Prescott, 0.5; Tucson, 0.1; Salt Lake City, 2 ; Los Angeles, 0.3 ; San Diego, 0.3 inch.

June. The month is *very dry* over California, Southern Nevada, Southern Utah, and Arizona, and is *dry* over Western Colorado. Over extreme Eastern Colorado it is *wet.* The normal rainfall-records for the principal cities are as follows for June : Denver, 1.4; Colorado Springs, 1.7 ; Santa Fé, 1.2; El Paso, 0.5; Prescott, 0.1 ; Tucson, 0.2; Salt Lake City, 1.2 ; Los Angeles, 0.1; San Diego, 0.1 inch.

July. A *very dry* month over California. A *very wet* month over Eastern Colorado, Western Texas, New Mexico, and the eastern part of Arizona. It is a *dry* month over Western Arizona (except in the extreme southwestern part) and Southern Utah. The normal rainfall-records for the principal cities are as follows for July : Denver, 1.6; Colorado Springs, 3.2 ; Santa Fé, 2.7 ; El Paso, 1.6 ; Prescott, 3 ; Tucson, 2.8 ; Salt Lake City, 0.9; Los Angeles, trace ; San Diego, 0.1 inch.

August. A *very dry* month over California (where it is practically rainless). It is *very wet* over Arizona, New Mexico, and the mountain region of Colorado and Southern Utah. The normal rainfall-records for the principal cities are as follows for August : Denver,

1.5; Colorado Springs, 2.2; Santa Fé, 2.7; El Paso, 0.2; Prescott, 2.8; Tucson, 2.3; Salt Lake City, 1.4; Los Angeles, 0.1; San Diego, 0.1 inch.

September. *Very dry* and nearly rainless in California, *dry* over Nevada and Southern Utah and Eastern Colorado. *Wet* over all of Texas. The normal rainfall-records for the principal cities are as follows for September: Denver, 0.8; Colorado Springs, 1; Santa Fé, 1.6; El Paso, 1.7; Prescott, 1.1; Tucson, 1.2; Salt Lake City, 1; Los Angeles, trace; San Diego, 0.1 inch.

October. The month is *dry* over Western Arizona, Southern California, and Southern Nevada. Elsewhere about the proportional amount of the annual rainfall, or a little less, occurs. The normal rainfall-records for the principal cities are as follows for October: Denver, 0.9; Colorado Springs, 0.6; Santa Fé, 1; El Paso, 0.8; Prescott, 0.6; Tucson, 0.6; Salt Lake City, 1.8; Los Angeles, 0.7; San Diego, 0.3 inch.

November. A *wet* month in Northern California. From the one hundredth meridian to the crest of the Rocky Mountains it is a *dry* month. The normal rainfall-records for the principal cities are as follows for November: Denver, 0.6; Colorado Springs, 0.3; Santa Fé, 0.9; El Paso, 0.5; Prescott, 0.8; Tucson, 0.5; Salt Lake City, 1.5; Los Angeles, 1.6; San Diego, 1 inch.

December. A *dry* month over Western Texas, Western New Mexico, and Eastern Colorado, with a tendency to be *very dry* in extreme Eastern New Mexico and extreme Southeastern Colorado. The month is *very wet* over Western Arizona and California. The normal rainfall-records for the principal cities are as follows for December: Denver, 0.7; Colorado Springs, 0.3; Santa Fé, 0.8; El Paso, 0.4; Prescott, 1.8; Tucson, 1.2; Salt Lake City, 2.1; Los Angeles, 3.7; San Diego, 2.1 inches.

"It must be clearly understood that these terms—*wet, very wet, dry,* and *very dry*—refer not to the absolute quantity of rainfall over the regions mentioned, but to the average monthly quantities with reference to the proportional part of the annual rainfall—that is, if equitably distributed, 8.33 per cent. of the year's rain would fall in each month."[1]

[1] The normal precipitation of cities is taken from the Report of the Chief of the Weather Bureau (1892), Report on the Climate of Arizona (1891), and the records of the United States Weather Bureau at Denver (1895), and are for the following years of record: Denver, 26 years; Colorado Springs, 16 years; Santa Fé, 33 years; El Paso, 28 years; Prescott, 18 years; Tucson 14 years; Salt Lake City, 21 years; Los Angeles, 21 years; and San Diego, 42 years.

Colorado.

For the purposes of health-resort stations the climate of Colorado may be divided into three groups: first, the prairie-plains, ranging from 4000 to 6000 feet in elevation; second, the foot-hills and adjoining valleys, varying from 6000 to 7000 feet; third, the natural parks, varying from 7000 to 10,000 feet.

It has been estimated that there are over 130 peaks of the Rocky Mountains in Colorado between 13,500 and 14,500 feet in height, of which Blanca Peak (14,483 feet) in the Sangre de Cristo range is the highest. Of this number but sixty or seventy have been named. Pike's Peak (14,134 feet)[1] is twenty-fourth in height on the list.

The rainfall of the State varies from 8 to 22 inches, with an average of 15 inches per annum, and is more copious near the mountain-peaks. It does not, however, necessarily increase with the altitude, as Gunnison, 7680 feet, with 10 inches, and Leadville, 10,200 feet, with 12.80 inches, have each less rain than Denver, 5300 feet, $14\frac{1}{2}$ inches.

The rainfall in Colorado does not reach great annual extremes. In Denver the maximum is $21\frac{1}{2}$ inches (1891) and the minimum 9.4 inches (1890). In Colorado Springs the extremes are $18\frac{1}{2}$ inches (1872) and $8\frac{1}{4}$ inches (1893).[2]

The influence of a high mountain-chain or peak in forcing the moisture out of warm air-currents through condensation is shown by the summer rains around Pike's Peak. The mornings in July and August are invariably fair ; for several hours after sunrise there is not a cloud in the sky ; then the hot air of the southern plains rises, and clouds begin to gather, until by noon the peak is likely to be quite obscured. When this heated air meets the colder currents its moisture is condensed, and a sharp fall of rain is the result.

There may be one or two showers during the afternoon, but the night is usually clear and the morning bright as before. These showers, which are apt to be of daily occurrence during a portion of July and August, form one of the pleasant characteristics of

---

[1] Elevation from the records of the United States Weather Bureau. A meteorological station was maintained by the government for several years on the summit of Pike's Peak. It was discontinued October 1, 1894.

[2] Compare with Los Angeles rainfall : extremes of $40\frac{1}{2}$ inches (1884) and $5\frac{1}{2}$ inches (1881).

the summer climate, as they materially cool the air, and the total precipitation is not very large during what is called the rainy season. The normal monthly rainfall at Boston or New York, for instance, is greater than at Colorado Springs during this "rainy" period.

The total yearly precipitation on Pike's Peak is less than one might expect, being barely 30 inches, and its local character is indicated by the fact that on the plains at its base the rainfall is but half as much—14 or 15 inches.[1] In Colorado, east of the Continental Divide, the rainfall of April and May frequently equals that of July and August, when about 30 per cent. of the annual precipitation occurs.

On the eastern plains snow falls occasionally during the winter and early spring, and more rarely in the late autumn; but it seldom lies for any length of time on account of the bright, hot sunshine.

The difference between the temperature in the sunlight and the temperature in the shade in these elevated regions is at all times very marked, being from 40° to 60° F. "The character of the sunlight of high altitudes is a nearer approach, if possible, to white light than at sea-level."[2]

After a cold night in winter, with the mercury perhaps down to zero, it will be warm enough during the day for an invalid to lie out in the sunshine. A table of sun-temperatures taken in 1886–'87 shows a range of maxima from December to March of 112° to 123°, and of minima during those months from 95° to 108°. The extreme maximum sun-temperature for the year was 155° in July.[3]

The average air-temperature for the month of January along the plains under the eastern foot-hills is 30°, which means the general average of twenty-four hours, and includes the extreme minima of the cold nights. The mean monthly temperature of Colorado Springs for January, 1887, taken at 2 P.M. (in the shade), was 38°, and for the winter months of 1893, taken at 12 M., 40°; while the mean monthly sun-temperature for January, 1894, was 60°. Air-temperature (January, 1894), monthly mean, 28°; per cent. of sunshine (January, 1894), 0.66. In Cañon City during January, 1887, the monthly mean, taken at 2 P.M., was 42°. In Denver the tempera-

[1] For details of rainfall, see Tables I. and V. to X.
[2] Rocky Mountain Health-resorts. C. Denison, M.D.
[3] An Invalid's Day in Colorado Springs. S. E. Solly, M.D., 1888.

ture of the air at 1 P.M. (mean of two years, 1884–'85 and 1886)
was 31° for January, and the mean for winter 37°. The solar
temperature for winter, taken at the same hour, was 99°, a difference of 62° between sun and shade. The mean monthly solar temperature for the three summer months (1886) was 144°; for the
year it was 120°.[1]

The unexpected mildness of the Colorado climate[2] is well described
by Dr. Carl Ruedi, of Denver, who formerly practised in Davos-
Platz, Switzerland. In October, 1891, he paid a visit to a small
place called Hygiea, situated forty miles north of Denver, at an
elevation of 5000 feet. When he left the house to look over the
grounds he first stepped into the orchard, where "the ripe apples
and pears were tumbling off the trees." Then he went through a
chestnut-grove, and saw at the further end of the garden watermelons
and sugarmelons sunning their backs and only waiting to be plucked.
Tomatoes and other vegetables, from cabbage to peppers, were growing everywhere. Referring to Davos, he adds: "No cherry or fruit
tree can bring forth its savory product, and even potatoes and barley
attain a very doubtful success."[3] One reason for this difference is
that the limit of vegetable growth is much higher in Colorado,
the "timber-line" being 11,000 feet above the sea, while in Switzerland it is 8500 feet.

In making his comparison Dr. Ruedi comes to the conclusion
that "Colorado, and probably also New Mexico and Arizona,
have in their mountains natural advantages and climatic conditions
which equal or surpass the best European health-resorts of this
character."

The great dryness of the elevated plains of Colorado is well
known. With a general mean of relative humidity during the
year of 50 per cent. and an average annual temperature of 50° F.,
the air shows but 2.04 grains of moisture to the cubic foot, which is
a small amount of absolute humidity for the average of the four
seasons. It is this quality which enables one to bear the changes
from an extremely high to a low temperature, or *vice versa*, with
comfort. A temperature of 56° or 58°, for instance, is not unpleas-

---

[1] See, also, articles on Denver and Colorado Springs.
[2] It may be interesting in this connection to call attention to the low average air-temperature of the New England coast during the month of January, as indicated by Boston 26°,
Portland 22°, and Eastport 20°, taking the government records for about twenty years. Davos-
Platz has for normal mean of January, from observation of eight years, a temperature of 18° F.
[3] Comparison of the Winter Health-resorts in the Alps and Rocky Mountains. Carl Ruedi,
M.D. See Estes Park.

ant if the air is dry, while the moist atmosphere of Florida, or even Southern California, is often chilly and disagreeable.

The wind-movement of the open plains is a marked feature of their climate and at times appears to be unpleasantly prominent, although an examination of its action will show the following surprising result. Taking Denver as an illustration, the monthly and annual wind-velocity is found to be less than that of New York, Boston, Philadelphia, Chicago, San Antonio, Key West, St. Paul, San Francisco, and numerous other representative localities (see Table V.).

A possible explanation of its apparent force in Colorado is that the wind blows with great irregularity, frequently in small squalls or gusts, as on the seacoast. Another reason why it is felt more strongly is that the towns are not closely built as in Eastern cities, and detached houses and wide streets offer a greater sweep for the wind. In the sheltered valleys of the foot-hills and in the smaller parks there is much less wind, the amount varying with the local conditions.

Along the plains east of the mountains the prevailing pleasant weather winds are from the south, southwest, and west, while their opposites, the north, northeast, and east, bring most of the stormy weather.

The advantages of continuous air-movement are well known. Perfect stillness of the air is only desirable in freezing weather. In other words, the warmer the atmosphere the more is a moderate air-movement desirable.

After studying the matter of the natural climate, the most important essentials of an invalid's knowledge are the advantages of particular localities as regards drainage, soil, water, accessibility, the quality of the hotels and boarding-houses, and the possibility of securing good food and cooking. Especial attention should always be given to avoiding, if possible, muddy or " adobe "[1] soil and hard or " alkali " water.

Colorado is more favored than some of her neighbors in having several resorts of considerable size where good accommodations can be relied on ; but even here the list is not extensive.

As most of these resorts are described later in detail, it will only

---

[1] Adobe (ay-do′bay). Throughout the southwestern portion of the United States the term *adobe* is applied indiscriminately to semi-dried bricks, to houses built of such bricks, or to the tenacious, clayey loam or soil of which they are made.

be necessary to refer in a general way to the advantages of the plains lying east of the protecting shelter of the mountains.

These health-stations may be said to extend from Estes Park, Longmont, and Boulder south to Cañon City or Pueblo, with Denver and Colorado Springs as places of refuge or main points of supply. The elevation of this country ranges from 4700 to 7400 feet. It is not easy in Colorado to get below an altitude of 4000 feet, certainly not without enduring great privations, which are not advisable for a delicate invalid. If a decidedly lower altitude is desired, it should be sought in New Mexico, in Western Texas, or in Arizona, where any elevation from 100 to 7000 feet or more can be found, or in Southern California, if the influence of a marine climate is not objectionable.

The belt of 45° to 50° F. mean annual temperature embraces Denver, Fort Collins, Greely, Boulder, Colorado Springs, and Trinidad. The Continental Divide runs irregularly north and south through Colorado, usually between the 106th and 107th degrees of longitude.

An attractive feature of the Rockies in Colorado is the natural parks—great tracts of land with meadows and pastures, brooks and trees, lying at high altitudes, sheltered on every side by huge ranges of mountains. Each enclosed valley has its own peculiar climate, and offers opportunities for hunting and fishing and the exploration of grand and beautiful scenery. During the summer and early autumn these regions are most attractive for visitors and those strong enough for camping out.

North Park lies at an elevation of 8000 feet and contains 2500 square miles.

Middle Park has an elevation of 7500 feet and an area of 3000 square miles. It is sixty-five by forty-five miles in extent. Its climate is milder than that of North Park. Within its boundaries are well-known sulphur springs.

South Park has an elevation of 9000 feet. It is sixty miles long by thirty miles wide, and contains 2200 square miles.

San Luis Park is the largest of all, containing 18,000 square miles. Its average elevation is a little under 7000 feet. It has a warm climate and is more thickly wooded than any of the others. All these parks contain game and trout, and fine feeding-ground for cattle.

Dr. Denison says : " We have every indication that these basins were once beds of immense bodies of water, which, breaking through their rocky barriers, cut deep, rugged gorges or cañons down which the rivers have flowed for centuries, depositing their débris below the foot-hills. That the plains are overlaid by mountain-washings is very evident to the intelligent observer."[1]

Egeria, Estes, Antelope, and Manitou Parks, and other smaller sheltered valleys are more or less known as summer-resorts, and in some of them good board can be found.

No fully equipped weather-stations have been established in the large parks, but there have been a few voluntary observers, from whose reports to the United States Weather Bureau the following details are taken:

San Luis, 7946 feet elevation. Temperature (average for two years, 1891–'92), autumn, 44°; winter, 20°; spring, 40°; summer, 62°. Annual mean, 41°; maximum, 91°; minimum, —25°. Average for January, 15°; July, 66°. Rainfall, 1891, 18.85 inches; 1892, 11.04 inches; 1893, 11 inches. Average for the three years, 13.6 inches.

Como, in Middle Park, shows a temperature even lower, as the annual average (for two years) is 33°, with the mean for July, 55°; maximum, 78°; minimum, —14°. Rainfall, 1891, 15.72 inches; 1892, 10.48 inches; 1893, 12.97 inches. An average of 13 inches for those years.[2]

In making a summary of the climate of Colorado, Dr. C. T. Williams, of London, says : " The chief elements appear to be—

" 1. Diminished barometric pressure, owing to the altitude, which throughout the greater part of the State does not fall below 5000 feet." Dr. Williams also quotes from General A. W. Greely's *American Weather* the interesting statement that " it has been found that the actual barometric pressure in the Rocky Mountains, generally at altitudes above 4000 feet, attains its minimum in January and its maximum in July and August, and that the barometric phases are of the same kind, in reference to the annual mean, as the temperature-phases at such stations. This phenomenon of atmospheric pressure is the reverse of that in parts of the United States at low elevation, and results, according to General Greely, from the lower average temperature of the winter months contracting the great body of air, so that much of it is brought below the summit of the mountains, while in summer the reverse conditions obtain.

[1] Rocky Mountain Health-resorts   C. Denison, M.D.
[2] Aërotherapeutics. The Lumleian Lectures for 1893. London.

" 2. Great atmospheric dryness, especially in winter and autumn, as shown by the small rainfall and the low percentage of relative humidity.

" 3. Clearness of atmosphere and absence of fog or cloud.

" 4. Abundant sunshine all the year round, but especially in winter and autumn.

" 5. Marked diathermancy of the atmosphere, or, as Dr. Denison expressed it, the 'increased facility by which the solar rays are transmitted through an attenuated air,' producing an increase in the difference of sun- and shade-temperatures varying with the elevation in the proportion of 1° F. for every rise of 235 feet.

" 6. Considerable air-movement, even in the middle of summer, which promotes evaporation and tempers the solar heat.

" 7. The presence of a large amount of atmospheric electricity.

"Thus the climate of Colorado is dry and sunny, with bracing and energizing qualities, permitting outdoor exercise every day, all the year round, the favorable results of which may be seen in the large number of former consumptives whom it has rescued from the life of invalidism and converted into healthy, active workers."

Estes Park (average elevation, 7200 feet).   Estes Park is situated sixty miles northwest of Denver, on a branch line of the Chicago and Burlington Railroad.   From Lyons, which is the present end of the railroad, it is necessary to stage twenty miles, usually in a light, covered wagon.   The road winds through the picturesque St. Vrain Cañon, which was once known as Muggin's Gulch.

Estes Park is a plateau about ten miles long and six miles wide, with a number of little side-valleys.   " In general contour it is not unlike the other valleys which make up the park system of Colorado, abounding in gentle slopes, dark pines, and beautiful winding trails leading from the open glades of the valley up dark cañons. Its clear brooks, fed by snowbanks high up on the mountain-side, and filled with speckled trout, unite in one large stream, the Big Thompson, which breaks its way from the hills to the plains below. From any of the neighboring mountains the view from this charming little valley is one of tranquil beauty, in marked contrast with the sublimity of its surroundings.'"

Dr. Carl Ruedi, of Denver (formerly of Davos), had weather-observations taken during the winters of 1891-'92 and 1892-'93, and compared them with similar ones taken at Davos, with a result exceedingly favorable to the Colorado resort.

The elevation of Davos is 5200 feet. A greater elevation was selected in Colorado because the timber-line is so much higher, and it was desired to choose a location similar in regard to the character of the fauna and flora. "A difference of 2000 feet between Colorado and Switzerland is required to put invalids under the same conditions."[1]

The barometric pressure in Colorado is remarkably uniform. The Estes Park record shows December, 1892, 22.06 inches; January, 1893, 22.11 inches; February, 22.01; March, 22.01.

The percentage of relative humidity is much less at Estes Park than at Davos.

Dr. Ruedi's short table is as follows:

Relative humidity.

	Denver.	Estes Park.	Davos.	Sentis.[2]
October,	0.53 per ct.	.... per ct.	0.77 per ct.	0.85 per ct.
November,	0.45 "	0.40 "	0.79 "	0.76 "
December,	0.60 "	0.47 "	0.81 "	0.77 "
January,	0.39 "	0.38 "	0.86 "	0.71 "
February,	0.54 "	0.49 "	0.82 "	0.79 "
March,	0.52 "	0.41 "	.... "	0.62 "

The following table is compiled from a record of the rainfall in Denver for ten years:

November	. . . .	0.577 inch
December	. . . .	0.370 "
January	. . . .	0.706 "
February	. . . . . .	0.563 "
March	. . . .	1.080 "

The total precipitation for ten years showed an average of 0.759 inch for each of the months given.

A comparison of the rainfall between two Swiss stations and two in Colorado is markedly in favor of Colorado:

DAYS OF PRECIPITATION.

Denver,	28 in 6 months (October to April) .	6.79 inches.
Estes Park,	19 in 6 months (October to April) .	6.64 "
Davos,	53 in 5 months (October to March)	11.24 "
Sentis,	103 in 6 months (October to April) . .	30.52 "

[1] A Comparison of the Winter Health-resorts of the Alps with Some Places in the Rocky Mountains. Carl Ruedi, M.D. Denver, 1894.

[2] The Sentis is an isolated peak 9500 feet high, and the observatory is 2000 feet above timber-line. It is the highest point in Switzerland where official observations have been taken.

The record of hours of sunshine during the winter of 1891-'92
shows a daily average from November to February, inclusive, of
2.85 hours for Davos and 6.80 hours for Denver. The hours of
possible sunshine during the winter of 1892-'93 were October to
February, inclusive, daily average: Davos, 3.26 hours; Denver,
7.85 hours.

The hours of possible sunshine by months for Estes Park and
Davos is recorded as follows :

	Davos.	Estes Park.
December	165 hours.	264 hours.
January	182 "	268 "
February	186 "	267 "
March	Not received.	325 "

(See Table VIII.)

The record of temperature for Estes Park is not reported by Dr.
Ruedi in detail, but he states that "for weeks (in winter) you read
20° to 40° F. at 7 A.M., 40° to 45° at 1 P.M., 25° to 30° at 9 P.M.
The only variations take place when sharp continental winds set in,
and then at once the temperature falls very low, stays low until
the snowfall has set in, to rise again to the usual point after three
days."

The greater amount of electricity in Colorado as compared with
Switzerland is noticed by Dr. Ruedi. In the important matter of
wind-velocity he considers the air-movement in the sheltered moun-
tain-parks very light. While the wind-movement for the year will
average from 7 to 9 miles per hour in Denver or Colorado Springs,
in Estes Park the air is usually still, owing to the protection of the
surrounding mountains.

The main basin of Estes Park rises from an elevation of 6800
feet, near the base of Mount Olympus, to about 7800 feet, near
Moraine, which is seven miles further up, on Divide Creek. Owing
to its irregularity of configuration, Estes Park cannot all be seen at one
time, even when viewed from one of the surrounding peaks. Numerous brooks come down from ravines and cañons and flow over the
meadows. Pine-groves are scattered over the park and on the slopes
of the hills. Several large cañons leading from the park will repay
exploration. They contain a number of picturesque waterfalls and
cascades.

Good trout-fishing is furnished by the numerous mountain-streams.
Hunting is usually sought "over the range" to the west. Long's
Peak (14,271 feet) is one of the highest elevations in Northern Colo-

rado. It can be climbed to a point above the timber-line on horseback, but the journey must be completed on foot.

There is a small but comfortable hotel in the park. Accommodations can also be secured at various ranches.

The following seasonal meteorological data for Moraine are based on observations for five years. Mean temperature: winter, 24°; spring, 37°; summer, 57°; autumn, 43°; annual, 40°. For January, monthly mean, 24°; minimum, —17°. For July, monthly mean, 60°; maximum, 88°.

Rainfall: winter, 3.56 inches; spring, 6.19 inches; summer, 4.25 inches; autumn, 2.31 inches. Mean annual, 16.31 inches.

Boulder (elevation, 5300 feet; population, 4000). The town of Boulder is thirty miles northwest from Denver, on a branch of the Union Pacific Railroad. It has fair hotels and has a local popularity as a summer-resort.

The town is situated close to the foot-hills, near the entrance to Boulder Cañon. The water-supply is drawn from a reservoir on Boulder Creek, five miles above the town. Water is also used from wells, from springs, and from the creek. There are no sewers in the town. Many of the streets are well shaded and bordered with handsome residences.

The Seltzer Springs, of Springdale (6500 feet), are ten miles northwest. From Boulder the Union Pacific Railroad has built a narrow-gauge railroad through Boulder Cañon to Sunset (7696 feet), about ten miles to westward.

The mean monthly temperature for Boulder, from observations for one and one-half years, is as follows: winter, 30°; spring, 49°; summer, 65°; mean for January, 30°; July, 69°. As the month of September is missing, the mean for autumn and the year cannot be given. They would each be probably not far from 49°.

No reports of the rainfall for a complete year nor records of the percentage of relative humidity and wind-velocity could be obtained. The average number of cloudy days during the winter is 13.

Denver (elevation, 5300 feet; population, 150,000; latitude, 39° 45' north). It is very nearly in the same latitude as Baltimore. Its altitude is easily remembered, as the business portion of the city is said to be exactly one mile (5280 feet) above the sea. The streets in this section of the city are wide, level, and paved with asphalt.

Julian Ralph, in an article on "Colorado and its Capital," first

published in *Harper's Magazine*, says : " I had supposed it to be a
mountain-city, so much does the Eastern man hear of its elevation,
its mountain-resorts, and its mountain-air. It surprised me to discover that it was a city of the plains."[1] It is fifteen miles due west
to the foot-hills and thirty or forty miles to the highest snow-clad
peaks of the range ; yet in the clear air they seem only distant a
comfortable walk or a ride of an hour or so. There are a number
of fine office and mercantile buildings in Denver. To quote Mr.
Ralph once more, " They are massive and beautiful, and they
possess an elegance without and a roominess and lightness within
that distinguish them as superior to the show buildings of most of
the cities of the country." There is a large and lively retail district. The animation on the streets usually impresses a visitor as
being greater than in other Western towns of about the same size
—Kansas City, Omaha, and St. Paul, for instance. This may be
partly due to the constant coming and going of a large number of
invalids, who are usually accompanied by friends. Many of these
visitors remain, nearly one-fifth of the population being credited
to the class who came for climate and health. The most desirable
residence-section of the city is on Capitol Hill, east of the new
capitol, on a *mesa* thirty or forty feet above the business streets.
Here are modern, well-designed dwellings, some of great size and
costliness, always standing alone, so that light and air have free
access, and each house is surrounded by its own grounds. This
generous method of building spreads the city over an unusual amount
of territory, but insures pleasanter homes and a healthier way of
living than the customary method adopted in large cities of building in crowded blocks.

Denver is unusually well provided with electric and cable street-
roads. The supply of water is ample, derived both from submerged
drains in the Platte River and from a reservoir in the mountains,
thirty-five miles distant. The water from the first-named of these
systems is alkaline. There are a number of artesian wells, a source
of supply upon which many of the hotels rely.

Denver possesses a number of good hotels, the principal one being
surpassed by few Eastern hotels. There are six national banks, four
theatres, two of them of the highest class, and well-built churches of
all denominations. Denver has a number of social clubs, three of

[1] The Great West. Julian Ralph, New York, 1894.

which—the Denver Club, the University Club, and the Athletic Club—have each a fine, modern club-house.

The visitor will find congenial society and better opportunities for business in a city of this size than in the smaller towns.

The markets are well supplied with the best beef from Kansas, and fresh fruit and vegetables from the fruit-growing centres of Colorado and from the warmer climates of Texas, Arizona, and California. The stores are large and well equipped and supplied.

Denver has about 315 sunny days during the year. It is a pleasant place in which to spend the winter. The weather is usually mild enough for a light overcoat, except during a week or so of snow in December and perhaps the same length of time in February. The direct rays of the sun usually melt the snow in a few hours, and the ground lies dry and unfrozen nearly all winter. The advantages of the Colorado plains as a winter-resort should be better known. The contrast between the storms and cold of the Eastern or Northwestern States and the dry ground, moderate temperature, and constant sunshine of the country sheltered by the Rocky Mountains is always a matter of astonishment to those who have had no previous knowledge of the climate. Reference to the records of the Weather Bureau will show the surprising fact that the climate of Denver greatly resembles that of Prescott—a town in the middle of Arizona—so much further south that it would be natural to expect a much milder climate. As Denver is now a city of sufficient size and resources to gratify any want, and is less than fifty hours by rail direct from New York, its desirability as a place of residence compared to the half-civilized adobe-towns of the southern alkali-country is obvious.

During the spring the wind—especially in the afternoons—becomes annoying, as there is frequently a great deal of dust flying about. The presence of the smelters north of the city is sometimes noticeable on account of the amount of objectionable smoke which, like a fine brown haze, fills the lower business streets. This condition of the atmosphere is usually due to a north wind, and to avoid it as much as possible a residence east or southeast of the capitol should be secured. "There is a feeling that the air at times is being contaminated by the smoke and gases from the massive smelting-works, containing, as they do, volatilized galena (lead), arsenic, antimony, etc., although an effort is being made to condense these from the smoke, as is done in other smelters. The same objection is increasing against Pueblo and El Paso."[1]

As before stated, the business streets have asphalt pavements. In the surrounding residence-districts the soil is usually sandy, but in some places it is very tenacious after a rain. There are occasional outcroppings of clay on Capitol Hill—the neighborhood of which should be avoided.

The monthly average temperature for January is 27° F.; for July, 73°; for the year, 50° (means for twenty years). Number of days in the year above 90°, 21; below 32°, 143; cloudy days, 57; stormy days, 73 (means for six years).

The hourly velocity of the wind in Denver from the Government records for ten years shows a yearly average of a little less than 7 miles per hour.

The total wind-movement for the year 1893 averaged, for the full day of twenty-four hours, 8 miles per hour or 193 miles per day. During the extreme length of what I have termed the "invalid's day," from 8 A.M. to 6 P.M., it blew at the rate of 9.2 miles an hour. The smallest wind-movement was from 2 A.M. to 9 A.M., when it averaged 6.6 miles an hour. The greatest velocity was usually in the afternoon between the hours of 1 and 5 P.M., when it averaged 10.5 miles per hour.[2]

Occasionally wind-storms come up during the night, and the velocity at that time is greatly increased. A record, however, of the number of days with gales (wind forty miles per hour or over) during the year 1892 shows 3 days for Denver as against 3 days for Santa Fé, 15 days for El Paso, 4 days for St. Paul, 6 days for New York, 8 days for Boston, and 59 days for Chicago. In Denver a record of the high winds (18 miles per hour or over) during the year 1886 was 23 days.

The record of sunshine for Denver for three years (1891, 1892, 1893), as reported by the United States Weather Bureau, is given herewith, adapted from the record for each month. For purposes of comparison the record for Davos-Platz for six years, adapted from the tables prepared by Arthur William Waters, Esq., is included:

[1] Health Journal. W. P. Roberts, M.D.
[2] Fourth Report of State Board of Health, Colorado. Denver, 1894. Weather-tables.

Sunshine-record.	Monthly means.				
	Winter.	Spring.	Summer.	Autumn.	Year.
Percentage of possible sunshine— Denver	Per cent. 62	Per cent. 60	Per cent. 66	Per cent. 71	Per cent. 65
Davos-Platz . .	57	52	55	52	54
Hours of actual sunshine— Denver	Hours. 188	Hours. 242	Hours. 290	Hours. 243	Hours. 240
Davos-Platz . .	100	166	196	126	147

The general weather-characteristics are much the same as at Colorado Springs, except that the rainfall at Denver is about 2 inches less during the summer season and the snowfall is two or three times greater in winter. There are also more hot days during the summer.

Eighteen miles northeast of Denver, on the Burlington Railroad, is the Colorado Carlsbad Spring, the water of which is bottled for commercial purposes. An analysis by Karl Langenbeck yielded:

		Grains per gallon.
Sodium sulphate	87.16
Sodium chloride	10.00
Calcium carbonate	19.70
Sodium carbonate	2.44
Magnesium carbonate	3.58
Potassium sulphate, iodide, and bromide . .	trace
Iron, aluminum, and silica	trace
Total solids . .	122.88

Colorado Springs (elevation, 6000 feet; population, 22,000). This well-known health-resort is seventy-five miles south of Denver. It is well located on gravel-soil which extends to a depth of sixty feet and insures dryness and good natural drainage. The town has a system of sewerage, electric lights, and electric street-cars. The supply of good water from mountain-sources is ample. The streets are wide, level, and well shaded by trees, usually cottonwoods. There are good stores, three banks, an opera-house, several first-class hotels, one of which ranks with the best hotels in the State, a fine social club-house, and remarkably fine churches for a place of its size. The residences are detached and surrounded by handsome lawns. There are a surprising number of large, costly dwellings, built in excellent architectural taste, giving the town the appearance of a long-settled, prosperous Eastern city. Five miles to the west are the foot-hills of the Rockies, while twelve miles distant, in an air-line, the summit of Pike's Peak—brown in summer and covered with snow during the rest of the year—rises to a height of 14,134 feet above the sea. On the peak there is a stone building, formerly occupied as a signal-service station, and now used as a hotel, and the cogwheel railroad carries up visitors daily during the summer season. From Colorado Springs to the southwest, west, and northwest the hills rise upward of 4000 feet above the town. To the north and northeast are high *mesas*, also affording protection from winds and storms, while to the east and south the country is an open plain, rolling, brown, and dry. Even so small a distance as separates the town from the foot-hills is of benefit during storms, for a heavy downpour is often seen on the mountains which does not reach Colorado Springs.

The rainy months are from April to August, inclusive, the most protracted season being usually a period of four or six weeks of daily rains, beginning during the last half of July. In the mountains these rains are more severe, but on the plains they frequently amount only to afternoon thundershowers.

The total yearly rainfall averages 14.46 inches; of this the normal precipitation is 11.18 inches during the five months from April to August, inclusive, leaving 3.28 inches to fall during the remaining seven months from September to March, inclusive. "It therefore follows that the fall of snow is infrequent and scanty through the winter.[1] By reason of the dryness and porosity of the soil and the dryness of the air, with the warmth of almost constant sunshine, the evaporation of snow is very rapid. In looking at the temperature-record it will be noticed that once or more during the winter the temperature drops below zero, and sometimes a long way below. Such temperatures rarely last throughout the day, as the sun seldom fails to shine. When such cold weather occurs during the day the wind, instead of taking its usual daily course from north to south through the eastern quarter, remains in the

---

[1] Owing to the near protection of the mountains on the west, and the presence of the Arkansas Divide, twenty-five miles north, which breaks the force of the northern storms, Colorado Springs has usually lighter winter snowfalls than her neighbors which have a more exposed position on the plains. The record for total rainfall and snowfall for three consecutive winters is as follows :

Denver.	*Pueblo.*	*Colorado Springs.*
1892-'93. Nov. to May, 3.73 inches.	Nov. to May, 2.74 inches.	Jan. to May, 0.06 inch.
1893-'94. " " " 5.98 "	" " " 3.21 "	Nov. to May, 1.46 "
1894-'95. " " " 4.09 "	" " " 3.57 "	" " " 2.25 inches.

north, but has a small velocity, a high wind and low temperature being rarely in conjunction. On these exceptional days the younger and more vigorous invalids go out to exercise with benefit, but those who are more delicate remain indoors, bearing their captivity with grace, as something that occurs to them only three or four days out of the whole winter. Driving is most agreeable in the morning, as there is usually very little wind; but after luncheon a strong breeze from the southeast is not uncommon, making walking or riding more pleasant."[1]

June is likely to be one of the most suitable of the summer-months for camping-out, as there are usually several weeks of dry, sunny weather before the coming of the cooling summer-rains.

. The mornings are invariably fine during the entire year, and are therefore the most favorable time of the day for being outdoors.

A few hot days occur during the summer, but the nights in these high altitudes grow rapidly cooler toward morning. "The absence of dew permits all but the most delicate invalids to sit on the porches in the evening with enjoyment."[2]

The autumn and winter are delightful, the sun usually shining strong and clear and warm, even after a night when the mercury has dropped pretty low. The air is so dry in winter that during the "invalid's day" lasting from 9 to 4 o'clock the temperature outdoors *in the sun* is rarely uncomfortable.

"After a night in which there have been a hard frost and a clear sky, with a light breeze from the north, and during which the invalid has usually slept soundly under several blankets, with his window partly open, he awakes to find the sun shining into his eastern window. And this is a feature which, whatever the weather may be later in the day, is rarely absent. After breakfast our invalid steps into the street, being then in an atmosphere in which the heat in the sun is 92° and in the shade 30° F. A gentle air is stirring from the northeast at the rate of six miles an hour. The mean dew-point is 18°.

"As the day proceeds the temperature rises to its highest point, being 100° in the sun and 40° in the shade between 2 and 3 P.M., while the wind, which has veered rapidly from the north to the south, blows with its highest daily velocity, thirteen miles an hour. After 2 P.M. the wind works back again toward the east, being at sundown northeast, and continuing as darkness falls to

<hr/>

[1] An Invalid's Day in Colorado Springs. S. E. Solly, M.D.   [2] Ibid.

shift back to the northern quarter, whence it blows from 8 P.M. to
9 A.M., its velocity dropping to between seven and eight miles an
hour,[1] and the temperature of the air at the same time falling from
three to four degrees.   The ground is usually bare of snow, no rain
falls from mid-September to mid-April, and the sun shines unob-
structed by clouds.   During the three winter months the number
of cloudy days does not average more than five a month.   The effect
of such an air is bracing and genial, and being so dry the cold in
the shade is very little felt, a medium-weight wrap being all that is
needed.   The roads are good and seldom obstructed by snow or
mud, and the neighboring hills and plains are full of interesting
points to visit, and pleasant, sheltered nooks where the invalid can
rest under the agreeable heat of the sun and eat his midday meal
without fear of taking cold."

The very cold days of winter (they are not numerous) and the
windy days of spring are the greatest trials for an invalid at Colorado Springs.   "The spring weather in Colorado, as in most
climates, is the least desirable; during late March and early April
the chief part of the snow falls and the wind that goes mainly to
swell the total of the annual movement occurs."[2]

The night-air is cold and dry all the year round, with very few
exceptions.   Windows can be left open on the coldest nights, as the
air in a room will never get as cold as the air outside, and the colder
the air is the less moisture it contains.   Immunity from fog is proved
by the low dew-point, the yearly average dew-point for Colorado
Springs for 1893 being 24°.

The mean daily range of temperature in Colorado Springs is 25°,
and it varies but little from that average for every month in the
year.

It is a question how adversely the constant and wasteful watering of the lawns in summer affects very delicate invalids who
spend a great part of the day on porches exposed to such influence.
A short series of experiments in the way of daily observations of
the relative humidity, taken during one summer by means of
Mason's hygrometer, on a front porch over a lawn that was not
sprinkled excessively, showed a usual variation of 13 per cent.
increase over the Government observations, which were taken
forty-nine feet above the ground, with no lawn around the building.

[1] See, also, Wind-movement in Denver.
[2] An Invalid's Day in Colorado Springs. S. E. Solly, M.D.

There is little doubt that the desire for fresh grass in a naturally arid country leads to a questionable freedom in the use of water.

Except for the short-lived objections of spring winds and rains, August rains, and winter cold—none of which periods is long continued—there are few objections to Colorado Springs as a health-resort for those who are not affected by long-continued residence at a considerable altitude.

On the other hand, it possesses advantages such as few other resorts can offer : a dry, porous gravel-soil ; unusually wide streets; town-sewerage ; remarkably pure, soft mountain-water, containing less than three grains of total solids to the gallon;[1] and good drives, beautiful natural scenery, and the absence of manufactories.

The markets are surprisingly good in Colorado Springs. The grocery, provision, and fruit-stores are numerous and unusually well supplied for a town of this size.

There are few opportunities for business, but Cripple Creek—twenty miles west in an air-line—has grown to be a bustling mining-town of over 20,000 inhabitants. It is quite dependent on Colorado Springs as the nearest town of any size, and the development of its mines has brought much wealth to the Springs.

Colorado Springs has sixteen trunk-lines of railway and direct daily communication with the East *via* the main line of the Chicago, Rock Island and Pacific Railroad to Chicago.[2]

The monthly mean temperature for January is 26° F.; for July, 69°; for the year, 47° (means for sixteen years). Average annual days

[1] The town-water of Colorado Springs is drawn from Lake Moraine, a natural basin on the shoulders of Pike's Peak, situated above the timber-line. It is fed by the melted snows of the summit, seeped through granite and gravel. The water is unusually pure and soft, and is probably as near an approach to distilled water as any public water-system could furnish. The supply is ample for a town several times the present size of Colorado Springs. The last analysis of this town-water made by Prof. William Strieby, of Colorado College, showed two and one-half grains of total solids to the gallon.

[2] As Colorado Springs is not a large place, it may be useful to travellers to explain its railroad connections :

Passengers from the East have a through sleeper from *Chicago* to *Colorado Springs* on the Santa Fé and the Rock Island Railroads. There is one change from *St. Louis* on those roads. The Missouri Pacific runs a through sleeper from *St. Louis*, and has one change from *Chicago*. Passengers from *New Orleans* or *Galveston* find a through Pullman running from Houston to Denver (Houston is one and one-half hours from Galveston) on Union Pacific, Denver and Gulf Railroad.

Passengers from *California* come over the Santa Fé, Colorado Midland, or Denver and Rio Grande Railroad with but one change.

From *Oregon*, with but one change, over Colorado Midland or Denver and Rio Grande Railroad.

Chicago to *Denver*, direct through sleeper on the Burlington Railroad, one change to Colorado Springs.

above 90°, 7 (means for seven years); below 32°, 155. Stormy days, precipitation, 0.01 or over, 69 (means for four years). Normal winter-rainfall, 0.7 inch. Number of cloudy days in winter, 13; in spring, 20; summer, 13; autumn, 11; year, 57 days (means for six years).

The annual wind-movement averages 9.1 miles per hour.

TABLE II.—COLORADO SPRINGS. MEANS FOR TWO YEARS, 1893 AND 1894.

Month.	Temperature. Monthly.			Extremes.		Relative humidity.	Absolute humidity.	Dew-point.	Total. Rain-fall.	Cloudy days.	Wind, hourly velocity.
	Mean	Max.	Min.	Max.	Min.						
	(1894)	(1894)				Per ct.	Grains		Inches.	(1893)	Miles.
January,	31°	42°	15°	65°	—2°	44	0.83	9°	0.01	2	11.1
February,	26	36	10	56	—3	53	0.80	8	0.36	9	10.7
March,	36	50	25	72	6	46	0.99	13	0.18	7	12.4
April,	44	60	32	73	11	41	1.16	17	0.38	8	12.5
May,	54	68	43	81	27	48	1.88	29	4.34	10	11.1
June,	63	75	48	91	38	43	2.27	34	2.17	3	10.0
July,	67	78	55	90	47	52	2.36	45	2.73	8	8.2
August,	64	78	52	85	42	59	2.45	46	2.22	7	7.2
September,	59	72	44	82	31	43	2.03	31	0.49	2	8.9
October,	48	...		76	14	51	1.61	25	0.45	1	9.1
November,	39	67	11	71	2	44	1.12	16	0.17	5	10.0
December,	32	58	2	69	—7	49	0.99	13	0.22	6	9.6
Annual,	47			91	—7	47	1.48	23	13.72	68	10.0
Winter,	29					48	0.87	10	0.59	17	10.4
Spring,	44					45	1.26	19	4.92	25	12.0
Summer,	64					51	2.95	41	7.12	18	8.4
Autumn,	48					46	1.55	24	1.11	8	9.0

NOTE.—The maximum temperature for 1893 was 93° on one day in June and one day in July. For 1894 it was 90° on one day in June. The number of days (or, more accurately speaking, nights) below 32° during 1893 was 168. The minimum temperature for 1893 was —3° in February, and for 1894 —10° in January.

The month of October is missing from the record for 1894, and the mean for six years is used, except in cases when unattainable, when the mean for October, 1893, is employed. In the seasonal averages no difference of importance will be perceptible.

The year 1893 was a dry year, and the year 1894 unusually wet, especially in May, when the precipitation was 7.34 inches. The mean of both years (13.72 inches) is nearly one inch below the normal annual precipitation for sixteen years.

The wind-movement for 1893 was unusually high, averaging 10.3 miles per hour for the year, which is about one mile per hour above the normal.

The year 1893 also contained an unusual number of cloudy days. (See Table V.)

The summers in Colorado Springs are usually cool. In 1892 there were 11 days when the thermometer rose above 90 , the highest point being 94°; in 1893 there were two days above 90°, the highest temperature being 93°; in 1894 a temperature of 90° was reached on one day only ; in 1895 there were no days over 90°, the highest temperature being 89°.

In the warm days that occur in Colorado Springs the air is invariably dry. When the temperature goes above 85° the relative humidity usually drops below 25 per cent., and the "sensible" temperature is about 60°. The mean· of the summer minima is 51°, showing cool nights.

There are severe wind- or dust-storms at infrequent intervals when the air is very dry and electrical. They are exceedingly disagreeable, but do not occur more than half a dozen times a year, and the greatest violence of the wind rarely lasts more than a few hours. The gravel-soil of the plain on which the town of Colorado Springs stands greatly mitigates this annoyance as compared with that experienced in towns built on adobe-soil.

Manitou (elevation, 6300 feet; population, 3000). Five miles west of the town of Colorado Springs, in a sheltered valley at the foot of the mountains, are located the mineral springs which give the name to Colorado Springs and form one of the principal attractions of the lively hotel-resort called Manitou.

Two railroads, a line of electric cars, and good roads for driving connect the two towns.

There are eight sparkling soda and iron springs, varying in temperature from 44° to 59° F.—all strongly charged with carbonic acid gas. The water of one of these—the Navajo—is bottled in immense quantities, and is known all over the country as a table-water of great purity and value. In composition and taste it closely resembles the imported Apollinaris Water.

A half-mile or so above the village, situated in a beautiful cañon, and on the banks of Ruxton Creek, is the summer-house over the iron Ute Spring, which is "highly effervescent, of the temperature of 44.3° F., and very agreeable in spite of its marked chalybeate taste."[1]

Professor O. Loew, Mineralogist and Chemist of the Wheeler Expedition, said that the springs of Manitou " resemble those of Ems and excel those of Spa."

There are three picturesque cañons leading into the mountains

[1] Manitou, Colorado ; its Mineral Waters and Climate. S. E. Solly, M.D., 1882.

from Manitou—Ruxton, or Engleman's Cañon, William's Cañon, and the Ute Pass, the latter of which contains the old Indian trail through the apparently impassable range.

Besides the cog-railway to the summit of Pike's Peak there is also a wagon-road.

Manitou has a number of good hotels, and during the "season," from June to September, they are usually lively with visitors. One or two keep open during the entire year.

The climate of Manitou differs slightly from that of Colorado Springs. The valley is more sheltered from the winds, but is somewhat damper in summer, owing partly to the effect of more frequent mountain-rains and partly to the presence of the brooks that flow through the town. The days in winter are short, on account of the height of the mountains on the west. During the month of December the sun sets in the town of Manitou a few minutes before 3 o'clock, which is about one hour and a quarter earlier than it sets in Colorado Springs.

ANALYSIS OF THE PRINCIPAL MINERAL SPRINGS IN MANITOU, MADE BY DR. WALLER.

In a pint of water are contained grains as follows :

	Ute Chief.	Navajo	Mani-tou.	Hiawa-tha.	Ute Iron.	Little Chief.	Sho-shone.	Minnie-haha.
Potassium sulphate . .	1.949	1.919	1.336	trace	0.979	0.501	0.333	trace
Sodium sulphate	0.932	1.367	1.268		1.880	3.601	2.601	0.750
Sodium chloride .	3.346	2.974	2.193	......	2.116	3.333	3.000	1.003
Sodium bicarbonate .	6.557	5.326	5.083	4.271	3.613	1.114	6.201	1.666
Calcium bicarbonate .	10.477	8.667	8.635	6.023	3.963	5.251	7.602	2.801
Magnesium bicarbonate .	2.235	2.005	2.085	0.092	1.005	1.001	......	0.501
Lithium bicarbonate .	0.141	0.089	0.077	0.092	trace	trace	trace	trace
Iron oxide . .	0.046	0.003	0.003	0.031	0.035	0.125	......	0.102
Silica . .	0.467	0.308	0.312	0.134	0.452	0.143	trace	trace
Aluminum . .	0.015	0.013	0.009		0.019			
Ferrous carbonate .				......	0.049			
Strontium carbonate .	trace		trace					
Potassium chloride .			0.213					
Magnesium sulphate .	......		1.123					
Calcium sulphate .	......	......	0.214					
Total .	26.165	22.671	21.801	12.193	14.211	15.069	19.751	6.823

The **Garden of the Gods** is situated a mile or two from Manitou and about four miles from Colorado Springs. This imaginative title is applied to a number of fantastic rock-formations. The red sandstone has been ground by a mighty agency in the past into strange forms resembling with an accuracy sometimes grotesque the shapes of animals and objects. The huge rocks at the gateway are over three hundred feet high.

Glen Eyrie and **Blair Athol**, a few miles further north, contain similar freaks of nature.

Broadmoor. Four miles southwest of Colorado Springs is Cheyenne Mountain, which guards the Broadmoor Casino, where, during the summer season, visitors can hear good music and cooling refreshments can be obtained. The Broadmoor Hotel, a well-equipped building fronting on the grounds of the Broadmoor Casino, has been recently finished.

In the Cheyenne Valley are a number of charming homes. There are also several summer-camps and a few boarding-houses, where quarters can be secured all the year round. The valley is sheltered from the wind by the higher land rising on either side.

Not far from Broadmoor are **South Cheyenne Cañon** and the **Seven Falls**, **North Cheyenne Cañon**, and still further north **Bear Creek Cañon**, through which the first Government trail to Pike's Peak was constructed.

A few miles west of Manitou are a number of attractive summer-resorts, rising higher in altitude as the railroad steadily climbs around the huge shoulders of Pike's Peak to reach the park-like country beyond.

Cascade (7240 feet), **Ute Park** (7500 feet), **Green Mountain Falls** (7730 feet), and **Woodland Park** (8480 feet) are stations on the Colorado Midland Railroad, between eleven and twenty miles from Colorado Springs.

They are picturesque resorts, situated in cañons and small natural parks, and are supplied with good hotels. Some of them have also accommodations for tent- and cottage-life.

Manitou Park (elevation, 7800 feet) is a green valley surrounded by beautiful mountains, the slopes of which are covered with pine-forests. The rides and drives are very fine. It is seven miles from Woodland Park, and is reached by a stage which makes daily trips between the two places during the season. There is a comfortable hotel surrounded by pleasant cottages, and the grounds contain golf-links, tennis-courts, and a lake which affords excellent trout-fishing.

There is a greater summer rainfall on the mountains than on the plains, the showers usually occurring during the afternoon.[1]

Pueblo (elevation, 4700 feet; population, 35,000). It is best known as a rising manufacturing-city, but the records of dryness and fair, mild winters entitle it to consideration among the health-resorts of the Colorado plains. The town is situated on both sides of the Arkansas River, which is a muddy, rapid-flowing stream. The soil in the lower portion of the city is adobe, caking to the hardness of brick under the hot summer sun, dusty under the influence of a strong wind, muddy and tenacious after heavy rain or snow.

From late September to March the weather is usually all that can be desired. The nearest mountains are twenty-five miles away, and the lower altitude of the Arkansas valley insures milder weather than that found further north. This portion of the year is a season of almost perpetual sunshine and moderate temperature. The spring months are more doubtful on account of occasional dust-storms and parching winds. The summers are very hot. The highest temperature is usually about 10° F. above that at Colorado Springs. The Pueblo Board of Trade report for 1893 contains, on page 96, the statement that "wind-storms are unknown, winter or summer." On June 14, 1893—one of the days this report was being sent through the mail—a dust-storm raged in Pueblo about sunset for an hour or more, at times too thick to see through it a church-steeple at a distance of two blocks. Such storms are infrequent, but this incident illustrates the constant allowance that must be made for local prejudice. Regarding the heat, the average number of days over 95° during five years is 16 for each year, showing a high general temperature for the impaired vitality of invalids to withstand. A visiting physician writes: "Pueblo is several degrees warmer than either of the other cities visited (Denver and Colorado Springs). In fact, it was hotter than any place in New Mexico when I was there. Except for the smelters already noticed, it is a good health-point for those with weak lungs."[2] If Pueblo is to be entirely a manufacturing-city, the smoke-nuisance should be carefully considered by delicate visitors, who should select a place of

---

[1] See Climate of Colorado.   [2] Dr. W. P. Roberts, in Health Journal.

residence as far from the noxious outpourings of chimneys as possible. The water-supply of the town is drawn from the Arkansas
River, the water of which is alkaline and contains considerable
muddy sediment. There are artesian springs of good quality, but
settling basins for the north side water-system should be provided
to clear and purify the regular river-supply for domestic use.

Pueblo has good stores, banks, and churches. There is one good
hotel, and others are contemplated. There is a pleasant social club,
and the Opera House is one of the finest buildings of its kind in the
State.

The average mean temperature for January is 29° F.; for July, 76°;
for the year, 52° (from records for fifteen years). Average number
of days above 90°, 41; below 32°, 146; cloudy days, 53; stormy,
52 (from records for four and six years). The average annual
velocity of the wind is 7.4 miles per hour.

In 1892 the number of days on which there were gales (wind
over 40 miles an hour) was (for Pueblo) 19 days.

Two of the artesian wells in Pueblo yield water of remarkable
qualities. In one the water contains a large amount of lithium
salts. Tub-baths are given at the hotel adjoining the spring,
which in every way offers good accommodations. The water flows
from a depth of 1200 feet. An analysis made by Professor A. A.
Cunningham, of the University of California, follows:

			Grains per U.S. gallon.
Lithium bicarbonate.	.	.	7.81
Lithium sulphate	.	.	1.92
Magnesium sulphate .	.	.	60.47
Magnesium carbonate	.	.	4.71
Iron carbonate .	.	.	4.14
Potassium chloride .		.	8.83
Calcium sulphate .		.	14.79
Calcium carbonate .		.	15.12
Sodium sulphate .			12.41
Sodium chloride .			19.72
Phosphoric acid .			trace
Organic matter .	.		none
Total solids		.	149.92

The other well furnishes a sulphurous chalybeate water, flowing
from a depth of 1400 feet, which is extensively used for bathing. It
is called a "Magnetic Mineral Spring." A circular issued by the

proprietor states that "a qualitative analysis of the water gives the following constituents :

"Sulphuretted hydrogen, iron (form titanic acid), bicarbonate of lime, sulphate of soda, sulphate of magnesia, manganese, potassium (trace), sulphuric acid, arsenious acid(?). One remarkable peculiarity of the water is that knives are readily magnetized by holding them in the water."

No quantitative analysis of this water has been published.

Beulah, twenty-eight miles from Pueblo, is a resort at an elevation of 5600 feet, situated in a basin along the foot-hills on the eastern slope of the Greenhorn range. It is reached by regular stage in about five hours. The village has a population of 200. There is a good hotel. Board can also be secured at some of the ranches. Water is obtained from the mountain-streams. There are cañons and other places of interest to visitors. The Beulah Springs are mild soda springs of excellent quality.

Fifteen miles west from Pueblo is the Carlile Soda Spring, which is carbonated and put on the market.

An analysis made by T. A. Stoddard, M.D., of Pueblo, is as follows :

		Grains in U. S. gallon.
Sodic carbonate	.	9.33
Calcic carbonate	.	18.56
Magnesium carbonate	.	11.60
Ferric carbonate	.	0.58
Potassic sulphate	.	1.16
Sodic sulphate	.	23.20
Sodic chloride	.	8.70
Calcic phosphate	.	1.50
Sodic phosphate		0.87
Silica	. .	trace
Total solids	.	75.50

Cañon City (elevation, 5300 feet; population, 3800). Cañon City is situated in a valley near the entrance to the Arkansas Cañon. The town lies mainly on the north side of the Arkansas River. It is forty-five miles northwest of Pueblo, on the main line of the Denver and Rio Grande Railroad and on a branch line of the Santa Fé Railroad from Pueblo. It has a milder winter climate than the cities on the plains. This is due largely to the shelter from the wind afforded by the surrounding mountains, which rise 2000 or 3000 feet above the town and protect it on the north and south. A westerly wind blows occasionally through the cañon, but its force is principally felt south of the river. To the east—or a little south of east—the valley opens to the plains. Snow seldom falls and quickly disappears.

The soil of the valley is adobe. There is a small residence-district south of the river where the soil is more sandy. The town is supplied with electric lights, waterworks, and a system of sewerage. The waterworks are artesian wells sunk near the Arkansas River, on the Holly system, into which the river-water filters. It is good water, but hard.

Cañon City has a number of comfortable-looking residences, each house being built usually of brick and standing alone on its own lot of land. The residence-streets are well shaded.

There are two or three fair hotels and a few boarding-houses.

About three-quarters of a mile west of the postoffice are the two cold soda springs, one of which has slight traces of iron. Half a mile further are the hot springs, where there are a hotel and a bath-house.

The analysis of these mineral springs by Professor Loew is appended :

ANALYSIS OF MINERAL SPRINGS, CAÑON CITY.

	Iron Duke. Cold.	Little Ute. Cold.	Ojocaliente. Hot, temp. 102°.
Sodium chloride.	83.0	118.0	18.2
Sodium sulphate	12.2	12.1	79.3
Sodium carbonate	76.8	76.4	73.2
Calcium carbonate	33.0	22.5	33.5
Magnesium carbonate	14.6	14.0	12.8
Iron	traces	traces	
Lithia	traces	traces	traces
Grains in one gallon	219.6	243.9	217.0

The valley in which the town of Cañon is situated is famed for its strawberries, grapes, and apples. The latter fruit is extensively cultivated with the aid of irrigation-ditches.

The attractions in the neighborhood are the Royal Gorge (which begins within two miles of the town, but which requires a twelve-mile drive to be fully appreciated) ; the Saurian bone-fields ; the Marble Caves (twelve miles) ; the Bottle Rocks ; the Bottomless Pit; and Grape Creek Cañon (distant about two miles).

The State Penitentiary is located on the edge of the town, about half a mile west of the postoffice.

Directly south of the town is a large zinc, lead, and copper smelter, the smoke from which is usually carried away from the business-centre and residence-district by the prevailing winds.

An abstract of the temperature-record for Cañon City for four years is as follows: mean temperature for January, 32° F.; for July, 74°; for the year, 53°. The minimum for January is —6° and the maxima for July and August 100°. By seasons the temperature and rainfall are as follows:

	Winter.	Spring.	Summer.	Autumn.
Temperature, four years,	34°	52°	72°	53°
Rainfall, six years,	1.80 in.	4.24 in.	4.21 in.	1.63 in.

The mean of the annual rainfall for six years is 11.38 inches. The year 1894 had an excessive rainfall, its total being 15.96 inches.

There are no records of the relative humidity, but it is about the same as Pueblo, perhaps a little higher in summer on account of the irrigation and the absence of strong winds.

Glenwood Springs (elevation, 5200 feet; population, 1500). A little west of the centre of the State of Colorado, not far beyond Leadville (10,200 feet), is the crest of the great Continental Divide. Glenwood Springs is on its western slope, and the waters of the Roaring Fork and the Grand River, which meet opposite the town, find their way eventually into the Pacific Ocean.

The town is reached by the Denver and Rio Grande and Colorado Midland Railroads.

The principal attraction is the famous Yampa Hot Spring, which flows the enormous quantity of 2000 gallons per minute, at a temperature of 124° F. A large and complete bath-house has been built to utilize this hot saline water under medical direction. There are also sulphurous vapor caves and an open swimming-pool 700 feet in length by 100 feet in width, where outdoor baths may be enjoyed during the entire year.

Glenwood Springs has one of the finest hotels in the West. It is a large building in the Italian style of architecture, constructed of Colorado peachblow colored stone and Roman brick. It surrounds an open court which is terraced, and has grass-plats, fountains, and beds of flowers. The hotel has 200 guest-rooms; it is well heated, liberally supplied with open fireplaces, and is lighted by electricity.

Glenwood Springs has an ample water-supply, and sewers in the populous section.

The meteorological record for Glenwood Springs is incomplete. It is only possible to obtain a record of the temperature for two full years, and of the rainfall for three full years, which, given by seasons, are as follows:

	Winter.	Spring.	Summer.	Autumn.	Year.
Monthly mean temperature,	27°	50°	69°	47°	48°
Rainfall (total),	4.90 in.	3.04 in.	3.80 in.	4.22 in.	15.96 in.

Monthly mean temperature for January (two years) . 22°
Monthly mean temperature for July (two years) . . 72
Mean yearly minimum (three years) . . . . —5
Mean yearly maximum (three years) . . . 100

ANALYSIS OF THE YAMPA SPRINGS (GLENWOOD SPRINGS, COLORADO) (made by C. F. Chandler, Ph.D., June, 1888). Temperature 124° F.

	Grains in U. S. gallon of 231 cubic inches of water.
Sodium chloride .	. . 1089.83
Magnesium chloride	. 13.09
Sodium bromide . .	. 0.56
Sodium iodide . . .	trace
Calcium fluoride . .	trace
Potassium sulphate . .	24.04
Calcium sulphate . .	92.38
Lithium bicarbonate .	0.22
Magnesium bicarbonate .	13.55
Calcium bicarbonate	24.37
Iron bicarbonate .	trace
Sodium biborate .	trace
Sodium phosphate .	trace
Alumina . . .	trace
Silica . . .	1.97
Organic matter . .	trace
Total	. 1250.04

Steamboat Springs (elevation, 6800 feet) is one hundred miles south of Rawlins, Wyoming, on the Union Pacific Railroad, and eighty-five miles northwest from Hot Sulphur Springs, Middle Park, Col. Leaving the Denver and Rio Grande Railroad at Wolcott, the tourist first goes by stage to McCoy's, distant twenty-two miles to the north. From there a stage can be taken the next morning at 6 A.M. for Steamboat Springs, sixty miles further north. This part of the ride occupies about twelve hours.

The road goes through Egeria Park, which is noted for its lovely scenery. Elk, deer, antelope, bear, and mountain-sheep are said to be fairly plentiful on the mountains in this vicinity, and the best trout and grayling fishing in the State is found between Zampa (formerly Egeria) and Steamboat Springs, on the Bear River.

The town of Steamboat Springs has about 400 inhabitants. It is situated in a narrow valley surrounded by mountains. Within a radius of two miles from the centre of the town there are over 300 springs, of which more than 60 are known to contain soda, magnesia, sulphur, and iron. Some of the cold, sparkling waters are very palatable, noticeably those of the Iron Spring. They vary in temperature from cold to 156° F. Many of the hot springs send off clouds of steam, which are particularly noticeable on a cold morning.

There is no detailed weather-record for Steamboat Springs. It is on the western slope of the Continental Divide, and has a larger rainfall and snowfall than many localities further south in the State. It is a beautiful place for camping-out and hunting during the summer and early autumn. The value of the hot springs is very great, and it would be considered a marvellous resort if it were more accessible and better known.

The mean record of rainfall for Steamboat Springs for three years is as follows: winter, 9.30 inches (including melted snow); spring, 6.03 inches; summer, 0.85 inch; autumn, 3.11 inches. Total annual, 19.29 inches.

The mean monthly temperature at Steamboat Springs in 1893 by seasons was as follows: winter, 19°; spring, 37°; summer, 59°; autumn, 44°; year, 40°. Monthly mean for January, 16°; for July, 61°. Maximum, 91°. The minimum temperature was below 32° every month in the year.

Grand Junction (elevation, 4500 feet; population, 2500). The town of Grand Junction, which is on the Denver and Rio Grande and Colorado Midland Railroads, lies in a valley fifteen miles long, opposite the junction of the Gunnison and Grand Rivers. It is a busy town in the centre of what will probably be in time the greatest fruit-growing region of Colorado—the moderate altitude, adobe soil, shelter from heavy winds and storms, and ample supply of water for irrigation all contributing to that result. All deciduous fruits flourish in the valley, but it is especially noted for its fine peaches.

Like Glenwood Springs, Grand Junction is on the Pacific slope, west of the Continental Divide, which is crossed at an elevation of over 10,000 feet on the Denver and Rio Grande Railroad, and on the Colorado Midland Railroad at an elevation of 11,500 feet. There is no complete weather-record. The mean temperature for two years (1893–'94) by seasons was as follows: winter, 31°; spring, 52°; summer, 75°; autumn, 52°; for the year 53°. The minimum temperature in 1894 was 3°; maximum, 100°.

The rainfall of the valley is very low, the annual mean for five years being 8.45 inches, divided as follows: winter, 1.88; spring, 2.38; summer, 1.97; autumn, 2.22 inches.

The annual snowfall is 34 inches (in Colorado equal to about 3 inches of rainfall), of which nearly 19 inches fall in February. Snow seldom remains on the ground for any length of time. The wind-movement is small, although an occasional lively wind-storm blows through the valley.

The town-supply of water for domestic purposes in Grand Junction is taken from Grand River, and is poor. An appropriation has recently been made, however, for the construction of a pipe-line from the mountains, which will furnish pure water. There are no sewers.

The valley has mild winters, the thermometer seldom reaching zero. The summers are hot. When it was an arid plain the humidity was low. How far the universal soaking of the soil by means of irrigation will affect it as a resort for pulmonary invalids remains to be seen. During the winter months, however, this objection does not exist.

Springs.

Colorado is favored with a great number of mineral springs, the value of which is known; but it may be said now, as it was said several years ago, " it is impossible to do full justice to the curative properties of the whole number, as but a small portion has as yet been subjected to quantitative analysis and few have been tested as to the effect of their medicinal qualities upon diseases."

The best known of these resorts are Manitou and Glenwood, where villages have sprung up around the springs, and there are first-class hotels.

The **Siloam Springs** (6000 feet) are in the centre of a beautiful valley on the Grand River above Glenwood Springs, on the line of the Denver and Rio Grande Railroad.

The springs range in temperature from 94° to 104°, and are said to resemble the combination of saline waters found at Saratoga.[1] The total flow is about 10,000 gallons per hour. They are similar to those at Glenwood, and can be used in the vapor-caves and in an open pool one hundred feet in diameter.

The **Seltzer Springs** at Springdale, ten miles from the town of Boulder, are situated near the edge of James Creek, surrounded by wooded mountains and in the midst of picturesque scenery. They are reached by stage from Boulder. A small hotel has been erected near the springs.

Idaho Springs (elevation, 7700 feet; population, 2000) is thirty-five miles west of Denver, on the Colorado Central Railroad, a branch of the Union Pacific Railroad. The town is situated in the Clear Creek Cañon, and is reached through picturesque scenery ; on account of its accessibility from the plains it promises to become eventually a desirable retreat for invalids.

The mountains on the north protect the town from inclement winds, "while the less precipitous mountains to the south allow a desirable long continuance of sun-exposure."

There are several hotels and boarding-houses.

There are a number of warm springs, and hot springs with a temperature of 110° to 120° F. A little further up the valley are the Cave and Tunnel Springs and bathing-pool.

The town of Idaho Springs is supplied with pure water from mountain-springs.

The soil is sandy and porous.

The mean temperature for winter for five years is 28°, with over 2 inches of melted snow, falling on 20 stormy days. Relative humidity for winter, 50 per cent. The minimum temperature in 1892 was —14°.

The mean temperature for spring is 40°. Total precipitation 3 or 4 inches, and cloudiness about the same as for winter, with perhaps more stormy days.

The mean temperature for summer is 62°; relative humidity, 50 per cent. Rainfall about 7 inches. About 36 cloudy days. The summer-showers are usually hard and brief. The water is quickly drained away.

The mean temperature for autumn is 44°. Rainfall from 2 to 3 inches on 15 days.

[1] Mineral Springs of Colorado. C. Denison, M.D., 1889.

The remarkable Steamboat Springs are referred to on page 267.

In Middle Park are the **Hot Sulphur Springs**, previously mentioned, situated on the banks of the Grand River. It is a delightful region for camping-out. In a basin, bounded by peaks of the snowy range, southwest, and high up among the mountains, are effervescing soda springs, not yet improved.

In South Park are Hartzell's **Hot Sulphur Springs** and the **South Park Springs**, saline alkaline waters, on the bank of the South Platte River. There are also other undeveloped chalybeate and alkaline or sulphur springs in the vicinity.

In the small valley occupied by Cañon City are several hot and cold springs, chalybeate, sulphur, and soda, some of which have been improved. (See Cañon City.)

South of Cañon City sixty miles, in Poncha Pass, are the **Poncha Springs** (7480 feet), a collection of hot and cold waters. There are thirteen hot springs at a temperature of 120° F., with sulphur and soda as the principal constituents.

Twenty or thirty miles north of Poncha Pass, situated in the Arkansas Valley, at an elevation of 8000 feet, are the **Chalk Creek Springs**, twelve or more in number, at various temperatures, the hottest being 150° F.

The **Pagosa Springs** (7100 feet) lie in a picturesque country on the northern bank of the San Juan River. These are purgative alkaline waters, with a large excess of sulphate of sodium, and they rise to a temperature of 140°.

In **Wagon Wheel Gap** (8500 feet) are purgative and alkaline hot springs of great local reputation, while near Cañon Creek and Uncompahgre are great numbers of mineral springs, varying in temperature from 120° to 140° F.

An analysis of the Colorado Carlsbad Spring will be found on page 253, and of two mineral artesian wells in Pueblo, the lithia water and the sulphurous chalybeate "magnetic" water, page 263.

Utah.

Salt Lake City (elevation, 4300 feet; population, 70,000). It is about ten miles from the northwest corner of the city to the shores of the Great Salt Lake.[1] Some of the popular bathing-beaches are

---

[1] The Great Salt Lake (elevation, 1200 feet) is about 80 miles long by 30 or 40 miles wide, covering an area of 2360 square miles. It has an average depth of 12 feet, with about 60 feet as its greatest depth. It has no outlet. The last analysis of the water showed 16.88 grains of mineral ingredients in one U. S. gallon, which is about five times as salt as the Atlantic Ocean.

a little more distant. Away to the east and north are the Wasatch, and to the west the Oquirrh range of mountains, distant about twelve miles.

The streets of Salt Lake City are wide, and, except in the active business-centre, well shaded with trees. In the business portions of the city most of the streets are paved. There are large, well-equipped stores and several good modern hotels. The city has an excellent service of electric street-cars.

Water for domestic purposes is brought from the mountains. It is pure, but hard. Ditches along the sides of the streets carry the water for irrigation. In the residence portion of the town are well-kept lawns. The soil is adobe.

There is a town-system of sewers.

Salt Lake City is one of the three Western cities of good size possible for the residence of those to whom a sunny climate is necessary and who desire to settle in an active business-centre. The other two large cities are Denver, which shares with Salt Lake City the advantage of altitude, and Los Angeles, which is equally sunny but exposed to ocean-influence.

Salt Lake City is accessible by rail. Its moderate elevation is a point worthy of consideration by a certain class of invalids.

On some of the higher benches or *mesas* near the town desirable situations for residence can be found. It is important for invalids in all large cities to be at a safe distance from any possible ill-effects from the chimneys of manufactories or smelters.

The proximity of the cañons and mountains, the hot springs, and the immense inland sea, with its bathing-resorts, give Salt Lake City unusual facilities either as a temporary resort during the summer and autumn or as an attractive place for a permanent residence.

Compared with some of the other high-altitude weather-stations, Salt Lake City has a somewhat capricious climate. This may be partly attributed to its being situated west of the Continental Divide, and therefore more directly subject to the far-reaching Pacific coast influence than localities like Denver and Colorado Springs, which are on the eastern slope and have the protection, on the west, of the great ranges of the Rocky Mountains.

While the Government-records show it to be very dry in summer, it is also exceedingly hot. In the autumn it is moist compared with other cities on the plains, and in the winter it is colder and more stormy, judged by the same standard. The lowest record of temperature is 20° below zero, in January, 1883. The mean temperature for the month of January is 27°; for July, 76°; for the year, 51°.

Average number of days above 90°, 30; below 32°, 109. The annual mean of cloudy days is 88; stormy days, 81 (means for six years). Rainfall 18.9 inches a year (from records for eighteen years), of which 5½ inches is the average rainfall for the winter.

The wind-movement throughout the year is moderate, being about five miles per hour.

New Mexico.

In going from Colorado into New Mexico by rail the change of climate is noticeable. The traveller passes from the cool air of the northern side of the Raton range, in Colorado, and after a glimpse of the Spanish peaks crosses the State-line, plunging at once into the blackness of a half-mile of tunnel, which pierces the mountains at an elevation of 7600 feet. From this he emerges on to the sunny, southern slope of the range, where it seems perceptibly warmer and brighter than on the other more shaded side. It is a milder air, certainly, and bears promise of long, sunny winter-days and freedom from snows.

Topographically, New Mexico is a high plain, called by the Spaniards "mesa" or table-land. At Santa Fé this plateau rises to 7000 feet, while in the Lower Pecos Valley it is depressed to about 3500 feet above sea-level.

The Mesilla Valley, forty miles above El Paso, has about the same elevation, from which the land rises to the northwest and north. The average elevation of the State is probably not far from 5000 feet. The great size of these far Western States has always to be borne in mind in dealing with their peculiarities. In this brief summary it is not proposed to treat of New Mexico's entire 390 miles of length, but only of those portions of the State suitable for the residence of invalids, so far as they are known.

The rainfall for the year 1891 (which was evidently a rainy year) varied from 4.55 inches at Deming to 32.83 inches at Chama. The general average for the State is 13¹ inches, with, of course, a quantity as large again falling on the mountain-summits, principally in the form of snow. In New Mexico there are several "islands of greater rainfall," where a much heavier precipitation may be ex-

[1] Rainfall and Snow of the United States. Prof. M. W. Harrington, Washington, 1891.

pected than in the country more than fifteen or twenty miles away
on either side. Two of these are especially noticeable—one in the
mountain-ranges north of Santa Fé and Las Vegas and west of
Springer, where the rainfall surpasses by 5 or 10 inches a year that
in the area immediately around it; and the other around Fort Stanton, in the mountains to the east of the Rio Grande. The rainy
season, as in Colorado, is in July and August. In the vicinity of
El Paso 40 per cent. of the rainfall of the year descends in these
two months.

Captain W. A. Glassford, of the Signal Corps, U. S. A., describes
the process of annual precipitation as follows :[1]

"In New Mexico the winter-precipitation does not begin until
the earlier days of January. By April the winter-rains are definitely ended, but in the eastern or Atlantic Divide section (the Pecos
Valley region) an area of considerable precipitation remains. This
follows the high summits of the Sangre de Cristo range until it
sinks into the tableland of 4000 feet, thence it tends southeasterly
across the Pecos head-waters region as far as Gallinas Springs, where
it sharply curves to the north and extends over the Raton range.
The winter-rains, according to the nomenclature of the meteorologist,
are marked by curves of from 1 to 7 inches of precipitation, running
nearly parallel. These curves are outside the mountain-lines and
indicate the diffuse and diverting influence of topography on the
aqueous currents borne to New Mexico from the South Pacific Ocean
across Arizona.

"The summer-rains are otherwise influenced, and the highest pre-
cipitations appear upon the levels west of the Canadian River, and
upon the cañon-course of the Pecos, which includes Las Vegas and
Fort Union ; at this point the fall reaches 17 inches. The lowest
summer-precipitation is found in sections most favorably influenced
by the winter-rains. The minimum is found in the southwest.

"Step by step the humid winds are drawn across over graduated
plateaus and extrusive summits, and at each higher step discharge so
much of their moisture as is a surplusage over the saturation-amount
of atmosphere of a given tenuity at a given temperature. . . .
There is nothing violent in these systematic draughts of the humid
air from the sea toward the continental cyclones or vortexes ; the
air is chilled by the seasonal causes which make the winter climate.

---

[1] New Mexico. Official Publication of the Bureau of Immigration, Max Frost, Secretary.
Santa Fé, 1894.

The earth-surfaces soon become largely covered with snow, and their radiating influence is thus mechanically obliterated; the air lies in practically even strata of uniform temperature. The humid wind is drawn along these ruling conditions; on every plateau it discharges down to the point of saturation; the diminution in actual amount of moisture is constant and large; by the time it overlies the Rio Grande trough its last available moisture has been condensed by the heights of the Continental Divide, and shifts down to leeward. Such precipitation as is induced appears as snow, which forms a storage whose supply is constantly utilized until July. With the vanishing screen of snow the conditions proportionately alter. The surface of the elevation, with its soil and rock-masses, ceases to reflect the incident heat-rays of the sun, but absorbs much of it; at the same time it radiates the heat which it receives, currents are formed in the surrounding air, and the mountain becomes a focus of activity, about which are currents rushing rapidly skyward and a lateral draft to supply the place of the air withdrawn by its action of convection; its excess of moisture and consequent precipitation therefore reach the maximum."

The greatly decreased altitude and more southern latitude of the lower Rio Grande Valley together combine to provide a place of residence for winter for those who seek refuge from the higher and cooler regions above.

In referring to this country El Paso will be considered as belonging, climatically, to New Mexico.

The climate of the elevated towns in the northern part of the State is similar to that of Colorado. There is a great resemblance, for instance, between the temperature of Santa Fé (7000 feet) and Las Vegas (6500 feet) and that of Denver. Silver City (5800 feet), in the southwest, has a milder climate and is sheltered from severe winds. Albuquerque (5000 feet) is a larger town, with better accommodations, but more exposed to the wind. Las Cruces, Eddy, and El Paso (3700 feet) offer each about the same elevation, but varying grades of accommodations.

The general mean of relative humidity during the year is about 40 per cent., which, with the average annual temperature of about 60°, shows for the year 2.30 grains of moisture to a cubic foot of air, or about the same dryness as Colorado—the higher temperature increasing slightly the amount of absolute humidity.

The fine climate of New Mexico is its greatest asset. Unfortunately, there are no cities of importance, and the comforts of civilization are not easy to find, except in two or three towns.

"Sun, silence, and adobe!" is the graphic summary of the country made by Charles F. Lummis in his delightful book, *The Land of Poco Tiempo.* The adobe Indian pueblos where he lived would, of course, be avoided by invalids ; but even Santa Fé will be found somewhat unsanitary, although quaint and interesting. 320 cloudless days a year can be claimed for any portion of this great domain, but gravel-soil, soft water, and good supplies are more difficult to rely on.

The climate is superb. It is in the details of living that the visitor is most liable to meet with disappointment. Among other testimony relating to the climate that of Ex-Surgeon-General W. A. Hammond is valuable. He says : "New Mexico is by far the most favorable residence in the United States for those predisposed to or afflicted with phthisis . . . In a service of three years in New Mexico, during which period I served at eight different stations, ranging from the extreme northern to the extreme southern part of the territory, I saw but three cases of phthisis, and those were in persons recently arrived from the United States. Inflammation of the lungs is also very infrequent, as are likewise pleurisy and bronchitis."

Dr. W. M. Yandell, of El Paso, asserts that if a mild climate during the cold season is desired, New Mexico and Arizona, south of the thirty-fifth parallel of latitude, furnish by far the best winter climate in the United States for consumptives.

The summers, however, in the southern portion are exceedingly hot. The heat begins to gain in strength after the middle of March and increases up to September and October. In spite of the low humidity the impaired vital forces of an invalid are apt to be weakened by the prolonged high temperature.

On the mountains a comfortable summer climate can be found, the nights in particular being usually cool. Santa Fé has very few hot days, and for those who enjoy camping-out the fine, wooded country of the Upper Pecos—where there is said to be plenty of game and where the streams "abound" in trout—or the top of the Black Range, will bring the visitor into a delightful summer climate and an interesting country.

The annual wind-movement in the elevated northern portions of the State is about the same as that of Colorado, but the wind-move ment of the lower and more southerly portions is less, as is shown by the tables. A disagreeable feature of all open, treeless plains is the occasional dust-storm, which is the more objectionable when the soil is adobe in character, producing a very fine dust, which penetrates every crack and cranny, no matter how obscure. Furious dust-storms are not frequent and are rarely of longer duration than a few hours.

Springer (elevation, 5700 feet; population, 900) is situated on the main line of the Atchison, Topeka and Santa Fé Railroad, about fifty miles south of the Raton Range, Springer is the centre of a rich and fertile country. Forty miles to the west are the Rocky Mountains, as represented by the Taos and Sangre de Cristo ranges, whose average elevation is 10,000 feet and whose peaks reach an altitude of 13,000 feet. The soil is adobe, the clay having a depth of several feet. The country is under an immense irrigation-system, and large lakes are projected in the foot-hills, thirty or forty miles northwest of the town.

Large ranches have been the rule previously throughout this region, which has been devoted mainly to stock-growing and the cultivation of alfalfa. As the irrigation-ditches are extended smaller fruit-ranches will become more numerous. The climate is fine, with apparently an uneven rainfall. The Pacific passage-winds which reach these ranges are relieved of their last moisture, and the precipitation even on the eastern slope is considerable. The snowfall is said to be from 15 to 20 feet on the level on the mountain-summits, or in water-measurement about 50 inches, an estimate that is probably too liberal.

In the valleys the mercury rarely falls as low as zero, and the summer-heat is not excessive. Good well-water is said to be obtainable, and there are some artesian wells.

There are no complete weather-records, but from the voluntary observers' report the following facts are obtained :

	Winter.	Spring.	Summer.	Autumn.	Year.
Temperature, 1892,	35°	53°	68°	51°	50°
Maximum,	99°		Minimum,	—5°	

Winter-estimate is made up of two Decembers and January, as the month of February is missing. Spring-estimate is for April and May only. March record missing.

	Inches.	Inches.	Inches.	Inches.	Inches.
Rainfall, 1891,	3.04	3.07	10.53	5 28	21.92
" 1892,	0.63	5.23	5.31	2.78	13.93

The precipitation in the summer of 1891 was unusually heavy.

The record for 1893 is incomplete. The mean monthly temperature for spring was 50° and for summer 71°. The total rainfall for spring was 11.93 inches and for summer 7.02 inches.

Las Vegas (elevation, 6500 feet; population, 6000). Las Vegas, like most of the Mexican towns in New Mexico, has a mixed population, consisting perhaps half of Americans and half of Mexicans and Indians. The last two live in low adobe-dwellings.

The town is hot and dusty in summer, while in winter the temperature occasionally goes to zero or below. During the day and in the sunshine, however, it will be found warm and comfortable. The town lies somewhat exposed to north winds.

Of the rainfall of 20 inches a year about half falls during July and August, with some rain during September and April. During the first four months of the year high winds blow, as is the case in most towns of the plains. Las Vegas is a healthy place, with the generally fine climate and bountiful sunshine of New Mexico.

Las Vegas Hot Springs (6700 feet) are seven miles from the town, lying in a position that is well sheltered from the winds. Here one can get mud-baths of great efficacy in rheumatism and kindred troubles. There are both hot and cold springs. A general analysis of the hot springs made by Dr. Walter S. Haines shows the temperature to be 140° F.

	Grains per gallon.
Calcium carbonate	0.89
Magnesium carbonate	0.15
Sodium carbonate	8.38
Potassium carbonate	0.28
Sodium sulphate	3.45
Sodium chloride	14.68
Silica	3.50
Alumina	0.10
Volatile and organic matter	0.32
Lithium carbonate	trace
Sodium bromide	trace
Total solids	31.65

"The Montezuma," built by the Santa Fé Railroad, is one of the largest and best hotels in the mountains, but is at present closed. It is about 7000 feet above the sea. There are picturesque cañons to explore and fine mountain-scenery.

Santa Fé (elevation, 7000 feet ; population, 7000). This still retains the appearance of a Mexican town, the outgrowth of an old Indian pueblo. The *plaza*, the winding streets, and the adobe-houses are interesting features to visitors.

The town lies on a treeless plain, at the base of one of the spurs of the Rocky Mountains. The general trend of the valley or plain is west southwest, the mountains affording protection to the north and east. Within thirty miles are peaks of 12,000 and 13,000 feet elevation. The low hills are covered with a growth of piñon-trees. The town has of late years improved its water-supply, but much is yet to be desired in the matter of drainage, and there are few shade-trees. The soil is light and sandy.

The climate is not very different from that of Denver. It is somewhat cooler during the summer, not quite so cold in winter, and a little less windy throughout the year.

The windiness of Santa Fé has been so frequently alluded to that it may be well to call attention to the fact that the hourly velocity of the wind during the year is much less than in New York, Boston, or Chicago. The Government record for ten years shows for Santa Fé an annual mean of 6.4 miles; for New York City, 9.4 miles; for Boston, 10.9 miles ; and for Chicago, 9.6 miles per hour.

The weather-record for Santa Fé is said to show for the past twenty years but eight days when the thermometer registered over 90°, and but fifteen times during the same period when it went below zero. Average number of days in a year below 32°, 134; cloudy days, 48; stormy days, 72 (means for six years). Average temperature for January, 28°; for July, 70°; for the year, 49° (means for seventeen years).

The maximum temperature in Santa Fé was, during the summer of 1891, 87°; 1892, 90° (one day); 1893, 89° (one day); 1894, 84°; 1895, 87°.

There is a good hotel in Santa Fé, and an excellent sanatorium, managed by Sisters of Charity.

Upper Pecos Valley and Timber Reservation, covering 702 square miles. The region is rugged and mountainous, with numerous small streams, which, flowing south, form the Pecos River.

In the beautiful upland valley, for twenty miles north of the Glorieta Mountains, are scattered stock-ranges and small farms. It is an attractive, park-like district, well-timbered and surrounded by mountains. The altitude ranges from 6000 to 8000 feet.

Not only is the location picturesque, but there are also good soil, water, and climate. Much of the country is wild, and game is found, including bear and deer, while the streams are well filled with trout. It has not been in any way developed as a health-resort, but its natural advantages are too great to allow it to remain long unknown.

Through the lower end of the valley the main line of the Santa Fé Railroad passes, between San Miguel and Glorieta.

The only way of learning to know the attractions of the country at present is by camping-out.

Albuquerque (elevation, 5000 feet; population, 8000). This town is situated in a wide valley, watered by the Rio Grande, and is on the Atlantic and Pacific division of the Santa Fé Railroad system. It is quite well known as a health-resort station. The spirit of the town is progressive, and it has street-cars, waterworks, and electric lights, gas, and sewers. There are churches of all denominations, and a hotel. Albuquerque has a large social club-building, but there are few diversions for visitors. In selecting a place of residence it is important for an invalid to live on the *mesa* or table-land, as far from the influence of the river as possible, for intermittent and malarial fevers are not unknown on the river bottom-land in the "old town".

There is no detailed weather-record. The temperature for 1892 by seasons was as follows: winter, 41°; spring (March missing), 60°; summer (June missing), 77°; autumn, 58°; maximum, 97°; minimum, 15°. Mean for January, 41°; for July, 78°. For the year 1893: winter, 38°; spring, 54°; summer, 72°; autumn, 53°; annual mean, 55°; maximum, 98°; minimum, 11°. Mean for January, 37°; for July, 77°.

The mean annual rainfall is about 8 inches, of which over 4 inches may be expected during the summer-months.[1] No report of the wind-movement is obtainable, but the situation of the town is open and exposed.

Deming (elevation, 4300 feet;[2] population, 2000). This town

[1] Rainfall and Snow of the United States. Prof. M. H. Harrington, 1894. Mean annual rainfall of Albuquerque for ten years, 7.7 inches. Buck's Reference Handbook of the Medical Sciences gives the mean rainfall for 17 years as 8.12 inches. Precipitation for June, July, and August, 4.35 inches.

[2] The engineers of the Santa Fé Railroad have estimated the elevation of Deming at 3600 feet. Barometric determinations of the signal Service give the approximate elevation at 4300 feet.

is situated on a plain about fifty miles square, surrounded by mountains. It is on the line of the Atchison, Topeka and Santa Fé and the Southern Pacific Railways.

The climate is mild, the temperature never going below 32° *during the day* in winter. The mean of the rainfall-record for nine years is 8.8 inches, the precipitation being principally in July, August, and September. The winds begin in February.

The town-water is brought from the Black Range—a spur of the Rockies. There are also a number of windmills for the purpose of raising water.

The following is the mean seasonal temperature for two years : winter, 44°; spring, 63°; summer, 87°; autumn, 64°. Mean for the month of January, 40°.

Silver City (elevation, 5800 feet; population, 3000; latitude, 32° 46' north) is situated on an elevated plateau in the Chihuahua Valley, south of the Pinos Altos hills. It lies at the end of a branch-line of the Santa Fé Railroad, running up forty-eight miles from Deming. Silver City is about ten and one-half hours by rail from El Paso.

The climate is considered very salubrious, being mild during the winter and spring and not subject to sudden changes, while during the heat of summer it is possible to go a little higher into the cooler regions of the pines.

In winter the frosts are said to be less severe than in the valleys of lower altitude along the Rio Grande.

The record of temperature for ten years[1] shows the mean for January to be 37°; for July, 72°; for the year, 54°. By seasons it is as follows : winter, 37°; spring, 53°; summer, 72°; autumn, 55°. The highest record (in June, 1871) is 100° and the lowest 1°. In ten years there were only 22 days above 90° and 6 days below 10°. The average annual precipitation—based on a record for twenty years—is 14.58 inches, of which 8.11 inches usually fall in July, August, and September.

The average number of cloudy days during the year is 37. Buck's *Reference Handbook of the Medical Sciences* states that the relative humidity showed an average of 50 per cent. for one year. The average during the winter was 49 per cent. The same authority

---

[1] Quoted by A. F. McKay, in American Climates and Resorts, November, 1891.

refers to Silver City as "enjoying the exceptional advantage of shelter from the wind." Actual wind-records were not obtainable, but the movement is small.

Silver City has been called an American town with a Mexican quarter, to distinguish it from many of the towns in New Mexico where the old Mexican influence still predominates. The buildings consist largely of brick, which is manufactured there. An ample supply of town-water is obtained from wells flowing into the reservoir above the town. An analysis made by Professor W. E. Waring shows an excellent quality of hard water. His report gave the number of grains of "total hardness" per gallon as 11.72, which would be equal to about 20 parts of "total hardness" per 100,000 parts. Softer water can be obtained, it is said, from some of the artesian wells.

The soil is of a sandy nature. The general rock-formation of the vicinity is principally slate and limestone.

Besides the usual public buildings there are four hotels, one of them a large three-story brick building, and a Sisters' Hospital. There is very little irrigation around Silver City; fresh vegetables and grapes are supplied from neighboring ranches.

This portion of New Mexico offers every advantage for outdoor life. The pine-forests on the mountains are available for camping, and small game is said to be abundant. The roads are good. The heat is more moderate during spring and summer than in the lower river-valleys, and there is more freedom from dust and insects.

Fort Bayard is a postoffice and United States Army post, lying at a lower altitude than Silver City and distant nine miles northeast. The extent of its accommodations is not known to the writer. Its population was over 500 by the census of 1890. The height of Fort Bayard is given in Toner's *Dictionary of Elevations* as 4450 feet above the sea. In 1893 the air-temperature by seasons was as follows: winter, 41°; spring, 52°; summer, 72°; autumn, 55°. Annual mean, 56°; maximum, 99°; minimum, 13°. Mean monthly temperature for January, 41°; for July, 74°. The rainfall during the year 1893 was slight in all the months except July, August, and September, when 12½ inches fell.

The **Valley of Mesilla** (see Las Cruces) is about seventy miles long and from one to six miles wide. Its elevation varies from 4000 feet at the Rio Grande, to 7000 feet among the foot-hills. The climate is mild, ranking high for dry air and continuance of sunshine. Snow is rarely seen.

The valley produces a fine variety of grape containing a large amount of sugar. Because of the fertile soil in the lower portion this district is one of the garden-spots of New Mexico. Mesilla was formerly noted for being one of the least windy of the Government weather-stations, but no official observations have been taken for several years.

The wind-movement for Mesilla, given in Dr. Denison's charts,[1] shows for the year 2.3 miles per hour and for winter 2.1 miles per hour.

The average temperature by seasons is : winter, 42°; spring, 61°; summer, 78°; autumn, 60°; year, 63°.

Buck's *Reference Handbook of the Medical Sciences* gives the mean relative humidity for one year as 41 per cent., and for three months of winter as 43 per cent.

Las Cruces (elevation, 3800 feet ; population, 3000). This is one of the Mexican towns above El Paso, distant forty-three miles —a little over an hour's ride by rail on the Atchison, Topeka and Santa Fé Railroad. The Mesilla Valley in which it lies is wide, and when irrigated the soil is fertile. Fine grapes are grown, and large fields of alfalfa. The principal supply of water for irrigation is drawn from the Rio Grande ; there are also a number of artesian wells, which furnish water for domestic use. The water, while alkaline, is said to be wholesome, and is best when obtained near the centre of the valley.

The houses in Las Cruces are usually one story in height, and are built of adobe. There are two small hotels. Good board in the town can be had at Mrs. Barker's, and in Doña Ana, one mile from the Las Cruces railroad station, at "The Alameda."

The mean temperature for January is 39°. Extreme yearly range from —2° to 106°. The average rainfall for twenty years has been at the rate of 7 inches a year. The number of cloudy days from August, 1892, to July, 1893, inclusive, was 20. The average wind-movement is said by one writer to be about 5½ miles per hour, which is probably an overestimate.

Winter visitors can remain until April without suffering from the heat.

---

[1] Climates of the United States in Colors. Charles Denison.

The **Van Patten Mountain Camp** is situated fifteen miles east of Las Cruces, at an elevation of 6000 feet, in a sheltered notch of the Organ Mountains. It is protected by rocky walls from all winds, and has an abundance of pure mountain-water.[1] Small game can be found in the vicinity. The greatest height of the Organ Mountains is about 8000 feet.

Twenty-five miles east of Las Cruces, on the eastern slope of the Organ Mountains, Dr. Petin has selected a site for a sanatorium as affording the greatest advantages.

Its claims to favorable consideration are " moderate altitude of 4800 feet, good mineral-water, no malaria or dust, porous soil, temperature quite even, with an annual average of 62° F. Scarcely any snowfall. Rainfall about 4(?) inches a year. But little wind. There is an abundance of game of all kinds and good fishing. Beautiful shade-trees grow at the foot of the mountains. There any patient can sleep out of doors eight months of the year without fear of catching cold."

As yet there are no accommodations for invalids. It is necessary to camp out and rough it.

Pecos Valley. Principal town Eddy (elevation, 3200 feet ; population, 2000). An extensive scheme of irrigation is transforming this valley in Southern New Mexico from an alkali-desert into a fruitful region. There are two or three large canals now in operation, the water being taken from the Pecos River, which not only carries a large amount of rainfall and snowfall from the Rockies, but is fed along its course by numerous living springs, flowing from fissures in the lime-tone which underlies the country. This river-water has been analyzed by Professor Precht, as follows:

	Grains per gallon.
Calcium and sodium chloride	69.23
Sodium sulphate (Glauber's salt) and magnesium sulphate (Epsom salt)	34.62
Calcium carbonate	39.00
Calcium sulphate	30.00
Total solids . .	172.85

The amount of alkali present is indicated in another way by the deposit of over two tons of salts yearly, per acre, on the irrigated land, which is claimed to be of benefit to the soil for fertilizing-

[1] A. F. McKay, in American Climates and Resorts.

purposes. Water can also be obtained by means of artesian wells.
Eddy secures its town-supply in this way.

There is no timber in the valley except mesquite,[1] but plenty of
timber is found fifty to one hundred miles west. The air is pure and
dry. There are from 320 to 340 days of sunshine during the year.

The winters are mild and the summers hot, the nights being
usually cool. The extreme range of temperature is reported to be
from 102° above zero to 7 ° below, but the mercury in winter
seldom drops so low. The annual average is 63°.

The rainfall is about 8 inches yearly.

The wind-movement is not obtainable. The most objectionable
feature is said to be the spring-winds, which are not dangerously
heavy, but tiresomely persistent and charged with light sand.

Eddy is a small town recently built. Fairly good accommodations can be secured at "Hotel Hagerman," or furnished ranch-
houses of moderate size can be hired.

The Eddy *Argus* for April, 1894, contained a review of the
weather for the preceding five months, which showed the following :

	Mean temperature.	
	6.30 A.M.	10 P.M.
November, 1893	36°	48°
December, 1893 . . .	35	43
January, 1894 . . .	31	32
February, 1894 . . .	32	41
	2 P.M.	
March, 1894 40	71°	53
Average temperature for the month of March .	. 59	
Cloudy days, November 1, 1893, to April 1, 1894 .	. 43	
Windy days, November 1, 1893, to April 1, 1894 .	. 52	

During the winter of 1894–'95 the monthly means at Eddy were
as follows (based on the reports of G. W. Lane, observer) :

	Temperature			Rainfall, inch.	Cloudy days.
	Mean.	Mean max.	Mean min.		
November, 1894,	54°	70°	37°	0.00	0
December, "	45	58	32	0.02	1
January, 1895,	42	57	27	0.65	2
February, "	37	50	23	0.19	9
March, "	55	72	38	0.00	1
April, "	64	80	47	0.10	1

[1] The *mesquite* is a sort of shrub sometimes growing to a height of thirty feet or more, but
usually scrubby, forming dense clumps of bushy thicket or chaparral.

Snowfall in February, 1895 .   .   .   .   . 2.10 inches.
Extreme minimum temperature in January, 1895   .   .   5°
   "   "   "   February,   "   .   . —2
   "   "   "   March,   "   .   . 22
   "   maximum   "   March,   "   .   . 88
   "   "   "   April,   "   .   . 93
Cloudy days from November 1st to May 1st   .   . 14 days.
Wind-velocity not given.

The temperature of Eddy by seasons is as follows (based on the records for the year 1894, except for March and April, which are the means of 1894 and 1895): winter, 40°; spring, 66°; summer, 79°; autumn, 65°; mean for the year, 62°. Extreme maxima for summer—June, 104°; July, 103°; August, 96°; all for the year 1894. Extreme minima for winter 1894 and 1895—December, 18°; January, 7°; February, 11°. During the cold wave of February, 1895, the extreme minimum temperature was 2° below zero. The mean daily range of temperature during the winter months is 34°, and the greatest daily range will average 41°.

During the summer of 1894 the temperature at Eddy rose above 90° 73 times, and above 100° 8 times.

Roswell. The railroad has been completed for seventy-five miles north of Eddy to Roswell, which is situated at an altitude of 4000 feet, and has a population of about 1000. The older ranches are at this end of the valley.

MINERAL SPRINGS.

Like all mountainous countries, New Mexico has a great number of mineral springs—both hot and cold—but only a few of them have been analyzed or developed. The best known are the Las Vegas Hot Springs (6700 feet), where the Atchison, Topeka and Santa Fé Railroad has built a fine hotel. An analysis of these waters will be found in the description of this resort.

The Folsom Hot Springs (6500 feet), near Alps, in Colfax County, are said to have medicinal properties, but of what nature is not stated. The Ojo Caliente Springs (6290 feet) are on the creek of the same name and twelve miles from Caliente station, in the southwestern portion of Taos County. They have a temperature of 108° to 114° F. The Jemez Hot Springs (6700 feet), in San Diego Cañon, Benalillo County, are in two groups, two miles

apart, and they vary from 70° to 168° in temperature. Some of
them flow from caves of carbonate of lime. Their ingredients are
principally sulphate and carbonates of sodium, calcium, and magnesium. At Cherryville, near the Black Range, southwest of Socorro,
are some hot springs, close by the falls of the Alamosa. The elevation
is 6500 feet and the temperature 130°. They can be reached from
Engle. In the southwestern part of Socorro County, on the west of
the Gila or Diamond Creek, is a group of springs (elevation, 5500
feet; temperature, 130°) similar to the Jemez Springs. They can
be reached by wagon from Silverton. Twenty-four miles north of
Deming and three miles from Hudson are several springs, situated
at an elevation of 5780 feet. They contain soda, lime, and magnesia, and the water has a temperature of 130°.

Arizona.

An inquiry into the climate of Arizona discloses the fact that it is
climatically distinct from each of its neighbors—New Mexico and
California—and has natural laws of its own, although these laws are
modified, in turn, by the climatic influences of both these Pacific coast
and the Rocky Mountains.

Captain W. A. Glassford, of the Signal Corps, U. S. A., prepared in 1890 an exhaustive "Report on the Climate of Arizona,
with particular reference to the Rainfall and Temperature," which
is of the greatest value to all students of the subject and from which
the author has drawn freely. Although the report was the result
of an investigation undertaken in order to learn the possibilities of
irrigation in the territory, the information gained is also of value
for the benefit of invalids for whom the curative power of climate
is necessary.

To understand the geographical position of Arizona attention
must first be directed to the presence of the great Continental
Divide, which passes through the western portion of New Mexico
in a fairly direct course north and south along the one hundred
and eighth degree of longitude.

West of this line the waters all flow toward the Pacific; and
this vast country, which decreases steadily in height in a series of
plateaus to the southwest corner, is influenced by the winter rainy
season of the California coast.

The northern half of the territory is also affected by another great divide, running southeast and northwest (from the point where latitude 34° crosses the New Mexican boundary to the San Francisco Mountains, north of Flagstaff). This ridge separates the watersheds of the Gila and Colorado Rivers, and may be roughly described as lying along latitude 35°.

"The axis of the mountain-system of Arizona," says Captain Glassford, "is remarkably well defined and appears with the utmost distinctness, not only in the general trend of the main mass of elevation, but also in minor ranges and notably in detached spurs, often widely separated from the plateau-system, to which, on the score of altitude, they may claim to belong. With sufficient accuracy to satisfy all legitimate demands of the present inquiry, the direction of the mountain-axis may be placed at northwest and southeast." In the valley of the Rio Santa Cruz, in Pima County, there are fifteen peaks rising to an altitude of more than 3000 feet from a *mesa* 2000 feet high. In the Gila valley are twelve members of a butte-system rising from 1000 to 2000 feet above benches 1000 to 3000 feet high. The tableland of 3000 feet is crowded with sierras of 5000 feet and upward, and on the high plateau of 5000 feet are six examples of mountain-masses rising from 7000 to 9000 feet, and culminating (in the group called the San Francisco Mountains) in three peaks—Humphrey, Agassiz, and Humboldt—of which the first two have an elevation of nearly 13,000 feet above the sea. In no case does the mountain-system diverge from the characteristic axial direction.

"It must be remembered that the Pacific Ocean is the reservoir of Arizona, and an important result of this uniformity of the mountain-axis, carried consistently over five hundred miles, is that the prevailing moisture-bearing wind, being from the southwest, comes at right-angles to the broadside of the mountains, and thus encounters the maximum bluff-surface.

"In other words, the passage of the rainy winds across Arizona is by no means an easy gliding over an inclined plane, but the laborious ascent of a flight of steps.'"[1]

For meteorological purposes Arizona is divided into three series of elevations :

1. THE PLAIN. This embraces about one-third of the territory which lies to the south and west below the level of 3000 feet, and

[1] Captain W. A. Glassford.

includes most of the desert-country that has given Arizona its unenviable reputation for heat and discomfort. On this low plain the rainfall is only from 2 to 6 inches during the year, and, including the desert on the California side of the Colorado River, the records approximate the absolute minimum of rainfall of the world.

It was, however, a careless estimate of the early emigrants to consider it all a desert, as this area—as great as that of Italy or the six New England States—has a rich alluvial soil along the river-bottoms, brought down from the hillsides by centuries of washing, which is found to be capable under irrigation of growing almost anything. Oranges, lemons, almonds, figs, apples, grapes, etc., all thrive and even mature earlier than in Southern California. The mountains above can furnish the needed supply of water if it is properly stored, and this rich land promises to be eventually one of the most productive agricultural sections of the United States. Its intense heat, unfortunately, forbids its consideration as a health-resort except during the winter months, when its cloudless skies and moderate temperature afford perfect immunity from the cold and snow which characterize that season in northern regions.

2. THE PRO-PLATEAU is a bench of from 3000 to 5000 feet elevation, which, from its geographical and physical relations to the high plateau covering fully half the territory, has been distinguished by this term. "It closely follows the axial inflection of the mountain-system, although its continuity is somewhat interrupted by more or less detached spurs of its higher neighbor. Across the central portion of the territory it preserves, with considerable uniformity, a mean width of less than one hundred miles. Widening at the cañon of the Gila, it covers the whole southeastern corner of the territory. . . . The pro-plateau is so narrow a strip for the greater part of its length, and so vestibular in its relation to the plateau, that in the absence of climatic data it should be provisionally included in the great plateau-mass which overshadows it. This may be well done with all that portion lying northwest of the Gila. The southeastern expansion of the pro-plateau, embracing portions of the counties of Graham, Pinal, and Pima and the whole of Cochise, is so marked by two systems of extrusive highlands, each composed of a considerable number of extensive masses of elevation reaching in every case the altitude of the plateau and in some cases 1000 or 2000 feet higher, that this region may be rationally included

19

in the discussion of the rain-making influence exerted by the extrusive summits of the plateau."

This country is that surrounding Tombstone, Huachuca, Crittenden, and Calabesas, as far as Nogales on the Mexican line. There are settlements of from 3000 to 6000 feet elevation, in or near cañons and valleys, affording picturesque scenery and claiming a delightful winter-climate. The accommodations are, however, crude and primitive. The principal industries are mining and fruit-raising.

3. THE PLATEAU. This is an approximately level *mesa* above the 5000-foot line, which embraces more than half the territory. It enjoys the greater portion of the rainfall, chiefly because of its two mountain-systems. These summits exert an important influence on the yearly rainfall.

The annual amount of rainfall on the pro-plateau approximates 10 inches and that on the plateau from 10 to 20 inches, the greatest precipitation falling to the south of the Great Divide—the San Francisco Mountains.

The curve of precipitation of 20 inches appears, *first,* southeast of Tucson, over the Santa Rita Mountains; *secondly,* over the Natanes mountain-group, between the valleys of the Upper Gila and Salt Rivers; *thirdly,* over the flanks of the Mogollan range and the San Francisco Mountains; *fourthly,* in a narrow region over the head-waters of the Hassayampa, the Agna Fria, and the Rio Verde, near Prescott; another region showing 20 inches is the high country east of Fort Apache, over the White Mountains and the head-waters of the Little Colorado, extending to the edge of New Mexico. A small curve of 25 inches appears south of the San Francisco Mountains. These are "islands of greater rainfall," where the rainfall decreases from the centre in every direction, until not more than half the amount of rain may be expected but a few miles away. One point to be borne in mind is that the actual maximum rainfall is not known, as the stations are for the most part in the valleys, where the gauges cannot record the heavy rains which are seen on the tops of the mountains. The record of the station on the summit of Pike's Peak in Colorado may be mentioned to illustrate the increased precipitation compared with the plains, although it shows an unexpectedly low average, being barely 30 inches per annum.

A noticeable feature of Arizona meteorology is that it has two plainly marked rainy seasons. The winter-rains begin usually some time in December, and terminate in February. "As in the case of the seasonal rains of California, so in Arizona the variability of the winter-rains in amount and frequency is in the ratio of the intensity and recurrence of barometric disturbances. . . . . These rains are caused by the proximity of approach of great storms in low-pressure areas which form a part of the storm-system of the country at large. . . . They are moderate in force and are interrupted by the anti-cyclonic types of high barometer and cloudless skies which are distinctive of the Pacific coast weather."[1] Much of the winter-precipitation occurs in the form of snow, which quickly melts on the plains, while on the mountain-tops it may be found from three to seven feet deep. This acts as a storage-reservoir to supply the streams through the dry season almost to the beginning of the summer-rains. In fact, it is a practice of the Indians and old settlers to calculate that when the last snow disappears on the mountain-summits the summer-rains commence.

These come on during July, August, and September, " being somewhat sharply defined from the preceding dry season, but shading off so indeterminately toward the beginning of the winter-rains that it becomes quite proper to say that while Arizona has two rainy seasons it has but one dry season."[2]

The rains of summer are local in character and due to mountain-influences. They are of almost daily occurrence on the high summits, falling invariably in the afternoon. Their intensity is remarkable. "From 30 to 40 per cent. of the entire precipitation occurs in heavy showers, where the rainfall is upward of 0.75 inch during a day of precipitation, and frequently more than an inch falls in a heavy shower."[3] The brief and violent summer-storms are felt almost entirely on the elevated *mesa* called the plateau. Captain Glassford is authority for the curious statement that in Arizona in summer, owing to local causes, "the greatest pluvial effort is registered on the *leeward* side of ranges."

Regarding the proportional rainfall during the two rainy seasons, a record of ten weather-stations[4] (excluding Yuma) gave a total precipitation for the winter months (December, January, and February) of 3.55 inches; for the summer season (July, August, and Septem-

---

[1] Captain W. A. Glassford.   [2] Captain W. A. Glassford.   [3] General A. W. Greely.
[4] The ten stations were Fort Apache, Calabesas, Crittenden, Fort Defiance, Fort Grant, Huachuca, Phœnix, Tucson, Prescott, and Verde. Their average elevation is 3900 feet.

ber), 7.71 inches. Total for the entire year, 14.67 inches. The mean annual rainfall for the whole territory is 11 inches.[1]

The belt of mean annual temperature of 60° follows nearly the line of 3000 feet elevation. It runs north of Fort Thomas, Globe, Gillette, and Signal—except for a narrow strip which runs up the valley of the Rio Verde, penetrating into the higher *mesa*. The belt of 50° to 55° mean annual temperature includes Prescott, Flagstaff, and nearly all of the *mesa* or plateau above 5000 feet.

The belt of 50° or less mean annual temperature includes the northeast corner of the territory above Fort Defiance. In the southern half of the territory the mean temperature of 60° or more shades into the heat of the desert, with a mean annual temperature of about 70° and a monthly mean for July of 90°.

The relative humidity will average about the same percentage as in New Mexico, giving a little over 2 grains of absolute humidity to the cubic foot.

The marked dry season of Arizona is in the spring and early summer before the coming of the cooling summer-rains; but the number of cloudy days during the autumn and during the year is fully as low—if not more so—on the high plateau as in New Mexico and Colorado, while on the almost rainless desert the number of cloudy days is exceedingly small.

The wind-movement at Prescott, Fort Grant, and Fort Apache shows an annual average velocity (for four and six years) of a little under 7 miles per hour. Phœnix has the smallest recorded annual wind-movement, averaging about $2\frac{1}{3}$ miles per hour.

In Buck's *Reference Handbook of the Medical Sciences* the article "Arizona," based on reports of surgeons of the United States Army, states that "the exceptional dryness and purity of its atmosphere, the wonderful clearness of its skies, and the very small number of rainy days there occurring, would all seem to point to the desirability of the territory of Arizona as a health-resort for patients suffering from pulmonary phthisis or from other diseases of the respiratory system. . . . The climatic advantages of Arizona are almost identical with those to be found in New Mexico."

Arizona is sadly destitute of large modern towns where satisfactory accommodations for invalids can be relied on. In all the vast country there are but two or three towns where it is possible to get

---

[1] Rainfall and Snow of the United States. Professor M. W. Harrington, Washington, 1894.

needful supplies. The largest, Phœnix (1100 feet) and Tucson (2400 feet), in spite of their fine winter climate, leave much to be desired in the matter of modern conveniences and accommodations for visitors. Prescott (5300 feet) is even more poorly equipped, as it has been losing ground for several years. The opening of the railroad connecting Phœnix through Prescott with the main line of the Santa Fé road at Ash Fork may be the m ans of restoring its commercial prosperity. Prescott is in the centre of a fine country, and if it were larger and more prosperous would make a suitable place of residence for the entire year.

Near Flagstaff (7000 feet), on both sides of the railroad, are great forests of tall pines. It is a park-like country, with grazing-land, sandy soil, and good water in the mountains. Photographs of the plateau show picturesque groups of pines and a general resemblance to Estes Park, Colorado. From the hills south of Prescott to the forest-plateau north of Flagstaff the altitude varies from 4000 to 7000 feet. In many of the cañons are found old villages of the cliff-dwellers. The pines rarely grow below 4000 feet.

It is a country with fine possibilities, which, so far, are utterly undeveloped. Besides its climatic advantages this locality is but a few miles north of the lower and warmer Phœnix fruit-country, a change to which may be advisable in the course of winter, while the high plateau stretching away north to the Grand Cañon of the Colorado makes a pleasant summer camping-ground; or it is but a couple of days' journey to the rainless, pine-clad slopes of the Sierra Nevada range in California.

It should be noted among Arizona's peculiarities that, in spite of her two rainy seasons, the annual total precipitation is no more than in Colorado or New Mexico, while the spring is usually dry and there are but few cloudy days in the autumn.

Prescott (elevation, 5300 feet; population, 3000; latitude, 34° 30' north; longitude, 112° 30' west from Greenwich). The town of Prescott is situated a little to the west of the centre of Arizona (on the Santa Fé, Prescott and Phœnix Railroad), sixty miles beyond Ash Fork. Its geographical position is in some respects unique. Seventy miles to the northeast is the San Francisco range of mountains, while the snow-clad peaks of Humphrey, Agassiz, and Humboldt rise to an altitude of nearly 13,000 feet. Away north are the Coconino hills and the great cañons of the Colorado River, where the surface of the rushing water is over

6000 feet below the level of the plateau. Through this country above Prescott, at a varying elevation of 6000 or 7000 feet, are great forests of pine and an open grazing-country. It is a land of summer-rainfall and winter-snows. Eighty miles to the south are the plains of the Gila and Salt Rivers, which, although naturally arid, have been so cultivated by means of irrigation that Phœnix, situated in their midst, at an elevation of 1100 feet, is surrounded by fertile fields and groves of oranges, olives, figs, and semitropical fruit. It may be said that Prescott is distant but an hour or two by rail from both the torrid and the temperate zones.

Although very much further south, Prescott has a climate resembling that of Denver, which has the same elevation, but exceeds Prescott slightly in the number both of very hot and very cold days.

The high plateau and mountains of Arizona, lying mostly north of latitude 34°, have "two plainly marked rainy seasons"[1]—i.e., the season of winter-rains between December and February and the summer-rains of July, August, and part of September. The mean annual rainfall at Prescott for eighteen years is 16 inches, and the record for that time shows but two years exceeding 20 inches.

The town is situated on Granite Creek. The town-supply of water is hard. There are wells of softer water for domestic use. The horizon is bounded on all sides by distant hills and mountains, which afford protection from cold northern storms in winter and the desert-heat in summer. Near by are the pines, which offer especial attraction to invalids. These trees are seldom found in Arizona below an elevation of 4000 feet. The pine-forests are most extensive to the east and south of Prescott.

Prescott is situated in a gray-granite country, and much of the soil surrounding the town is of a sandy nature, drying quickly after a rain. The town itself is on adobe-soil, although Whipple Barracks, one mile from the *plaza*, has a sandy soil. There are windy and dusty days in the spring. The summers are hot, but are rendered more endurable by the cooling influences of afternoon rains. Snow falls occasionally during the winter, but the direct rays of the sun rarely allow it to remain long, although sleighing has been known even in the streets of Prescott. The climate during the autumn and winter is delightful. It is an American town, with few adobe-dwellings.

---

[1] Climate of Arizona. Captain W. A. Glassford.

Accommodations are poor. The hotel is principally for the accommodation of miners, and the best way of living is to go to housekeeping.

The average temperature for January is 31°; for July, 74°; year, 53° (means for thirteen years). Average number of days above 90°, 21; below 32°, 115.

Cloudy days, 51; stormy, 69 (means for three years).

Average annual velocity of wind 7 miles per hour.

Phœnix (elevation, 1100 feet; population, 10,000). This town has become favorably known as a winter-resort of low altitude. It is situated near the centre of the great Salt River Valley, about two miles north of that stream and twenty-eight miles (in a direct line) north of Maricopa Station, on the Southern Pacific Railroad, with which it is connected by rail.[1] Direct railroad communication with the Santa Fé Railroad at Ash Fork is now open.[2] Phœnix is the present capital of Arizona. Its growth during the past few years has been phenomenal. It has waterworks, street-railways, gas and electric lights, seven churches, six banks, and three daily newspapers. Its accommodations for visitors are not yet of a high standard, and its best hotel is greatly overcrowded in winter, especially when the Legislature is in session. However, two new hotels are being built.

The streets of Phœnix are broad and level, shaded on each side by rows of cottonwoods and willows. Streams of water flow in the ditches, and the houses—which until lately were usually built of adobe—are well shaded and surrounded by grass lawns.

The Salt River Valley is a favored region for the cultivation of fruit, as snow seldom or never falls, and the rich alluvial bottom-lands yield largely under irrigation. Fruits mature one or two months earlier here than in Southern California, and a great number of small ranches are being developed since the completion of the Arizona Canal. In order to show the character of the climate it may be stated that an exhibit of the valley's products sent to Chicago, December 15, 1894, consisted of oranges, lemons, olives, grapefruit, peaches, pears, strawberries, figs, watermelons, muskmelons, ripe tomatoes, green pease, green corn, pomegranates, almonds, etc.; also alfalfa, broom-corn, sorghum, sweet potatoes, squashes, pumpkins, etc. These were gathered outdoors in the month of December.

---

[1] The Resources of Arizona. P. Hamilton.
[2] The Santa Fé, Prescott and Phœnix Railroad.

In the matter of summer-heat this country rivals the famous Yuma Valley.

The average temperature (from records for twelve years) is for January, 49°; for July, 90°; for the year, 69°. By seasons it is as follows: winter, 51°; spring, 67°; summer, 87°; autumn, 69°. The mercury frequently rises to 95° in March and sometimes to 100° in April, while during the summer months the heat is intense. Records of the humidity were not obtainable, but the atmosphere is considered dry.

The average annual rainfall is 7 inches. The amount falling in each month is light, the heaviest rainfall occurring in August and December. The average annual wind-movement (taken from observations for three years) is $2\frac{1}{3}$ miles per hour.[1] General A. W. Greely says that "Phœnix is the locality where the wind is perhaps the feeblest of any point in the arid regions."[2]

Tucson (elevation, 2400 feet; population, 6000). An old Spanish grant is said to show a Spanish town here in 1553, built on the site of an Indian pueblo which is lost in the mists of tradition. Another account dates the Spanish Mission of San Xavier from 1694. At any rate, in the words of a thoughtful writer, "it is certainly of sufficient age to promise permanence."[3]

The name is said to be from an Indian word pronounced "Chookson," meaning "Black Water" or "Black Creek".

With the exception of one or two public buildings and mercantile structures and a few residences of better style, the houses are built of adobe, one story in height, and arranged after the usual Mexican fashion. The town is situated on the Santa Clara plateau, and is surrounded by mountains. On the north are the Santa Catalinas, east the Rincons, south the Santa Ritas, and on the west the Tucson range, with its most prominent peak—Tucson Mountain. The plateau is about twenty miles wide. Water is brought from a point on the Santa Cruz River, seven miles distant. The winter climate has been highly praised for its warmth and sunniness.

The average temperature for January is 50°; for July, 88°; for the year, 69° (means for fourteen years). "The mercury is never below 90° at 2 p.m. in July, and the mean minimum for that month is 78°, indicating nights too hot for comfort. . . . On twenty-

[1] Report on the Climate of Arizona. Captain W. A. Glassford.
[2] Report on the Climatology of the Arid Regions. General A. W. Greely.
[3] Arizona. Honorable John A. Black.

five of the thirty-one days in January, 1878, the mercury fell to the
freezing-point, the lowest point being 24°."[1] In 1892 the lowest
points touched were : January, 17°; February, 30°; December, 16°.
These low night-temperatures preclude the successful cultivation of
semitropical fruit, which is so large an industry in Phœnix and
Yuma, but the winters are considered mild and desirable for invalids. The best part of the season is from December to April, as
after that it grows too hot for comfort. For the year 1892 the
highest temperatures were: in May, 100°; June, 107°; July, 106°;
August, 107°; September, 102°. The average number of days
above 90° is 128 ; below 32°, 33; cloudy days, 57; stormy days,
42 (means for two years).

The record by season is shown in the following table, the figures
for temperature and rainfall being based on the reports of the army-
posts for fourteen years :[2]

	Temperature.	Rainfall.	Relative humidity. 1892.
Winter . . .	52°	3.01 inches.	52 per cent.
Spring . . .	67	1.22 "	34 "
Summer .	87	5.47 "	27 "
Autumn .	70	2.41 "	32 "
Year .	69	12.11 "	36 "

The accommodations for invalids are not very desirable.

Tucson is on the main line of the Southern Pacific Railroad, and
there are through trains east and west daily.

See meteorological data for Fort Grant (4860 feet), sixty-odd miles
northeast of Tucson, page 299.

Oracle. Elevation, 4500 feet. This settlement is forty miles
north of Tucson, by a good road. It is less than one day's journey
by stage. The population numbers twenty-five or thirty persons.

Oracle contains three or four ranches, two of which take boarders.
The accommodations are good for the Southwest.

The soil is a granitic detritus. There is no adobe.

Good, soft drinking-water is obtained from wells. There are live-
oak trees, and pines grow further up the mountains, eight or ten miles
away. The great mountain-ranges afford shelter from northern
storms.

---

[1] Handbook of the Pacific Coast. Dr. J. S. Hittell.
[2] Report on the Climate of Arizona. Captain W. A. Glassford.

There are no springs or streams of water immediately around Oracle, but within convenient reach for exploring are said to be interesting cañons and attractive scenery.

The winter-climate is dry, warm, and mild. There is little or no dust.

Yuma (elevation, 140 feet; population, 1200). During the winter months there is probably no place near the sea-level where as mild and dry a climate can be found as in Yuma, which is situated in the Great Arizona Desert, on the banks of the Colorado River, fifty or sixty miles above the salt water of the Gulf of California.

Early semitropical fruit can be grown on the soil watered by the Colorado River, and this industry is one that will probably be extensively developed, as the fruit ripens earlier here than in California.

Yuma is famed for sunshine and heat, the latter being a delicate subject to refer to local opinion. The mean monthly temperature for January is 53°; for July, 92°; for the year, 72° (from observations for fourteen years). The mean of the minima for three years (1891–'93) for July is 75° and for August 77°, indicating hot nights. The average number of days during the year above 90° is 163; below 32°, 4; cloudy days, 21 (mean for six years).

The mean monthly winter-temperature for Yuma is 56°. The maximum temperature is likely to exceed 100° during the months of May to October, inclusive. Frequently it exceeds 100° in April, and it may rise to 90° during the month of March. In 1893, from April to October, inclusive, out of 214 days 162 days were over 90°. The maximum temperature for the year was 111°.

Physical suffering is not so great under this intense heat as it would be if it were not for the great dryness of the desert-air and the consequent rapid evaporation. The sensible temperature (wet-bulb thermometer) is, however, frequently over 80°, which would anywhere be considered *hot*.

The yearly mean of the relative humidity is 46 per cent.

The present hotel-accommodations are ordinary. The Southern Pacific Railroad has for several years been reported to have in contemplation the erection of a good-sized hotel.

There is little amusement in Yuma, as there are few objects of interest in the vicinity.

An army-post—Fort Yuma—is situated on the opposite side of the Colorado River, in the State of California.

Army-posts and Other Stations.[1] Fort Defiance (elevation, 6500 feet) is situated in the northeastern portion of Arizona, near the line of New Mexico. It is in the belt of mean annual temperature of 45° to 50°, and is an elevated country with forests and springs and a climate resembling the Colorado plains. Due north of Fort Defiance are the Moqui and Navajo Indian Reservations.

	Autumn.	Winter.	Spring.	Summer.	Year.
Temperature[2] (8 years)	46°	27°	46°	68°	47°
Rainfall (7 years)	3.72 in.	2.55 in.	2.03 in.	5.89 in.	14.19 in.

Fort Apache (elevation, 5000 feet). This post, like Fort Grant, is near the old Apache Indian Reservation. The country generally, though sparsely settled, is devoted to sheep- and cattle-raising and agriculture. About thirty miles northeast of Fort Apache are the White Mountains, which are said to possess fine natural scenery. They lie within one of the rainfall-belts of 20 inches for the year, of which there are four or five in Arizona. There are pine-forests and beautiful parks and valleys. Game and fish are plentiful.

	Autumn.	Winter.	Spring.	Summer.	Year.
Temperature (18 yrs.)	54°	37°	52°	73°	54°
Rainfall (15 yrs.)	3.05 in.	5.07 in.	2.96 in.	8.96 in.	20.04 in.

Wind-movement, annual average, 6½ miles an hour.

Relative humidity, annual average, 46 per cent.

Fort Grant. Elevation, 4860 feet. Sixty or seventy miles northeast of Tucson, in the foot-hills of the Graham Mountains. It is a military post, with commodious quarters for officers and men, hospital, waterworks, store, postoffice, ice-machine, etc. There is telegraphic communication with Fort Thomas.

	Autumn.	Winter.	Spring.	Summer.	Year.
Temperature (17 yrs.)	62°	45°	59°	77°	61°
Rainfall (16 yrs.)	3.64 in.	3.68 in.	1.80 in.	7.73 in.	16.85 in.

Wind-movement, annual average, 7 miles per hour.

Average number of days above 90°, 40; below 32°, 29 (for two years). Mean monthly temperature—January, 44°; July, 79°.

---

[1] These army-posts are mentioned on account of their fine climate, and the illustration they afford of the natural resources of the hill-country of Arizona. It would be important for a casual visitor to be provided with proper letters of introduction to some of the higher army officers, before venturing so far away from ordinary hotel-accommodations, as there is usually no settlement of any size near the post. It may be added that on the frontier it is often difficult to supply two of an invalid's greatest needs—fresh meat and milk.

[2] From Report on Climate of Arizona, etc., by Captain W. A. Glassford, Signal Corps, U.S.A.

Mean annual relative humidity, 45 per cent. Absolute humidity, 2.38 grains. Average number of cloudy days per annum, 45. (These last three records for ten years.)

Fort Verde. Elevation, 3200 feet ; population about 50(?). Forty-five miles east of Prescott. Situated near the Rio Verde. Camp Verde has a pretty location. It now consists largely of abandoned adobes. Cottonwood-trees grow near the river.

Hittell's *Handbook of Pacific Coast Travel* divides the year into two periods of six months each, and reports the relative humidity of Camp Verde as follows: cold semester, 41 per cent.; warm semester, 38 per cent. Average for the year, 40 per cent.

	Autumn.	Winter.	Spring.	Summer.	Year.
Temperature (21 yrs.)	61°	43°	61°	80°	61°
Rainfall (21 yrs.)	2.82 in.	3.45 in.	1.98 in.	4.88 in.	13.13 in.

Tombstone (elevation, 2300(?) feet ; population about 2500(?)). The town is built on an elevated *mesa* on the San Pedro River, twenty-eight miles south of Benson, on the Southern Pacific Railroad, and eight miles from Fairbanks, on the New Mexico and Arizona Railroad. The buildings are largely adobe, one story in height, with arcades shading the sidewalks. The natural facilities for town-drainage are said to be good. An ample supply of pure water is brought twenty-two miles from the Huachuca Mountains.

	Autumn.	Winter.	Spring.	Summer.	Year.
Temperature,	62°	46°	65°	77°	62°

Rainfall, for the year, about 16 inches (see Fort Huachuca).

Fort Huachuca (elevation, 4780 feet) is one of the most important military posts in Arizona. It is situated in the Huachuca Mountains, at an altitude where, as a rule, the summers are cool and pleasant. There are picturesque mountain-views and forests of pine-timber in the vicinity. Fort Huachuca—and also the settlement of Huachuca, ten miles north—are near the influence of the belt of 20 inches annual rainfall, which lies over the Santa Rita Mountains.

	Autumn.	Winter.	Spring.	Summer.	Year.
Temperature (4 yrs.)	61°	44°	59°	76°	60°
Rainfall (5 yrs.)	4.57 in.	3.02 in.	1.20 in.	7.60 in.	16.39 in.

In this corner of the territory are several other well-situated settlements of high altitude on the line of the New Mexico and Arizona Railroad, which runs into Mexico. In the mountains are a number of picturesque cañons. The country, unlike the low, desert portion of Arizona, is green and fertile. There are fine mountain-scenery and pine-timber on the ridges. It is not thickly settled; the principal occupations, aside from mining, are grazing and fruit-raising.

The rivers rise in the higher mountains in Mexico and flow north to the Gila.

Crittenden (4100 feet) has fine natural scenery and productive fruit-farms. **Old Camp Crittenden** (2000 feet) has an annual average temperature of 60° and an annual rainfall of 16 inches. It is near the Senorita Cañon, in which are hot springs. At **Calabasas** (about 4000(?) feet) the views of the mountains are extensive. There is said to be hunting in the vicinity. There is a brick hotel at Calabasas, with some modern conveniences. **Nogales** (4000 feet) is on the Mexican line. Transfer is here made to the Sonora Railroad, running to Homosillo and Guaymas. Nogales is an American town with some Mexican characteristics. It claims a population of 2000. The buildings are mainly adobe. It lacks many facilities for comfortable living, and is not exempt from dust-storms. The soil is sandy loam, which dries rapidly after rains. The climate is fine, sunny, and mild, though somewhat warm. The rainfall is about 10 inches a year.

SPRINGS.

The foot-hills and valleys of Arizona, in or near the " rain-islands " before alluded to, are supplied with mineral springs—both hot and cold—but they have not been developed, and so little is known of them that any reference to the springs of Arizona, at this day, is necessarily meagre and incomplete. In the hilly country between Prescott and Phœnix are numerous springs. The Rio Verde is formed from a series of springs in what is known as Chino valley, in the Great Colorado plateau, between the Jupiter range and Bill Williams's Mountain. In the Verde valley is the Montezuma well, near which is a warm soda spring. The **Castle Creek Hot Springs**, in the southern part of Yavapai County, have been much visited by invalids. Twenty miles or so below Phœnix, on the Gila River, are hot medicinal springs, well known to the Indians. The streams rising in the mountains in the eastern portion of Arizona are fed by numerous cold springs. Near Wilcox (Cochise County) are **Hooker's Hot Springs**, consisting of a cold sulphur spring and six hot springs, in which magnesia and iron are said to predominate.

CHAPTER XIV.

PACIFIC SLOPE REGION.

The Climate of Southern California.

"Two influences dominate the climate of California, radically dissimilar in every particular, combining in ever-varying forces to produce the resultant which is recorded by observers of the weather. One is the sea, tending always to charge the air with moisture; the other is the mountain-mass, tending always to discharge the moisture from the air. The combination of these two activities in varying proportions is responsible for the variation in the amount of precipitation, including months of drought."[1]

The important mountain factors are the great chain of the Sierra Nevada and the Coast Range. In the southern portion of the State is a third series of elevations which may belong to either of the other northern systems, and which has been distinguished as the Southern Coast Range. It extends from the Tehachapi region southeasterly to San Bernardino Peak and Grayback, the altitudes of which approximate to 12,000 feet. South of these peaks the range runs parallel to the coast into Lower California.

It is with Southern California that we have principally to do. That portion of the State south of latitude 35° lies more open to the sea. The hills no longer form a barrier to the wind directly over the shore. They are here further inland, and leave large and beautiful fertile valleys accessible to the daily ocean-wind. Among these are the Santa Inez and Santa Monica and the rich, broad valleys of Los Angeles; the fertile district extends through the valley of San Gabriel and as far as that of the Santa Ana. It includes Riverside, Colton, Pomona, and the long plains of the San Jacinto, stretching southward to the valleys of San Diego, while facing directly upon the sea are the valleys of the Santa Clara and Santa Buena Ventura and the Santa Barbara plains. These inland tracts of country are

---

[1] Climate of California and Nevada, with Particular Reference to their Rainfall and Temperature, etc., by Captain W. A. Glassford, Signal Corps, U. S. A.

irregular in outline, branching out in many directions and often merging into rolling upland *mesas*.

Dr. J. P. Widney says, in his able and entertaining *Climatology of the Pacific Coast :*[1] "The area of the plains of Southern California is really largely increased over their apparent size by the rolling, hilly uplands into which, in many directions, they merge. This is especially the case in the country which lies between the San Fernando valley and the lower Santa Clara valley, and also in the great upland which rises from San Jacinto toward the south in San Diego County. These uplands have a rich, deep soil, and are well watered by numerous small streams."

Directly north and east of the highest part of the Southern Coast Range are the great deserts. The Mohavé Desert has an average elevation of about 2000 feet. In the southeasterly corner of the State, bordering on the Colorado River, is the low desert-land called the Colorado Desert, which in some places is depressed nearly 300 feet below the level of the sea.

The influence of these vast desert-areas on the country to the windward of the mountains—that is, on the fertile plains south and west of the ranges—is plainly perceptible.

Besides the influences of the mountain-masses in aiding to discharge the moisture from the atmosphere, another influence has been referred to as constantly exerted to charge the atmosphere with moisture, namely, the Pacific Ocean influence.

In his elaborate and comprehensive memoir of the California climate, which is full of scientific detail, Captain Glassford has placed bounds around that part of the North Pacific Ocean which may be considered as modifying the climate of California, as follows: " To the west it is bounded by the extreme Orient, the islands of Japan, with their northern projection over the Kuriles to the coast of Kamtchatka and their southerly connections with the Philippines. The northern limit is drawn by the Aleutian Islands, and the eastern border is the shore of North America. To the south no consistent mass of land appears to hem this ocean in, yet the barrier is none the less strong because it may be measured only with the instruments of the meteorologist. It exists at the thirtieth parallel of north latitude. Below this bounding-line is the region of the northeast tradewind and the westward drift of the equatorial current, and these

[1] California of the South, by Walter Lindley and J. P. Widney, 1888.

two serve sufficiently to bound in wind and water the great basin above."

It is a basin within these limits—a rough ellipse having a major axis of 100° of longitude and a minor axis of 25° of latitude. It has its characteristic systems of circulation, both of atmosphere and sea.

The strongly individualized ocean-current of the region is the Kurosiwo. Developed from the culminative process of the equatorial drift and directed by the rapid alteration in the plane of the sea-bottom and the trend of the Asiatic coast, this warm stream moves across the whole Northern Pacific. Flowing through a broader sea than does the Gulf-stream, it shows several important differences—it has a slower motion, its warmth is not in such strong contrast to the water through which it flows, and the wind blowing counter to its course frequently avails to deflect it or even to check it entirely. Its eastern development and dispersion have been for years a battle-ground for theorists, and even now it is impossible to say definitely that it reaches any part of the Californian coast.

The winds upon this basin are of the system of the passage-winds which are developed upon the surface of the earth by the descent from high altitudes of upper currents. In general these winds vary with the latitude from southwest, westerly, to northwest. It should be noted that these winds begin to appear about the parallel of 30° north, and that at first they are practically dry winds, but presenting all the best conditions for absorption. The sea is warm and in the best condition for giving off moisture, and the wind is most receptive.[1]

---

[1] In order to gain a broader view of the distinctive climate of California—which differs from that of any district in this country—it may be well to quote here Captain Glassford's technical review of the climatic characteristics which dominate the whole region :

"The distinguishing characteristic of the climate of the region is that varieties of weather endure practically unaltered for days at a time, and even when supplanted by others return again and again, and on each such recurrence are symmetrical with their former appearance, even when they are not practically identical. In this regard there is a wide variation from the conditions which obtain elsewhere in the United States. Nor is this the only difference. Another notable one is that the storms of the Pacific are, with comparative infrequency, traced across into the Central Valley and the Atlantic Slope. Another is that the storm-area frequently increases rapidly toward the north.

"When the area of low barometer of considerable depth overlies Oregon and Washington, and probably is central far to seaward, and the cyclonic type appears, its translation eastward is checked if not prohibited by the barrier of the Cascade Range and the Rocky Mountains, which here begin to fuse. Held back by the mountain-wall and the equally potent barrier of high pressure eastward, the low is kept beating against these obstacles'and the high remains steadfast over the Great Basin* and the Northern Plateau. While this condition endures gales

---

* The Great Basin, a high plateau lying between the Sierra Nevada and the Wasatch Mountains. It contains 208,500 square miles, embracing all of Nevada and Western Utah, and portions of the adjoining States.

are felt upon the Californian coast as far down as Cape Mendocino, and rain occurs in the Great Valley* and down the coast to San Luis Obispo. These storms leave the southern part of the State untouched, except when a subsidiary low is developed over the Colorado Desert, when the brief 'Sonora storms' occur."

---

* The Great Valley. This term includes both the Sacramento and San Joaquin Valleys. It has a length of about 450 miles by about 40 miles average width, taking in the lower foot-hills.

"When this lower area is shallower, and can be plainly seen to have its centre not far out upon the sea, but over Washington, and the high is plainly marked upon the Great Basin, then occur light showers from San Francisco northward, with strong gales at Cape Mendocino; the temperature over the dry area is usually high and occasionally of steep gradients, and in the Los Angeles region the warm Santa Ana winds occur. The rain rarely passes south of San Francisco, except in cases where the definition of the high is so strong toward the south of the Great Basin as to condition a low advancing over the Southern Coast Ranges and back of the Sierras to meet it; then light showers may occur between San Luis Obispo and San Diego.

"These two cases have presented the conditions of low pressure over Washington and Oregon, accompanied by rains, which, for the most part, occur in California only in the region north of the southern inosculation of the Coast Range and the Sierra Nevada. When, on the other hand, a high area rests upon the two Northern States and the low type is permanent over Southern California, it conditions for California a climatic manifestation of extremely unstable equilibrium, and while this arrangement of the meteoric elements is of frequent occurrence it is often of short duration. When the low is in the north rain falls upon California; when the high is in the north fair weather is a marked concomitant.

"During the perfection and greatest intensity in the prevalence of this arrangement, and while the isobars are perpendicular to the general trend of the coast-line and the axial inflection of the Coast Ranges and the Sierra Nevada, the Great Valley is exposed to 'northers,' marked with disastrous desiccating influences. The day-temperature is usually high, increasing proportionally to the duration of this climatic type, but at night frosts are of characteristically frequent occurrence. The winds increase toward the south, being light and variable on the Oregon coast, but high gales on the Californian coast. When this type occurs in spring, and is accompanied in Southern California by high winds and sandstorms, rain is almost certain to follow. In general, the breaking up of this type is heralded by frosts of more or less severity.

"The most severe and general rains of the region occur in co-ordination with a general climatic disturbance over the whole country. To the eastward there is a series of waves of abnormally high pressure over the eastern guiding-planes of the Cordilleran system, reaching thence across the Central Valley and the Appalachian system to the Atlantic seaboard, and everywhere accompanied by severe storms and intense cold. Upon the Pacific coast, in correlation with this eastern disorder, the barometer drops very low, and exhibits rapid fluctuations, with remarkable gradients between the coast and the interior, the rain-area overspreads all sections, gales are marked with the greatest violence, the rivers attain their high levels and tend to floods; in general, the condition is that of an extensive cyclonic disturbance, which, proving unable to scale the Sierra Nevada, is forced to spread out over the entire length of the coast region, until it gradually wears out with the restoration of climatic equilibrium beyond the range, or if it does move eastward, does so at some extreme point beyond the sphere of observation. In this condition of the weather the rain is precipitated with practical impartiality from Siskiyou to San Diego.

"Another rainy condition is found when a diffuse and moderate high exists upon the southwestern coast, accompanied by unusually low temperature, and apparently unaffected either by the presence or the absence of a faint and shallow low on the Northern coast. With this arrangement of climatic factors the isobars are somewhat perpendicular to the coast, a condition almost certain to bring rain, while if these curves of pressure assume a parallelism with the coast-line fair weather soon follows. During the prevalence of this condition there are rains upon the Los Angeles country and the Great Valley, and the winds above San Francisco are feeble, except in the rare instances when the barometer sinks excessively. Should the absolute general pressure fall considerably below the normal, yet retaining the relative high upon the southwestern coast, gales rage in Southern California, with occasional storms of thunder and hail. This condition determines very suddenly by the movement of the high up the coast, and its obliteration as a distinct feature in its progress.

"A condition which leads to rains of local character, yet impartially distributed as to occurrence within geographical limits, is marked by a moderate low continued through a succession of days and below the normal over a large area. The isobars are then diffusely disrupted;

The amount of annual rainfall along the southern coast and on the valleys of the southern slope is from 10 to 20 inches, diminishing to 3 or 4 inches over the Colorado Desert and increasing to about 40 inches on the highest summits.[1]

The period of the winter-rains is seldom established before November, and is over usually by April. The months of the heaviest rainfall are December, January, and February, when two or three inches during each month may be expected, with intervening rainless periods, perhaps several weeks in length, of fine, sunny weather. In Los Angeles over 60 per cent. of the rainfall for the season occurs during these three winter months.[2]

The normal rainfall is well illustrated by the following table from the Government records for each of the months of the rainy season separately and the remaining six months of the year together.

The average rainfall for these three towns for the months from November to April, inclusive, is 14 inches; for May to October, inclusive, 1.1 inch.

Rainy season, Southern California.	Nov.	Dec.	Jan.	Feb.	March.	April.	May-Oct. inclusive.	Length of record.
Los Angeles	1.6	3.7	3.9	4.0	2.2	1.3	1.2	To 1892, 21 years.
Santa Barbara	1.6	3.9	3.7	3.8	2.1	1.4	1.3	" 1891, 24 "
San Diego	1.0	2.1	1.6	2.1	1.0	1.0	1.0	" 1892, 42 "

It is interesting to note that no rainy season ever runs along on normal lines in California. The rain may hold off for weeks at a time, and then it may pour for days and days, until from 3 to 10 inches of water have fallen before fair weather is again definitely restored. It is of great value to know the annual average precipi-

---

[1] Irrigation and Rainfall Maps. William Ham Hall, C.E.
[2] Rainfall and Snow of the United States. Professor M. W. Harrington, Washington, 1894.

they are wavy, or enclose several subsidiary lows, occurring over mountain-basins with a marked absence of any decided gradients. The winds are variable, the temperature declines, the sky is cloudy, rain comes at intervals, rising under favorable conditions to a gale, which, while quite local in character, sometimes does considerable damage within its narrow limits.

"The dry season shows little variation from beginning to end. Rain is almost entirely absent, and the light showers which sometimes occur on the Washington coast only rarely drop down upon a limited district of the Californian shore. Another feature of the season is the development and persistence of marked intensity of the high in Oregon, accompanied with a corresponding fixity of a slight low area over Southern California, creating the characteristic northerly winds which blow down the Great Valley."

tation for twenty years, but it does not in the least indicate what the rainfall will be for the coming year. Santa Barbara, with a normal precipitation of 17.6 inches, had a minimum record of 8.01 inches for the year 1881 and a maximum of 38.8 inches for 1884. Los Angeles, normal 18 inches, has varied in different years from 5.6 inches (1881) to 40.5 inches (1884). Riverside, normal 10 inches, has had minimum 2.46, maximum 25.32 inches. San Bernardino, normal 16 inches, has had minimum 8.98, maximum 37.51 inches. San Diego, normal 10 inches, has had a minimum of 3.71 and a maximum of 25.97 inches; while Julian, with a normal of 38 inches and no record below 25 inches, has known 61.62 inches during a season. One of the heaviest seasonal rainfalls in Southern California was during the winter of 1883-'84 when previous records were everywhere exceeded.

Southern California has been divided into three belts of mean annual temperature: (1) A belt of 60° to 68° F. along the seacoast and extending inland from thirty to fifty miles. (2) A belt of 52° to 60°, running up the sides of the Southern Coast Range and extending from Tehachapi to San Bernardino Peak and from San Jacinto Mountain south to Lower California. (3) A wide belt of 68° to 78°, beginning at the base of the mountains, stretching east and northeast to the Colorado River, and including the Colorado and Mohave deserts.

These arbitrary divisions cannot be followed closely, however, as every valley and hillside has variations both of temperature and rainfall, owing to its position with relation to the prevailing wind. On the Coast Range this variation is often experienced. Dr. Widney says: "While upon the ocean side of the range are great forests where the giant redwood is bathed nightly in the dense, cool fog which seems to be essential to its growth, just across the summit are warm mountain-slopes facing off toward the morning sun, their rolling hills green to the very crest with the olive and the vine; and yet from their sheltered warmth one may pass on for a few miles to some pass or gap in the range that is swept during all the summer months by the great, cool ocean-wind as it rushes through to the heated interior."[1]

The relative humidity of the coast is usually in the neighborhood of 70 per cent., which, taking a mean annual temperature of 60°,

[1] California of the South. J. P. Widney, M.D.

would show the amount of absolute humidity to equal 4.03 grains of vapor to each cubic foot of air. This humidity decreases steadily on going inland until the desert is reached, where, east of the mountains, the atmosphere is perceptibly drier. It is here, directly within the desert-influence, that any dryness in California is to be found, for within the sweep of the sea-breeze great moisture will always be present. During the so-called " dry " or rainless summer season the relative humidity in the air is equal in amount to that of the winter season, as shown by the Government weather-reports.

The Pacific coast has, during the spring and summer, a frequent night-fog, which becomes visible about sunset or a few hours later and disappears during the early forenoon.

Many observers who have lived directly on the coast believe the fogs to be more frequent during the winter, especially north of Point Concepcion ; but the record of night and morning fogs at Los Angeles (fourteen miles from the sea) for thirteen years shows that the number of days on which fogs occur increases steadily from the beginning of spring through July, continues large in August and September, and decreases to a marked degree between October and March.

By seasons the average number of foggy nights at Los Angeles is as follows (means of thirteen years) :[1]

| Winter | . | . 7.3 days. | Summer . | . | . 23.5 days. |
| Spring. | . | . 12.1 " | Autumn . | . | . 13.9 " |

Annual average, 57 days.

Owing undoubtedly to local conditions Los Angeles is more subject to fog than San Diego.

In 1891 there were 22 fogs at San Diego, the greatest number in one month being 7, which occurred in November. In the same year there were 62 fogs at Los Angeles, the maximum number per month being reached in June and July, in each of which there were 12 fogs.

In 1893 there were 12 fogs at San Diego, the greatest number in one month being 3, occurring in October; while there were 46 fogs at Los Angeles, 11 of which, the maximum monthly number, come to July.

In 1894 there were 26 fogs at San Diego, the greatest monthly

[1] From records of Los Angeles office of U. S. Weather Bureau.

number being 9 in November. At Los Angeles there were 46 fogs, of which 9 were in the month of July, 8 in August, 7 in September, and 6 in November. Santa Barbara had 73 fogs during this year, July and November, during which months they were most frequent, having 16 and 12 respectively. Records for other years were not obtainable.

In Southern California a fog lasting through the entire day is infrequent, the usual course being for it to appear in the night and vanish in the course of the morning. This fog is a virtual prolongation of the rainy season for the immediate coast-district, and its humidity and freshness help to make the day cooler and more bracing. It is said rarely to rise above an altitude of 2000 feet.

Dr. Widney writes: "The heat of the summer is not felt along the coast within reach of the sea-breeze—a midday temperature of 65° to 80° being the rule, varying with localities. Back from the coast, in the interior valleys, where the fog does not penetrate, the midday temperature may, in exceptional cases, during a hot spell, reach 90° or 100° or even 105°; but it is a dry heat, without the discomfort or danger attending a like temperature in the Atlantic or Mississippi States. These hot spells, as they are called, may occur several times during the course of the summer, generally lasting for three days, when the mercury drops and the normal coolness returns. Even during these hot spells, however, the night is generally marked by a rapid fall in temperature, so that sleep is restful and refreshing. The heating of the interior valleys gives rise during the afterpart of each summer-day to a strong surface-current sea-breeze, which dies away toward sunset."

As the cooler air from the sea flows in upon the warmer air over the land the moisture in the warmer air is condensed into visible vapor, and the formation of fog goes rapidly on. Along the immediate coast there is seldom or never fog on the land unless there is also fog on the ocean, and while there is little wind at night the sea-influence is very marked.[1] In some of the interior valleys, however, the light, cool evening-wind which blows off the sides of the moun-

[1] From Report of Chief of U. S. Weather Bureau, 1893.
San Diego. Frequency of winds, 1893.

Hours winds were blowing from the	W.	2904 ;	E.	190
" " " " " "	N.W.	1677 ;	S.E.	256
" " " " " "	S.W.	1522 ;	N.E.	704

Total sea-winds, 6103 hours ; total land-winds, 1150 hours.

As is usual on a seacoast, the time when the wind is blowing from the land toward the sea is during the night.

tains performs the office of condensing the vapor in the warmer strata of air lying over the plains. Thus a fog is formed which remains until dissipated by the rays of the sun during the early forenoon. But, after all, the source of the fog is of less importance than its frequent presence and the fact that the amount of moisture in the air increases with the approach and advance of night, as shown by the instruments registering that factor.[1]

Dr. Widney continues: "It is the coolness and nightly moisture of these summer-fogs which draw the frost-line well down the coast in Northern California. To persons of delicate constitution, those who do not make blood and bodily heat rapidly, these keen sea-breezes and the chill fog are very trying."[2]

During the months of June, July, and August, which are the driest months of the year in California away from the coast, any precipitation is rare except on the highest mountains. This is the time of year for camping-out in the pines on the sides of the Sierras, 4000 or 5000 feet or more above the sea. In well-protected spots there is little if any dew, and the relative humidity during the middle of the day is sometimes as low as 20 per cent.; between the 1st of June and the 1st of September there is little danger of frost.

In the fertile valleys during the rainless season the irrigation of groves, orchards, and vineyards is constant and increases with the more extended cultivation of the soil. How great an effect this practice has had in augmenting the rainfall and the atmospheric humidity is yet an unsettled question. The following evidence is submitted as suggestive, the first illustration especially:

In the official record of the rainfall at Riverside for fourteen years, from 1880 to 1893, inclusive, the average precipitation of the first seven years is 8.02 inches and of the last seven years 11.75 inches, an increase of 46 per cent. Looking at it in another way, the annual rainfall at Riverside for the four years 1880–'83, inclusive, averaged 4.42 inches; while for the four years 1890–'93 it was 10.69 inches, an increase of 140 per cent. in ten years.

The record of Santa Barbara for twenty-four years, from 1867 to 1891, gives the average precipitation of the first twelve years as 15.78 inches and of the last twelve years as 20.50 inches, an increase of 32 per cent. The question of irrigation does not, however, enter into the discussion, so far as Santa Barbara is concerned.

[1] See p. 313.   [2] California of the South.   J. P. Widney.

The cloudy days in Southern California number about the same as on the high plains of Arizona, New Mexico, and Colorado. There are usually from forty to sixty cloudy days during the year, except perhaps in particular locations on or near the coast, where the effect from fog is greatly felt. In the inland desert-country in Southeastern California and Southwestern Arizona the total of cloudy days numbers barely twenty during the year.

The average annual wind-movement of San Diego is equal to a velocity of 5.6 miles per hour; at Los Angeles, 5.1 miles; and at Santa Barbara about 4 miles per hour. The general movement in California is usually small. "The feature which most impresses the observer upon the Pacific coast in his study of the winds is their regularity. . . . He knows that at certain seasons there will be a prevalence of wind from a certain quarter, and that at a certain time of each day the wind will rise. He knows that a persistence of the wind from a certain quarter will bring a very moist atmosphere and rain, while the current from another quarter as surely means clear, cool weather, with a moderately humid atmosphere; and from yet another quarter means an exceedingly dry atmosphere, cold in winter, hot in summer. . . . While the whole Pacific coast has much less really calm weather than the Atlantic coast, yet the records of the Signal Service show that the total wind-movement is less; in other words, in a given length of time there are more hours of wind but of less velocity. It is a region of more continuous wind-currents, but of milder character. The brisk sea-breeze is diurnal; the gale rare; the hurricane and the cyclone are unknown."[1]

In an entertaining discussion of the climatic characteristics of Southern California by Dr. C. P. Remondino, of San Diego, in a magazine article,[2] he refers to the great range of the rainfall over that country from the lowest yearly record to the highest twelve-hour record for the United States, as indicating a "land of seeming and incomprehensible climatic contradictions." The arctic current that passes the coast with its waters never below 60° F. in winter, nor over 66° in summer, he regards as the "great equalizer of Southern California temperature."

Before summing up the advantages of California as a health-resort a frequent source of error in estimating the character of the

[1] Dr. J. P. Widney.
[2] California as a Health-resort. P. C. Remondino, in the California Magazine, October, 1893.

climate should be particularly pointed out. The great number of days of sunshine and the small amount of annual rainfall should never be confused with the actual degree of atmospheric humidity.

In connection with any discussion of the climate of California this important fact should always be kept in mind.

When writers speak of its "dryness" they mean—if they mean anything—to refer to the interior desert-country, beyond the mountains, where the wind has lost its coolness and moisture, and where it is dry, but for most of the year excessively hot during the day, although somewhat cooler at night. When they speak of its "coolness" they refer to the coast-districts, where high temperatures are rare and the cool sea-breeze is full of moisture. There are found the features of an ocean climate, and this is the secret of its equability.

In order to have a general knowledge of the climate of Southern California this relative distinction is important: the coast is *cool* and *moist;* the interior is *hot* and *dry.*

Or, to put it a little differently, it should be thoroughly understood by the Eastern visitor in search of health that if he seeks more days of sunshine and opportunities for outdoor life, with a more equable temperature and an average *humidity a little greater* than that of New York or Boston,[1] he can find what he desires at Santa Barbara or San Diego.

If he needs the element of absolute dryness with low altitude and sunshine, he will hardly find them together, except along the low plains of Arizona and New Mexico; that is, while the barren inland desert-country of California is dry, it lacks the conveniences of civilization, which cannot be obtained short of the towns of Phœnix, Tucson, or El Paso.

---

[1] The accompanying table, taken from records of U. S. Weather Bureau, shows that the number of grains of moisture to each cubic foot of air averages during the year *more by one-third* along the coast of Southern California than in Boston or New York. This is partly owing to a higher average of temperature.

From Table V. :

	Mean annual		
	Tempera-ture.	Relative humidity.	Absolute humidity.
New York	52°	73 per cent.	3.19 grains.
Boston . .	. 49	72 "	2.81 "
Santa Barbara	. 60	73 "	4.20 "
Los Angeles .	. 62	72 "	4.42 "
San Diego	61	73 "	4.34 "

There is, however, a good country along the foot-hills from Sierra Madre to Beaumont, where on the "benches" of from 1500 to 3000 feet elevation the air is moderately dry. The winters throughout this region are mild, but except at a considerable elevation the summers away from the sea are cloudless and quite hot.

It may be well to call attention to the fact that the climate only is under consideration here, the matter of accommodations being referred to elsewhere.

For a cool, dry air there are the elevated plains of Arizona, New Mexico, and Colorado.

The marked changes of temperature during the day in California are frequently trying, and the humidity is sometimes greatly increased by fogs and in winter by protracted rains. It is a point worth noting that even when the atmosphere has been fairly dry from 11 A.M. to 5 P.M., it is *always damp at night*. The author has noticed this at Redlands—one of the most favorably situated of the inland towns. On one occasion when at 4.30 P.M. the relative humidity, as indicated by the hygrometer, was 55 per cent., at 6 P.M. it had increased to 80 per cent., and a light haze was visible in the valley.

While the minimum relative humidity at Redlands is sometimes as low as 30 per cent. for a limited period taken during the middle of the day, it always reads high during the night, and will probably be somewhere from 70 to 80 per cent. or over, with a night-temperature ranging usually during the year from 44° to 60° and in the winter from 44° to 52°; freezing weather is very rare. This means to an invalid a climate possessing perhaps six hours in the day of moderate dryness and eighteen hours of positive dampness.

This peculiarity of the climate of Southern California explains many of its apparent contradictions. The great difference between the character of the atmosphere during the day and during the night usually escapes observation. It is of importance because it shows the danger of making an estimate of the climate without considering the preponderating night-influence.

The weather-records are not so complete for the night as for the day, but they are sufficiently so to establish this fact—that, in spite of the great amount of sunshine during the day in California, the foggy and damp nights and mornings take up a great part of the twenty-four hours.

At Los Angeles observations taken at 8.15 and 8.07 P.M. (local time) for six years (January, 1882, to December, 1887) show the following results by seasons (for office of United States Weather Bureau) :

	Night-temperature.	Relative humidity.
Winter	. 52°	76 per cent.
Spring	57	81 "
Summer	. 65	82 "
Autumn .	. 60	82 "
Yearly mean .	. 58	80 "

Observations showing the humidity at its greatest, between midnight and 4 A.M., have never been published.

One more example of night-humidity in California may be quoted, taken in the Ojai Valley during ten clear nights in January and February, 1895 : Nordhoff, average hour for observations, 10 P.M.: mean temperature, 47°; mean relative humidity, 79 per cent.

The observers in the United States Weather Bureau stations take synchronous observations at 8 P.M., Washington or 75th meridian time. A comparison of the tables of relative humidity for that hour for several health-resort stations, on the elevated inland plains and in Southern California, may be of value in this connection.

MEAN MONTHLY RELATIVE HUMIDITY.

		Winter.	Spring.	Summer.	Autumn.	Year.
		Per cent.	Per cent.	Per cent.	Per cent.	Per cent.
Denver (6 P.M. local time) .	1892	55	41	32	31	40
	1893	42	40	33	31	37
Colorado Springs (6 P.M. local time) No complete record for 1892.	1893	41	34	39	36	38
Santa Fé (6 P.M. local time) .	1892	52	31	33	31	34
	1893	36	17	28	33	29
El Paso (6 P.M. local time) .	1892	35	12	19	26	23
	1893	23	10	26	25	21
San Diego (5 P.M. local time) .	1892	69	69	71	70	71
	1893	67	68	70	72	70
Los Angeles (5 P.M. local time) .	1892	68	65	61	63	65
	1893	68	66	61	69	66

In reading the preceding summary it should be remembered that the observations for the first four stations were taken at 6 P.M., mountain or 105th meridian time, while the last two stations have the advantage of having had the observations taken still earlier, at 5 P.M., Pacific or 120th meridian time, only a few hours after the driest portion of the day.

If this record of relative humidity for Los Angeles, taken at 5 P.M., is compared with the record of the same station taken at 8 P.M., which was given on page 313, the steady increase of moisture with the approach of night at all seasons becomes most evident.

On the elevated plains of Colorado or Northern New Mexico the night-air can be admitted into sleeping-rooms on the coldest nights in winter because it is so dry. The temperature of a room with the window partly open seldom falls below 40° F., even if the air outside is 20°. At that temperature, even with 70 per cent. of relative humidity, the amount of absolute humidity the air can hold is less than one grain (0.91) to the cubic foot.

In California, on the contrary, as in the Riviera, the night-air is usually damp and frequently saturated with fog.

To those to whom the presence of dry air is not important California offers many attractions.

From Monterey to San Diego are wooded ranges, fertile valleys, vineyards, orange and lemon groves, tropical and semitropical fruit in abundance, plenty of sunshine, well-built and prosperous towns, pleasant society, and for the transient visitor, what is most important, good hotel and boarding accommodations.

The winter or rainy season is the favorite time for visiting California. The monthly rainfall at that time is not usually more than the normal monthly rainfall for Eastern cities, and there are longer periods of fine weather. The welcome rains bring a vivid green to the brown valleys and hills, and the beautiful "procession of the flowers," which continues from December to May, enraptures Eastern visitors.

In the protection from cold northern or eastern winds afforded by the mountains (with the additional climatic advantage of the miles of desert beyond), its soft, balmy air, bright days, and equability of temperature from one season to another, this southern coast resembles the Riviera and may even claim superiority over that resort.

Some of these points of comparison are alluded to in the more detailed descriptions of the towns themselves (see Santa Barbara).

The attention of invalids or delicate persons should be directed to a marked feature of the Californian climate—that is, its peculiar chilliness in the shade and when the wind blows, even on a summer-day. It is always chilly at nightfall, at which time a light overcoat or wrap is needed all the year round.   Rooms with a sunny exposure are much to be desired.

It is greatly to be regretted that no suitable health-resort station has been developed in the highlands of Southern California, where the natural advantages of sandy soil, soft water, moderate elevation, and an atmosphere containing less moisture than is to be found directly on the coast could be supplemented with good accommodations.   Such a resort is possible.   Beaumont (2500 feet) and the higher "benches" above may be suggested as possessing some of these advantages.   The Sospe Valley, above Ventura, which is moderately elevated and similar in character, has been strongly recommended, and other locations can be found equally good if not superior.

In San Diego County, thirty or forty miles from the sea, are several places which might be developed into health-resort stations, such as Alpine (2200 feet); Ballena (2500 feet); the Santa Isabel Valley (3200 feet); Julian (4300 feet); Cuyamaca (4700 feet); and the Palomar and Coleman valleys (each 4000 feet).   Above an altitude of 2500 feet the trees grow more thickly, and fir, pine, and oak trees are found up to the crest of the mountains.   On the west side of the mountains the rainfall also increases.

In the desert east of Jacinto are stations on the Southern Pacific Railroad where the experiment of living a little below the sea-level may be tried, viz., at Salton (260 feet depression) and Volcano Springs (220 feet depression); good accommodations are lacking. Indio (50 feet depression) has better facilities for visitors, but is not so low.

San Diego.   Population, 18,000.   The bay of San Diego is a beautiful land-locked body of water twelve miles long by a mile wide.   It is the best natural harbor south of San Francisco; but, although the settlement dates from 1542, when it was first discovered by Cabrillo, its commercial supremacy is still in the future.

The completion of the Nicaragua Canal will be of the greatest benefit to this port, which, from the strength of its geographical position, will receive a large share of the enormous trade that will then be developed.

The town of San Diego lies on the main land, facing the Pacific. Dr. Walter Lindley thus describes it : "Situated on one of the most perfect harbors in the world, with vessels unloading at its wharves from all the chief ports of civilization, the culmination of the Santa

Fé Railway system, that brings it into intimate relations with Chicago, New York, and Boston; planted on a series of hills that gently slope to the ocean, with a soil that produces almost everything desirable from a pumpkin to an olive; with business blocks which, for elegance, solidity, and size, are rarely surpassed; with a climate that is enjoyable and healthful both summer and winter; with every facility for boating, fishing, and hunting; with a population noted for culture and refinement; with schools, churches, and hotels that would be creditable to much larger cities; with commercial prospects of dazzling brilliancy—with all these attributes the visitor does not wonder when he finds that every one of San Diego's fifteen thousand inhabitants, from the infant just beginning to prattle to the great-grandmother who dozes away the sunny Christmas-day in her arm-chair on the veranda—has learned to sing her praises loud and long.''[1]

The soil in the highest portions of the town is sandy. The shores of the bay are high at Point Loma—the entrance—but low and marshy around the southern end. The *mesa* or tablelands rise higher going from the bay until they reach the mountains, distant about forty miles. In the fertile valleys and on the hillsides a great variety of soil can be found.

San Diego is now supplied with pure, soft water, brought from the mountains. There are also wells.

The natural facilities for drainage are good. Forty miles of sewers emptying into the bay have already (1894) been built.

San Diego has several lines of street-cars, mostly electric. There are a number of good hotels.

One of the best portions of the town for residence is Florence Heights, which lies at an elevation of 180 feet above the sea.

The climate of San Diego is noted for its equability, the range of the mercury between winter and summer being usually under 15°. The monthly mean temperature for January is 53°; for July, 68°; annual mean, 61°.[2] The average number of days during the year above 90°, 1; below 32°, none. During twenty-three years the records show that the mercury three times rose to 100° (going once to 101°). It went to 95° but three times, and to 90° but seventeen times. It does not usually reach 86° more than twice a year. The highest temperatures are likely to be recorded in September.

1 California of the South. Walter Lindley, M.D., 1888.
2 Government record for twenty-three years—to 1894.

Summer-nights are invariably cool. Frost is said to be unknown
on the coast, although 32° has been recorded five times in twenty-
three years. In the foot-hills a thin skim is sometimes formed
on the lowest lands when the slopes of the uplands are exempt.
The temperature of the winter-days on the high *mesa* near the
mountains is said to be usually about 45° at daylight, running to
65° or 70° at noon. It is seldom as low as 55° at noon, and sometimes it is as high as 75°.

San Diego has an annual rainfall of 10 inches. During a " dry "
winter it has been known to be as low as 4 inches. The heaviest
rainfall was during the season of 1883–'84, when 25.97 inches fell.
There are few stormy days during the year (37), but a large number
of cloudy days for Southern California (69), and the degree of
humidity is high.

Fogs are less prevalent at San Diego than at Los Angeles.[1] Four
or five miles inland from San Diego the amount of fog is still less
than directly on the bay.

The wind-movement is moderate, the yearly average being 5.6
miles per hour.

The prevailing sea-wind is frequently alluded to by local writers
as " dry". This is an error, as will appear if the movement of the
winds is considered. The wind *is* dry when it first starts from the
heated deserts of Southeastern California and Southwestern Arizona,
and rising to a great elevation passes out to sea over the incoming currents ; but when it has to descend to the ocean about latitude 30°
and, reversing its course, becomes in turn the southwest or west
wind blowing landward, it presents such conditions for the absorption
of moisture that long before it reaches the coast of California its
original dryness has been completely lost. This is the daily wind
from the sea.

What its characteristics are, then, is shown by the meteorological
analysis, which can be better illustrated by instituting a comparison
with a locality of admitted dryness.

El Paso will afford an excellent parallel,[2] although it has a little
higher temperature during the year—a fact slightly to the advantage of San Diego, as the warmer the air is the greater the amount
of moisture it can contain.

---

[1] See reference to night and morning fogs, in article on the Climate of California.
[2] Latitude of San Diego, 32° 43′ N. Latitude of El Paso, 36° 47′ N.

	Tempera-ture.	Relative humidity.	Absolute humidity.	Dew-point.
San Diego	61°	73 per ct.	4.34 grains.	52°
El Paso	. 64	48 "	3.16 "	40

We thus find that San Diego, at nearly the same temperature, exhibits over El Paso 58 per cent. increase in relative humidity, 37 per cent. increase in absolute humidity, and 30 per cent. increase in the dew-point. The only conclusion possible to reach is that air which at 61° of temperature shows an annual mean of 73 per cent. of relative humidity, 4.34 grains of absolute humidity to the cubic foot, and 52° for the dew-point (or point of saturation), cannot be properly described except as a *warm, moist* climate.

Particular emphasis is given to this point, because so many pulmonary invalids—to whom the amount of moisture in the air is sometimes of the greatest importance—have suffered sadly from lack of early knowledge regarding this detail.

It is surprising to notice how constantly one meets references to the "dryness" of this coast, not only in guide-books and in editorial utterances, but even in articles on the climate by medical writers. On the part of strangers such a misconception would probably be due to mistaken inferences based on the large number of sunny days during the year and on the small amount of the annual rainfall. It is unfortunate that it should be kept up by local writers to whom the actual dampness of the country within the ocean-influence ought to be well known.[1]

Admitting, then, as we must, the presence of the important factor of humidity in this as in all coast climates, it can be said that the climate is delightful, equable, and healthful. The ocean-current that brings so much fog and chill to the coast north of Monterey passes by so far out to sea that its influence is greatly modified. Curiously enough, it is said that one hundred and fifty miles further south in Lower California, where it again approaches the shore, it brings to San Quintin a climate cooler than that of San Diego.

Opposite San Diego, on the ocean-side of the promontory that

[1] An article by Mr. John D Parker on "California Electrical Storms," in the *American Meteorological Journal* for June, 1895, suggests that "one cause for the infrequency of electrical storms is probably found in the humidity of the atmosphere. . . . The humidity of the atmosphere in California, so contiguous to the Pacific Ocean, is naturally much greater than that found at points more remote from large bodies of water. At San Diego the mean humidity of the air in 1891 was 74 per cent. of complete saturation ; in 1892, 76 per cent. ; in 1893, 74 per cent.; and the mean humidity for eleven years, from 1884 to 1894 inclusive, is 77 per cent." This percentage of humidity for San Diego, which is for different years than those used for the computations for this book, is even more excessive.

forms the harbor, is **Coronado Beach**, the whole development of which—from the great hotel of 750 rooms to the tropical garden— has taken place since 1887. It is an attractive resort with an equable marine climate, and in many respects is unequalled in the country.[1]

Coronado Beach is easily reached by ferry from San Diego. There are a great many cottages near the hotel, and quite a settlement is growing up, with shade- and fruit-trees, which are developing rapidly under the stimulus of irrigation.

The Coronado Springs are worthy of notice. They are thirty feet above tide-water and flow 50,000 gallons per hour. The water is pure, soft, and sparkling, resembling the well-known Waukesha Springs in Wisconsin. There are about 26 grains of total solids to the gallon.[2] It is carbonated and bottled for commercial uses, making a fine table-water.

Miss Kate Field, in a letter which appeared in the San Francisco *Chronicle*, in June, 1893, made a strong and characteristic plea for better water for domestic purposes in California. Her summary of the general quality of the water in the State is somewhat severe, as many towns are supplied with excellent water from either the mountains or from artesian wells; but what she says is well worth reading, especially for its bearing on the climate.

"California has herself to thank for her national reputation as the purveyor of bad water. Nature produces no bane without an antidote. Though mountain and desert send water freighted with lime or alkali, heaven sends rain pure, soft, and health-giving. If every house in Southern and Lower California should have its cistern, nobody need drink hard or poisonous water. Were it known to tourists that soft water abounded in hotels and boarding-houses many an invalid would hasten to breathe the balmiest air in this country.

"Rain, however, is not the only means by which soft water can

---

[1] For meteorological data, see records for San Diego, Tables V.–IX.

[2] Coronado Springs. Light alkalo-carbonated water. Analysis by C. Gilbert Wheeler, of Chicago.

	Grains in U. S. gallon.		Grains in U. S. gallon.
Sodium chloride	10.16	Ferrous sesquioxide	0.04
Potassium chloride	0.91	Silica	1.08
Potassium sulphate	0.55	Organic matter	0.99
Magnesium	4.72		
Calcium carbonate	6.48	Total solids	26.25 -
Calcium sulphate	1.32		

be obtained. It is a remarkable, but little recognized fact that the dew or fog of the Pacific coast would, if caught in cisterns, supply every family with soft water. Seeing is believing, and I have seen. An acquaintance of mine, living near San Diego, has never dug a well, though his house is half a mile from the sea. His cistern supplies all the water necessary for domestic purposes and for a horse and cow. Twenty-four hundred square feet of roofs produce 140 gallons of water in twenty-four hours. Multiply 140 by 365 and you are amazed to learn that this householder obtains an annual supply of 51,100 gallons of soft water without counting the rainfall. Another acquaintance, while building on Coronado Beach a cottage with a roof measuring 1000 square feet, saw a stream of water as large as a small pen-handle running from this roof as late as 9 o'clock in the morning. Here is a beneficent nature coming to the rescue of that beautiful coast, yet purblind residents rarely accept the blessing falling upon them nightly.

" ' Irrigation ? No, indeed,' said a ranchero, as we sat in the shadow of his cottage on the border of Mexico. ' We don't need irrigation for farming, as the rainfall suffices to raise crops in our region. If it did not, dew would supply the deficiency. Here we are twelve miles from the sea, yet the nightly moisture almost equals rain. We don't irrigate, but, as we have no well, we send sixty miles for drinking-water.'

" ' Why don't you utilize your wonderful fog ?' I asked. ' Why don't you put gutters and spouts on your roofs and improvise a cistern ?'

" ' By Jove ! I never thought of that. That's an idea.'

"Acting on this suggestion all the water needed for domestic purposes was readily obtained.

"Many a morning I've risen at Ensenada, Lower California, fully persuaded from the dripping on the roof that rain had come, and I've found a fog almost thick enough to cut with a knife. Yet not one drop of soft water could be had at the hotel for love or money. Santa Barbara abounds in fogs so heavy as frequently to lay the dust in summer, despite which the fastidious traveller must pay ten cents a quart for distilled water if he objects to coating his mucous membrane with lime.

"South America has learned the wealth of its dews. In his memoirs General Sherman refers to a very interesting conversation in which Henry A. Wise, then our Minister to Brazil, ' enlarged

21

on the fact that Rio was supplied from the "dews of heaven," for it rarely rains there, and the water comes from the mists and fogs which hang around the Corcovado, drips from the leaves of the trees, and is conducted to the madre fountain by miles of tiled gutters.'

"What can be done in Rio Janeiro can be done from one end of California's coast to the other; the peninsula itself can be redeemed from desolation. Yet though nature failed to furnish this welcome moisture, there are simple methods of distilling water that come within the ken of all.

" ' For heaven's sake don't refer to our fogs,' exclaimed a boomer.

" 'Why not? They are a boon to the coast. Moreover, was there ever seaboard without fog?'

" 'True; but if people knew we had fogs they'd be frightened away.'

" ' What kind of people?'

" ' Invalids.'

"So I'm not to make known a valuable fact because it may keep a few sick people away from the coast. That's the reason it should be advertised far and wide   California gains nothing by concealing the truth. It is big enough and wonderful enough and varied enough in climate to be honest about every locality."

Julian (elevation, 4300 feet). East of San Diego is a chain of mountains rising higher in successive ranges until, about forty miles from the coast, they culminate in pine-clad summits with an altitude of from 4000 to 6000 feet.

The rainfall on the western slope of these mountains is said to amount to about 20 inches per annum at 3000 feet altitude and 30 to 34 inches on the top of the range. It is probably somewhat more, as General A. W. Greely, in his *Report on the Climatology of the Arid Regions of the United States,* 1891, credits Julian with about the same annual precipitation as Pittsburg, and by means of an interesting comparison shows the difference in the methods of distribution.

The following is by seasons :

	Winter.	Spring.	Summer.	Autumn.	Year.	Length of record.
Julian,	18.34 in.	16.56 in.	0.00 in.	2.78 in.	37.68 in.	6 years.
Pittsburg,	8.58 "	8.53 "	12.00 "	7.60 "	36.71	19 "

The California rainy season is made still more prominent when the year is divided into two periods, the rainfall at Julian being for

November to April 36.40 inches; for May to October, 1.28 inch. General Greely remarks: "At Julian only 9 per cent. falls from the 1st of May to the last of November, while nearly one-half (48 per cent.) of the entire precipitation of the year falls during the months of February and March." The greatest seasonal rainfall was in 1883 and 1884, when it amounted to 61.62 inches. The smallest record is 25.89 inches.

"The soil on the hillsides is more or less gravelly, resulting from the decomposition of granitic rock. In the valleys it is finer and often of a brownish or reddish color, and strong and fertile. In the lowest part of the valleys it is a dark alluvium."

Below 4000 feet the timber is largely live-oak; above that elevation there are pine, fir, cedar, and deciduous oak groves. Fruits and cereals can be successfully grown on the western slope of the mountains, where springs and small streams are abundant. There are fruit-ranches near Julian, which is also the centre of mining-interests.

The eastern slope pitches more steeply down to the desert, which lies nearly at the level of the sea. On this slope little rain falls, and the increased dryness and temperature produce a very different climate.

There is no meteorological record for the eastern side of these mountains at 4000 or 5000 feet elevation, but the influence of the ocean must have almost vanished, while the influence of the desert is marked.

The description of the climate of Hesperia and Daggett, on the Mojave Desert (pages 333 and 334), may be of value in this connection.

If water could be impounded and brought from the summits, or around from the western slope, this arid land would undoubtedly produce luxuriantly and trees and verdure would reward the cultivator.

The winter climate is probably fine. The degree of heat of blasts from the desert in summer at that altitude is not yet a matter of record.

The hill country back of San Diego has been proved by individual experience to possess most valuable climatic characteristics, and it is to be hoped that the necessary detailed information will soon be obtainable.[1]

---

[1] See Climate of California.

Los Angeles (elevation, 330 feet; population, 80,000). The full name bestowed by its pious founders was *La Puebla de Nuestra Senora la Reina de Los Angeles*—The Town of Our Lady the Queen of the Angels—eight-tenths of which we leave off nowadays, and usually mispronounce the remainder. The town is on a rolling plain, fourteen miles from the sea and about the same distance from the mountains. It is a beautiful and interesting place, full of architectural and social contrasts. Several elements go to make up the city, the Southern 'or Spanish, and the American; and brown faces, betraying Castilian and Indian ancestry, mingle on the busy streets with those of the fairer-skinned Yankee type. Low adobe-quarters and American country houses are found near each other, within a few minutes' walk, although the old-fashioned "adobe" is growing more rare. Modern office-buildings appear within sound of the bells of the early Missions.

The near presence of the mountains insures a good supply of water both for domestic and irrigating uses. Much of the sewage has been used for the purpose of fertilizing the plains below the city. Large sewers have also been constructed to the sea.

Los Angeles is well supplied with hotels and churches. The hotels are, however, not worthy of this attractive city. Cable and electric cars reach all points.

South of latitude 35°, that portion of the State called Southern California, is more free from cold summer-fogs and strong winds than the country near San Francisco. Los Angeles is far enough from the sea to have a temperature perceptibly higher in summer and lower in winter than directly on the coast. While its average annual rainfall is but 18 inches, it is subject to great extremes. For instance, during the summers of 1882 and 1883 there were 12.67 inches of rain. During the seasons of 1883 and 1884 there were 32.16 inches. In 1892 and 1893 about 30 inches fell. The mean annual relative humidity is 72 per cent.

There are frequent fogs both night and morning during the spring and summer. They usually clear up during the forenoon.[1] The average number of foggy nights and mornings for the year at Los Angeles is 57 days.[2]

The percentage of cloudiness and sunshine and the number of clear days for Los Angeles and San Diego by seasons and by the year are as follows:

[1] See Climate of California.
[2] From records of thirteen years at Los Angeles office of U. S. Weather Bureau.

Cloudiness and sunshine.	Winter.			Spring.			Summer.			Autumn.			Year.			Length of record.
	Mean Cloudiness.	Mean sunshine.	Average No. clear days.	Mean Cloudiness.	Mean sunshine.	Average No. clear days.	Mean Cloudiness.	Mean sunshine.	Average No. clear days.	Mean Cloudiness.	Mean sunshine.	Average No. clear days.	Mean Cloudiness.	Mean sunshine.	Average No. clear days.	
	p.c.	p.c.	p.c.		p.c.	p.c.		p.c.	p.c.		p.c.	p.c.		p.c.	p.c.	
Los Angeles,	34	66	49	41	56	36	31	69	39	25	75	54	33	67	178	11 years.
San Diego,	39	61	37	49	51	26	46	54	24	37	63	38	43	57	125	17 "

Total cloudiness 100 per cent.

The highest recorded temperature is 108°; lowest, 28°. The mean monthly temperature for January is 53°; for July, 72°; for the year, 62°.

Average annual number of days above 90°, 20; below 32°, none (one day in 1893). Cloudy days, 45; stormy, 36 (means for six years).

Normal wind-movement for the year, 5.1 miles per hour.

Pasadena, a suburb of Los Angeles, distant nine miles, has an elevation of 900 feet. It is twenty miles from the sea and five from the mountains. Population about 9000. Ample supply of water. Soil is gray gravel mixed with brown loam; it is light and porous. Some of the streets are paved.

Pasadena is a charming little city of attractive homes. As a place of residence its social advantages are well known.

There have not been any records published of moisture, but the percentage of relative humidity is believed to be about 60 for the year. Pasadena is usually a few degrees hotter at all seasons than Los Angeles.

There are large, well-equipped hotels and good accommodations for visitors.

Sierra Madre is a settlement twelve miles northeast of Los Angeles, situated at the base of the foot-hills, 1700 feet above the sea. There is less frost at this elevation than in the lower valleys, and also less fog. The climate is considered particularly healthful, and was selected by the State Board of Health in 1880 as the most desirable for consumptives in Southern California.

Along the foot of the mountains are innumerable desirable locations for country homes and fruit-farms, where an invalid can avail himself of the best there is in California outdoor life.

Echo Mountain (elevation, 3500 feet). Four miles from Pasadena is Rubio Pavilion, at an elevation of 2200 feet, from which the cable-road ascends 1300 feet further to the top of Echo Mountain.

The Echo Mountain House is a new hotel of seventy rooms, commanding wide views over the San Gabriel and Los Angeles valleys and the Pacific Ocean beyond.

It is claimed that the climate on the summit "is equable and delightful during the entire year. When clouds and fog obstruct the vision and render residence somewhat uncomfortable in the valley, the mountain is invariably bathed in sunshine." One of the sights from the hotel is to see the fog or clouds resting over the valley below.

There is no record of the weather on the summit to compare with that of the upper valley, but it is said to be warmer in winter and cooler in summer than on the plain. There is always a breeze on the veranda of the hotel.

Mount Lowe (elevation, 6000 feet). The summit of Mount Lowe, still higher up, is to be provided with a stone hotel. A large observatory has been built. The views are very extensive. Bridle-paths have been laid out through the mountain-paths and cañons.

Wilson's Peak (5500 feet) is four miles east of Echo Mountain. The view is extensive and beautiful. There are a hotel and camp on the mountain. Mount Harvard (5000 feet), an adjoining peak, is easily reached from Pasadena and has also a camp for the accommodation of visitors.

Riverside (elevation, 850 feet ; population, 10,000) is situated ten miles south of San Bernardino and sixty miles east of Los Angeles. It is famous for its orange-groves. The roads are hard and well shaded, but sometimes dusty. There are a number of handsome residences. The town is supplied with good hotels. Water for domestic use is furnished by a pipe-line from twenty-three wells (300 to 600 feet in depth), located nine miles above the city in an artesian belt, and the Waring system of drainage is being introduced.

A local physician, writing of the climate, leniently says it is "warm, but not hot, reaching in the summer months a maximum of 108° to 110° and in the winter from 78° to 80°. . . . In addition to the precipitation in rain, occasional and very infrequent fogs add a trifle to the total moisture. They drift into the valley from the seaward, coming up in the early morning and vanishing by 9 or 10 o'clock in the forenoon. They occur more often in the fall and winter months, but come so seldom and are so light that their effect upon atmospheric moisture is insignificant. From July, 1885, to July, 1886, there were 280 absolutely clear days, 38 days of rain, in many of which there was simply a shower with a precipitation of one-tenth of an inch or less, the balance of the time being clear, and 47 days in which there was a longer or shorter interval of trifling fog in the early morning."[1]

At Riverside these fogs are less felt than nearer the coast or mountains.

Irrigation is extensively used on the groves and orchards. The Santa Ana River—a small stream—flows by the town. A curious fact regarding increase of rainfall is shown by comparing the precipitation of ten years ago with that of recent years. During the four years from 1880 to 1883, inclusive, the annual average precipitation was 4.42 inches; while during the four years 1890 to 1893 it was 10.69 inches, an increase of 140 per cent.

The average yearly rainfall for the past fourteen years was 10 inches. The rainiest months are usually February and March. The percentage of relative humidity during the year 1888 was: spring, 67; summer, 56; autumn, 63; winter, 76. From December, 1890, to September, 1891, inclusive, the average monthly relative humidity was 67 per cent.

The mean monthly temperature for the seasons from a record for twelve years is: spring, 60°; summer, 74°; autumn, 64°; winter, 51°. The mean for July is 76°; maximum, 106°; mean for August, 76°; maximum 104°; mean for January, 50°; minimum, 29°. The average for the year is 62°.[2]

Redlands (elevation, 1350 feet; population, 3500). From the end of the platform of the railroad station at San Bernardino, facing to the east, one sees, away to the left, the high range of the San Bernardino Mountains, terminating in its two great peaks of Grayback and San Bernardino, each rising 12,000 feet above the level of the sea. Twenty miles straight ahead, due east, the huge shoulders of San Jacinto lift their snow-clad crest 11,000 feet into the air, while filling the middle ground of the picture, eight miles away, the homes and orange-groves of Redlands dot the hillsides at varying elevations of 1300 to 1500 feet.

The delightfully situated town of Redlands is four or five hundred

1 A Study of Riverside Climate. W. B. Sawyer, M.D.

2 Meteorological record from pamphlet issued by the Riverside Board of Trade, 1894.

feet higher than its neighbor, San Bernardino, and about six hundred feet higher than Riverside, which is distant fifteen miles to the southeast. It is over fifty miles from the Pacific Ocean.

The settlements of Lugonia and Mentone are included in this description of Redlands.

The hotels are not large, but are well kept and furnish good accommodations. The water-supply of Redlands is unsurpassed, coming as it does from the great Bear Valley reservoir—eight miles distant—which is situated in the mountains at an altitude of 6500 feet. The records of the water corporation are said to show an annual rainfall of 40 inches on the high peaks, which illustrates how the mountains " milk " the moist sea-winds. The irrigating-pipes throughout the valley are of steel and are buried underground, a method considered less subject to evaporation and generally superior to carrying the water in open ditches.[1]

The soil of the hillsides is red adobe, most fertile, and, of course, muddy after a rain. In the lowest portions of the valley it is a sandy " wash".

The annual rainfall at Redlands is slightly less than at Los Angeles. It averages 15.32 inches, divided as follows: winter, 6.55; spring, 7.45; summer, 0.48; autumn, 0.94 inch.

The sea-breeze blows usually from 10 o'clock to sunset, increasing in force during the middle of the afternoon. The presence of the high range on the north protects the valley from desert-winds, except for an occasional norther that sweeps around by way of the Cajon Pass.

There is frequently an early morning-fog, but it rarely remains after 9 o'clock. It clears here an hour or two earlier than at Los Angeles.

Like all of the California towns within the influence of the coast climate, Redlands cannot claim extreme or long-continued dryness. It is, however, one of the most favorable locations in Southern California for invalids where comfortable quarters can be secured. During the summer the temperature rises pretty high, but the town is conveniently near to the cool, tree-covered slopes of the mountains.

There are no official or complete records of the humidity, the only

---

[1] Regarding the possible influence of the use of irrigation in Redlands, it should be noted that irrigation is not used during the winter and early spring when the annual rains usually occur, and during the irrigating season each grove is irrigated for three days at a time but once in four or six weeks—not continuously. The readings of the hygrometer, as taken by Mr. E. N. Peirce, showed that Redlands had a drier air, during the day than the valleys nearer the sea-coast.

figures being furnished by voluntary observers. It is probably safe to say that the general average of the relative humidity will be at least 5 per cent. less than at Los Angeles. A record of the relative humidity taken at Terracina Heights for two weeks in April, 1893, showed at 8 A.M. 70 per cent.; at 12 M. 66 per cent.; at 6 P.M. 67 per cent. During the night the amount of atmospheric moisture was greatly increased.[1]

A record of the temperature taken by a voluntary observer during the years 1892 and 1893 is as follows:

	Winter.	Spring.	Summer.	Autumn.	Year.
Monthly mean for two years,	59°	63°	76°	66°	66°

Average minimum, January .		37°
Average maximum, July .	.	105

For the winter of 1893–1894 a record of temperature and relative humidity taken at the Hotel Terracina at midday—which is just before the warmest and driest period of the twenty-four hours— is as follows:

Observations taken at 12 M.	Temperature.				Relative humidity. Per cent.
	Monthly mean.	Extreme max.	Extreme min.	Monthly max.	
December, 1893 .	55	76°	35°	63°	51
January, 1894 .	47	69	26	58	46
February .	51	70	32	55	43
Winter mean	. 51	...	...	...	46
March .	. 55	80	31	66	44
April .	62	85	40	72	39

The monthly mean temperature for 1893 for Redlands, taken by the Southern Pacific Railroad, arranged by seasons, was as follows: · winter, 52°; spring, 58°; summer, 75°; autumn, 58°; annual, 61°; maximum, 103°; minimum, 34°. Monthly mean for January, 54°; for July, 77°.

Rialto (elevation, 1300 feet) is a colony four miles west of San Bernardino and twelve or fourteen miles from Redlands. The following imperfect weather-record is given as a basis of comparison. It is for the winter of 1892–1893, and was taken from *The Orange Belt:*

Temperature, November, 63°; winter (December, January, and February), 56°; spring, 59°. Relative humidity, November, 46 per cent.; winter, 49 per cent.; spring, 60 per cent.

[1] See the Climate of California.

The wind-movement during November and December averaged 5 miles an hour. April and May averaged 6 miles.

The Yucaipe Valley, eight miles northeast of Redlands, is worth the attention of invalids who can live in California. Its elevation is from 2000 to 4000 feet, a little too high for orange-culture, but well adapted for deciduous fruits. It is slightly more protected from the influence of the sea-winds and has also less fog than Redlands. Its rainfall and humidity are probably about the same.

Arrowhead Hot Springs. Among the many hot springs in San Bernardino County one of the best known and most resorted to is Arrowhead, six miles northeast of the city of San Bernardino, where the rude outlines of an arrowhead, a quarter of a mile in length and 350 feet across, are conspicuous on the side of the mountain. Here are twenty-five springs, at an elevation of nearly 2000 feet, with temperatures varying from 140° to 193° F. An analysis of the principal spring (temperature 193°) made by Professor E. W. Hilgard, University of California, is as follows:

	Grains per gallon.
Potassium sulphate	4.00
Sodium sulphate	42.47
Sodium chloride	8.17
Lithium	strong test
Calcium sulphate	1.34
Calcium carbonate	1.34
Barium	a faint test
Strontium	well marked
Magnesium sulphate	0.14
Magnesium carbonate	0.32
Silica	4.94
Organic matter	trace
Total solid contents	63.39
Free sulphuretted hydrogen	644 cub. in.

The hotel is situated at the entrance to a cañon, 2000 feet above the sea, and commands enchanting views. Baths are given in vapor, hot mineral water, and the mud poultice.

Seven Oaks, known also as the Lewis Ranch, is a camp in the Santa Ana Cañon, about twenty-four miles from Redlands, reached by a mule-train. The elevation is 4800 feet. Season extends from May to October. There are four log-cabins and several tents. Food plain but wholesome. Some brook-trout and small game can be found. There is a good cold-water mineral spring, the ingredients being principally iron salts.

Beaumont (elevation, 2560 feet; population, 200). Sixteen miles east of Redlands, on the Southern Pacific Railroad, is an almost deserted "boom" town of the era of 1888. At that time twenty-seven miles of streets were laid out, trees were planted, excellent water was brought from a cañon in the foot-hills four miles west of the town, a good hotel built, and other improvements projected; but it was started on a falling market and "busted" early in its career. Beaumont has decided natural advantages, particularly for invalids. It is situated on the southern side of a valley twelve miles long by six miles wide. It has warmer summers and colder winters than the coast-region, but is drier and freer from fog than the lower valleys at the foot of the mountains. In the summer the sea-wind blows over sixty miles of land before reaching Beaumont, thus losing much of its moisture and coolness. In winter the prevailing wind is from the east, blowing from the distant desert through the San Gorgonio Pass. This wind is dry and healthful, but tiresomely persistent. It can fortunately be avoided to a great extent by getting in the shelter of the foot-hills that form the eastern boundary of the valley. The air is quite dry and pure and invigorating, and at night fairly free from dampness.

The annual rainfall is 18 inches, divided as follows: winter, 9.87 inches; spring, 6.84 inches; summer, 0.12 inch; autumn, 1.22 inch. There are no records of relative humidity; but as there is almost entire immunity from fog the air is perceptibly less moist than at Los Angeles. A yearly average would probably be near 60 per cent. or below it. The highest record of temperature at the Highland Home, four miles northeast of the postoffice, is 104° and the lowest 36°. At the railroad station the extremes in 1888 were 104° and 24°. The mean for January was 41°; for July, 77°. Average for the year, 62°.

For 1893 the mean monthly temperature, taken at the Southern Pacific Railroad station, was as follows, by seasons: winter, 53°; spring, 57°; summer, 78°; autumn, 63°; annual, 63°; maximum, 102°; minimum, 30°. The monthly mean for January was 58°, and for July 80°.

The accommodations for visitors are poor, the hotel having been closed for years.

A few miles west of Beaumont are benches of tableland at an

altitude of 4000 and 5000 feet, and even higher. On the side of San Jacinto Mountain, fourteen miles southeast of Beaumont, is the **Strawberry Valley**, a well-known summer-resort, with an elevation of 6000 feet. It has fair accommodations.

Banning (elevation, 2300 feet; population, 300). Six miles east of Beaumont is the town of Banning, which has been resorted to by invalids, although its position in the San Gorgonio Pass renders it difficult to escape the trying desert-winds that blow in from the east. There are a number of cañons north of the town where the wind would be less violent.

Banning has a resident physician, a fairly good hotel, and several stores.

Palm Springs. On the eastern slope of the San Jacinto range, and on the edge of the desert, five miles from the Southern Pacific Railroad and thirty-five miles southeast of Banning, is the winter-resort of Palm Springs. The elevation is about 580 feet above the sea. The sheltered lands of Palm Valley produce, under irrigation, fruits and vegetables several weeks earlier than is possible directly within the sweep of the Pacific Ocean winds. Palms, figs, oranges, lemons, and dates grow luxuriantly. The winter climate is warm and mild. A copious, hot sulphurous mineral spring flows here, whence comes the name of the settlement. Palm Springs has warm winters and scorching summers.

Indio is a station on the Southern Pacific Railroad one hundred and twenty-seven miles east of Los Angeles. It is situated on the edge of the great depression in the Colorado Desert, which reaches in one place a depth of 360 feet below the level of the sea.[1] This basin is 130 miles in length by 30 miles in average width. At Salton, on the Southern Pacific Railroad, the surface of the earth is covered with a crust of salt four inches thick for nearly ten miles square.

Indio is near the northern rim of the basin, and is 50 feet below the level of the sea. This is the most arid portion of America, the annual rainfall being about $2\frac{1}{2}$ inches. The temperature for a portion of the year 1894 was as follows:

---

[1] Indio. Walter Lindley, in New York Medical Record, 1888.

	Extreme max.	Extreme min.	Monthly mean.
November . .	104°	46°	70°
December .	79	40	55
January	83	23	54
February	90	32	45
March	102	36	64
April .	100	53	75

The highest temperature during the year was 117° and the lowest 23°.

Dr. Lindley gives the mean temperature taken for the month of January, 1893, as follows : 7 A.M., 45°; 2 P.M., 83°; 9 P.M., 53°. No records of the humidity at Indio have been published, but the air is very dry and warm.

Indio is an oasis in the desert. Water is obtained from surface-wells, and also in great quantity from artesian wells sunk only to a depth of 115 feet. By means of irrigation trees and plants and flowers are grown. Near the mountains are date-palms, growing sometimes to a height of eighty feet.

There is a good small hotel, and a limited number of cottages have been built for visitors, but the accommodations are not yet very extensive.

Hesperia (elevation, 3100 feet) ; Victor (2700 feet) on the line of the Southern California Railroad. The climate of the country in the foot-hills on the leeward side of the mountains is worthy of notice.

The Mojave Desert is a vast plain lying north of the San Bernardino range of mountains, and extending from the Colorado River on the east to Ventura County on the west. It is 300 miles east and west and from 20 to 200 miles north and south—a *mesa* or table-land, most of it 2000 feet above the sea, except the northeastern portion, which is below sea-level and is known as the Death Valley. Close up to the mountains the soil is fertile under the application of water, which can be obtained by means of reservoirs on the mountains and by using the Mojave River. This watershed is 3000 to 6000 feet above the sea, and has a rainfall on the mountain-tops of about 40 inches per annum.

The climate on the north and east slopes of the mountains in Southern California is the driest, as it is protected from the trade-wind coming from the west or southwest, which is a moisture-bearing wind. The rainfall is about 15 inches. The humidity is less on that slope, and at an elevation above 2000 feet the nights are usually cool. There is occasionally a parching, dry north wind from the desert that is very trying, although healthful. It is intensely hot in summer, but drier both in winter and summer than localities on the coast-side of the mountains.

Artesian wells are already used to a limited extent at Hesperia to furnish a supply of water, and the number of fruit-ranches on this upland edge of the desert is increasing.

At Daggett (2000 feet) the record of temperature is: monthly mean, winter, 47°; spring, 57°; summer, 84°; autumn, 71°; average for the year, 65°; maximum, 104°; minimum, 20°. Annual rainfall, 4 inches.

The Antelope Valley, fifty miles west of Hesperia and Victor, has much the same characteristics. Lancaster (2300 feet) is the principal town.

There are no accommodations for visitors in these places. They are referred to here in order to preserve a record of the climate of this region.

Ojai Valley (average elevation, 900 to 1200 feet). Nordhoff (elevation, 1200 feet; population, 800) is the principal town.

This retreat in the mountains is in Ventura County, fifteen miles north of the town of Santa Buena Ventura, with which it is connected by a daily stage.

The valley is a "pocket" in the mountains, varying in width from two to four miles and entirely surrounded by the San Rafael and Santa Inez ranges, which rise on the east to a height of 6000 feet. It is thus well sheltered from harsh winds and partly also from the sea-fogs. The southern portion of the valley is about five miles long, and the northern portion, in which Nordhoff is situated, about ten miles long.

During the spring of 1895 the fogs in Nordhoff averaged ten for each month, clearing usually before 9 A.M. and always by 11 o'clock. During the winter months the fogs were not so prevalent. Directly on the coast the fogs are usually more severe and protracted than they are inland. The mean temperature during the winter is a little higher in the valley than it is on the seashore, although there is an occasional lower minimum. During the year 1892 the temperatures during the months of winter and spring were as follows ·

1892.	Mean.	Max.	Min.
January . . .	52°	78°	27°
February . .	53	79	...
March . .	54	85	28
April . . .	56	82	31
May . . .	62	100	34
December . .	50	77	25

During the year 1892 the maximum temperature was 110° in August and the minimum 25° in December.

During the year 1893 the maximum was 100° in August and the minimum 25° in December. The mean annual temperature was 58°. Monthly mean for January, 52°; for July, 70°. The temperature by seasons for 1893 was as follows : winter, 51°; spring, 54°; summer, 68°; autumn, 58°.

From a detailed report of the weather taken at the Gally Cottages, one mile from the centre of Nordhoff, during the year 1895, for the months of winter (December, January, and February) and spring (March, April, and May), the following particulars were obtained :

	Temperature.			Mean relative humidty. Per ct.	Fogs. Mornings.	Rainy days.
	Mean max.	Mean Min.	Monthly mean.			
Winter,	67°	38°	52°	66	4	14
Spring,	71	44	57	65	29	8

The relative humidity was based on three daily observations taken at 9, 1, and 6 o'clock. For winter the humidity is for two months only, January and February. The extreme maximum temperature was for winter, 85°; spring, 95°. Extreme minimum temperature for winter, 26°; spring, 32°. The number of days over 80° was 7 in winter and 15 in spring. Number of days below 32°, 17 in winter and 2 in spring.

The mean annual rainfall at Nordhoff, taken from a report published several years ago by General A. W. Greely, then Chief Signal Officer, entitled " Rainfall in California, etc., for from 2 to 40 years," is as follows :

Rainfall, annual mean .	. .	27.84 inches.
" winter " .	. .	12.82 "
" spring " .	. .	9.23 "

This shows the amount of annual rainfall to be about 10 inches greater than at Santa Barbara.

In the Ojai Valley are fruit-ranches and mountain walks and rides. There are pleasant boarding-houses and cottages. Trout-fishing can be found in the vicinity. There are several beautiful cañons, with fine oak trees and a wealth of flowers and vines. It has an agreeable winter climate, but the summers are hot and night-fogs are frequent at that season.

In Waterfall Cañon, five miles from Nordhoff, are the **Ojai Hot Springs**, situated at an elevation of 1000 feet. They are soda springs, carbonated and sulphurated, and flow about 50,000 gallons per hour at temperatures varying from 60° to 104° F.

Dr. Winslow Anderson, of San Francisco, states that these waters contain sodium, potassium, and magnesium carbonates and sulphates, calcium and ferrous carbonates, silicates, carbonic anhydride, and sulphuretted hydrogen gases.

In Matilija Cañon, six miles from Nordhoff, are twenty-eight springs, known as the **Matilija Hot Springs**. They are mostly sulphuretted, and flow about 5000 gallons per hour at temperatures ranging from 35° to 160° F. There is a resort at the springs.

Other cañons are the San Antonio Cañon and the Santa Paula Cañon to Santa Paula, fourteen miles distant.

Sespe Valley (elevation about 2000 feet) is fifteen miles northeast of Nordhoff. This valley is reached by a mountain-trail, suitable only for animals and foot-passengers, which crosses the range ten miles north of Nordhoff at an elevation of 5000 feet. The Sespe River shrinks to almost a brook during the rainless summer season. The water is alkaline. Water can also be obtained from springs. It is a rough country, without conveniences, there being but few ranches. It has a good climate, hotter, but with fewer foggy days than on the coast. It is cooler than the Ojai Valley. This region is well adapted for camping-out, and is a change from the climate of the seacoast.

Pine Mountain (elevation, 6000 feet) twelve miles north of Sespe, is also said to offer attractions for campers.

Santa Barbara (population, 6000). The coast of California, just south of latitude 35°, suddenly makes a sharp turn at Point Conception, and runs east and then southeast until the Mexican frontier is reached. The great Pacific drift-current, which is by many writers held responsible for so many peculiarities of the Pacific coast climate, not being prepared for such a sudden change, shoots straight on south and washes the Californian shore no more. Along these sheltered waters are situated the coast-resorts of Southern California, those facing on the Santa Barbara Channel being especially famous for a climate rivalling or surpassing that of the Riviera, while the interior resorts, on the border of the desert, are said to have a climate "similar in winter to that of Egypt". Rain falls only during the winter months, and the mildness and equability of the temperature on the coast render it desirable for invalids to whom the moisture is not objectionable.

The town of Santa Barbara (latitude, 34° 28′ north) is forty miles east of Point Concepcion, situated on a plain that rises three or four hundred feet from the ocean to the foot-hills. This whole plain is eighty miles long, with a width of about four miles. To the north the Santa Inez range reaches a height of 3000 feet, affording protection from cold northern and western winds. The Channel Islands, twenty-five miles south, afford a slight break against strong southwest winds. "Here is our Mediterranean. Here is our Italy. It is a Mediterranean without marshes and without malaria. It is a Mediterranean with a more equable climate, warmer winters, and cooler summers than the north Mediterranean shore can offer. It is an Italy whose mountains and valleys give almost every variety of elevation and temperature. But it is our commercial Mediterranean. The time is not distant when this corner of the United States will produce in abundance, and year after year without failure, all the fruits and nuts which for a thousand years the civilized world of Europe has looked to the Mediterranean to supply."[1]

There is little or no sand in the soil around Santa Barbara; it is alluvial or adobe, and extremely fertile and moist. Crops are grown in this portion of the State without requiring irrigation.

The town is supplied with water from Mission Cañon and from tunnels run into the mountains. The water contains lime, and should be boiled before being used. There are street-car lines, electric lights, and sewers in the principal streets. Santa Barbara has a good beach for bathing or driving. The fishing is said to be excellent around the islands.

Sergeant James A. Barwick, of the United States Signal Corps, the Director of the California State Weather Service, has prepared a series of "climatic comparisons of Santa Barbara with San Remo and Mentone," based on the averages of several years. A brief

---

[1] Our Italy. Charles Dudley Warner.

extract shows the temperature for January to be: San Remo, 47°; Mentone, 48°; Santa Barbara, 54°. For July: San Remo, 74°; Mentone, 75°; Santa Barbara, 65°. A further examination of the annual records shows that the resorts on the Mediterranean Riviera have also more cloudy days during the year, twice as many stormy days, and about twice as much annual rainfall, as well as more wind than Santa Barbara; but they are comparatively drier, as indicated by an almost complete absence of fog.

Along the coast of California the daily variation[1] of temperature during the winter months is exceedingly small. Comparing the winter climate of Nice with that of the Californian coast, General A. W. Greely wrote, a few years ago, in an article in *Scribner's Magazine*, that while Nice generally excelled California in equability of temperature, yet at irregular intervals cold, dry, piercing winds swept over the place, bringing sharp and sudden changes of 20° or more in a day. He considered Nice inferior as having lower temperature, higher winds, and occasionally snow and ice, which are unknown on the coast of Southern California.[2]

The equability of the temperature at Santa Barbara during the year is shown by the record for the seasons: mean monthly temperature for winter, 54°; spring, 58°; summer, 65°; autumn, 63°. There is an average number of 26 days above 80° and 6 days below 40°. The yearly mean of relative humidity is 73 per cent. Average number of cloudy days, 73; stormy days, 28 (means for eight years). Annual rainfall averages 18 inches. The record for twenty-four years shows an increase of 32 per cent. in the last twelve years over the first twelve. It rains partly at night and occasionally for nine or ten days in succession, giving about half a year's precipitation during that time.[3]

There are rarely during the year more than two or three unpleasant days of wind- and dust-storms. These may come in March. The least windy period of the year is from October into February. The wind-movement is very moderate, the annual average for seven years, recorded by the local observer, being nearly 4 miles per hour.

There are a number of foggy days, the damp and depressing influence of which is bad for invalids.

---

[1] See Tables V. to IX. and XIV.

[2] Where Shall We Spend Our Winters. General A. W. Greely, in Scribner's Magazine, November, 1888.

[3] For further consideration of the coast climate, see the Climate of California, and San Diego.

At Santa Barbara during the year 1894 there were recorded by the local observer 73 foggy mornings, the greatest number being in November (12) and July (16). The number was small from the end of November to April.

By seasons, Santa Barbara, 1894, foggy mornings :

Winter .	5 days.	Summer .	. 26 days.
Spring .	. 14 "	Autumn .	. 28 "

The nights are so cool as to make blankets necessary. There are usually about thirteen nights during the year above 60°.

Dr. C. B. Bates, of Santa Barbara, referring to the climate, warns delicate persons to beware of the chill. "Just before sunset the temperature rapidly falls, and the invalid at this time should remain in the house, or, if out of doors and not briskly exercising, should put on an overcoat. Indeed, although the climate of Santa Barbara is warm, it is not hot ; flannels next the skin, with moderately warm clothing, can and should be worn throughout the year."

This Pacific coast climate is damp, and presents its claims to sufferers on the grounds of equable temperature and sunshine. It lacks the dry air and tonic, stimulating qualities of the elevated inland plains, but offers less shock to the system from rapid changes.[1]

Santa Barbara is well supplied with good hotel and boarding accommodations.

The temperature of the sea-water off Santa Barbara varies from 60° to 66° (the warmest record being made in September), with a yearly mean of 62°. This is about 10° cooler than the sea-water off the coast of Southern Florida and the Bahamas.

Santa Barbara Hot Springs. A little over six miles northeast of the town of Santa Barbara are the Santa Inez Hot Sulphurous and Soda Springs, which are picturesquely situated 1450 feet above the level of the sea. There are about thirty mineral springs in all —sulphurous, saline, and chalybeate—ranging in temperature from 99° to 122° F. The water of several of the principal springs is used

---

[1] Memorandum showing the highest and lowest yearly records of temperature for several years at Santa Barbara. From the published records of Hugh D. Vail, Esq.

	Max.	Min.		Max.	Min.
1885	. 85°	35°	1892 .	. 97.5°	37.5°
1889	107	33	1893 .	. 88	38
1890	98	33.5	1894 .	. 94	33
1891	96	33			

for drinking and bathing purposes. An analysis of the Sulphur Springs Nos. 1 and 2, made by Dr. Winslow Anderson, of San Francisco, is as follows :

	Grains in one gallon.
Sodium chloride	1.74
Sodium carbonate	2.17
Sodium sulphate	14.92
Magnesium sulphate	7.75
Calcium sulphate	6.03
Aluminum sulphate	2.90
Arsenic	trace
Silica	1.18
Sulphuric acid	trace
Organic matter	trace
Total solids	35.95
Carbonic anhydride	19.14 cub. in.
Sulphuretted hydrogen	9.16 "

There are a hotel and a bath-house at the Springs.

Five miles from Santa Barbara are the Montecito Hot Sulphur Springs, situated in a ravine in the mountains at an elevation of 1460 feet. The temperature of the water is 120° F.

Paso Robles (elevation, 800 feet; population, 1000). The annual " Meteorological Review" of the State of California for the year 1891 contains the following statement relating to the climate of this town, taken from the *Pacific Rural Press:*

" This portion of the Coast Range is subject to greater fluctuations than has been heretofore supposed. The mean monthly temperature can never be used as conclusive evidence of any climate. Nothing could be more misleading in reference to the climate under consideration—a climate representative of many higher valleys in the California Coast Ranges. The station is situated eighty feet above the Salinas River, about 800 feet above the sea-level and forty miles due east of Estero Bay, with a high mountain-chain— the Santa Lucia—between.

" There are many places in the district where the thermometric variations are even more sudden than at the station. . . . The greatest atmospheric dryness, in October and December, occurred when a north wind was blowing. The hygrometer was exposed on the north side of the house, so that the conditions were favorable to extreme indications ; but in any case the record is an extraordinary one."

The table given is for each month in the year (1890), and is full of details. An abstract shows the mean daily range of temperature to be by seasons : winter, 21°; spring, 26° summer, 43°; autumn, 35°. The mean range of humidity was : winter, 44 ; spring, 30 ; summer, 46 ; autumn, 64 per cent. The greatest range of temperature was in January, 54°; and of humidity in December, 80 per cent.

The monthly mean temperature for January was 44°; for July, 75°. No reference was made in the report to cloudy or stormy days or to the wind-movement.

By seasons (1890) :

	Winter.	Spring.	Summer.	Autumn.	Year.
Mean monthly temperature,	45°	58°	74°	57°	59°
Mean relative humidity,	73 p.c.	80 p.c.	50 p.c.	58 p.c.	65 p.c.
Total rainfall,	10.89 in.	3.09 in.	0.00 in.	0.00 in.	13.98 in.

The mean monthly temperature for Paso Robles for 1893, arranged by seasons, was as follows : winter, 43°; spring, 55°; summer, 72°; autumn, 55°; annual, 56.25°; maximum, 107°; minimum, 20°. Monthly mean for January, 44°; for July, 75°.

This portion of the Salinas Valley has long been noted for its mineral springs, which are sulphurous and alkaline and both hot and cold, ranging in temperature from 59° to 122° F.

Near the hotel is the main hot, sulphuretted spring, which flows 5000 gallons per hour, at a temperature of 108° F. An analysis made by Dr. Winslow Anderson in 1889 is as follows :

	Grains in one gallon.
Sodium chloride	25.73
Sodium bicarbonate	41.19
Sodium carbonate	7.62
Sodium sulphate	7.25
Sodium iodide	trace
Sodium bromide	trace
Potassium chloride	1.57
Potassium carbonate	2.05
Potassium iodide	trace
Potassium sulphate	trace
Magnesium bicarbonate	......
Magnesium carbonate	2.15
Magnesium sulphate	5.11
Calcium carbonate	1.23
Calcium sulphate	2.94
Ferric peroxide	0.73

Grains in one gallon.

Borates	. trace
Lithates. .	. trace
Alumina .	. 0.25
Silica . . .	. 1.75
Iodides and bromides	. .....
Organic matter .	. 1.90
Total solids	. 101.47
Free sulphuretted hydrogen	3.75 cub. in.
Free carbonic-acid gas	8.90 "

The next most important spring is that used for the mud-baths. It is sulphurous, and is one and one-half miles from the hotel, situated on the edge of the Salinas River. There is a well-kept road, shaded with oaks. The mud-springs flow collectively about 6000 gallons per hour. The temperature varies from 107° to 122° F. A bath-house with necessary facilities has been erected.

An analysis made by Dr. Anderson is as follows:

Grains in one gallon.

Sodium chloride .	. 83.72
Sodium carbonate .	. 7.41
Sodium sulphate . .	. 36.97
Sodium iodide . . .	. trace
Potassium chloride . . .	. 3.19
Potassium iodide . . .	. trace
Potassium sulphate . . .	. 0.82
Magnesium carbonate . .	. 4.25
Magnesium sulphate . . .	. 1.13
Calcium carbonate . . .	. 2.10
Calcium sulphate . .	. 15.75
Ferrous sulphate . .	. 0.23
Alumina . . .	. 0.80
Manganese salts .	. trace
Silica . . .	. 0.25
Lithium salts .	. trace
Organic matter .	. 7.14
Total solids	. 166.02
Ammonia and nitrogen .	. trace
Free sulphuretted hydrogen	. 4.16 cub. in.
Free carbonic-acid gas .	. 42.50 "

About 200 yards north of the mud-baths is the "soda spring". Its temperature is 77°. Total solids, 84 grains in one gallon. The

"Garden Spring" is carbonated water, containing 76 grains of total solids to one gallon.

Beyond the "Garden Spring" is the "Sand Spring," an alkalo-sulphurous, lightly carbonated water, containing 190 grains of total solids. Temperature, 79° F.

A quarter of a mile or more southeast of the hotel is the "Iron Spring". Temperature, 64°. Total solids, 79 grains in one gallon.

Two and one-half miles southeast of Paso Robles are the Santa Isabel Hot and Cold Sulphur Springs, which are situated in a small cañon about one mile east of the Salinas River, at an elevation of 1000 feet above the sea. There are three warm sulphur springs, almost identical in composition. An analysis by Dr. Anderson of Spring No. 1 is as follows:

Temperature, 96° F. Flows 20,000 gallons per hour. The water is tonic, antacid, diuretic, aperient, and alterative. It resembles the Arkansas Springs.

	Grains in one gallon
Sodium chloride	18.10
Sodium bicarbonate	29.04
Sodium carbonate	6.91
Sodium sulphate	7.25
Sodium iodide	trace
Potassium bromide	trace
Potassium iodide	trace
Potassium chloride	trace
Potassium carbonate	0.83
Magnesium carbonate	6.16
Magnesium sulphate	4.85
Calcium carbonate	2.45
Calcium sulphate	2.32
Manganese carbonate	0.13
Ferrous carbonate	0.98
Borates	trace
Alumina	0.73
Barium salts	trace
Silica	1.68
Organic matter	trace
Total solids	81.43
Free sulphuretted hydrogen	4.65 cub. in.
Free carbonic-acid gas	11.76 "

A quarter of a mile further up the little cañon are the cold or White Sulphur Springs. They vary in temperature from 56° to 60° F., and are much lighter than the warm sulphurous water.

Monterey is one hundred and twenty-five miles south from San Francisco by rail and eighty-five miles by sea. It is one of the most interesting of the old Spanish settlements on the Pacific coast. There is a fine drive around the peninsula on which the town lies, affording views of the beautiful Bay of Monterey and of the Pacific Ocean. There is said to be good fishing in the bay and in the Carmelo River.

The Hotel del Monte is one of the largest and most attractive hotels in America. In appearance it resembles an immense country mansion placed in a large park. The main portion of the hotel is 340 feet in length by 110 feet in width. The entire structure contains 430 rooms, and is surrounded by wide verandas.

The grounds around the hotel consist of a plat of 126 acres, devoted to lawns and flower-gardens, while the entire park contains 7000 acres, containing groves of oaks, pines, and cedars.

The roads are well kept, and boating and bathing may be enjoyed.

The climate of Monterey is exceedingly humid, and therefore quite equable. The mean temperature for several years for the month of January is 50° and for July 65°. In 1894 it was: January, 46°; July, 60°; annual mean, 56°; maximum, 88°; minimum, 26°.

The mean annual rainfall is 14.4 inches. The highest record is 21.4 inches and the lowest 9 inches.

Fogs are more prevalent than on the coast south of Point Concepcion.

Napa Valley : Atlas Peak (elevation, 1500 feet). It is curious to find the best record for dryness in California credited to a portion of the State only from forty to sixty miles north of San Francisco, on the slope of the Napa Valley, which is noted for its wine-producing qualities. Dr. J. S. Hittell states that the average percentage of relative humidity at Atlas Peak, east of Napa Ridge, about fifty miles from the sea, is for summer (half-year), 39 per cent.; winter (half-year), 51 per cent.; year, 45 per cent., or about the same as Denver. The temperature, however, is usually higher than in Denver, especially in winter. Dr. Hittell credits Los Angeles with an annual relative humidity of 65 per cent. (which is low), and Santa Barbara and San Diego are rated still higher. He says: " The figures here given show that Atlas Peak and Blake's, in the coast mountains of California, are unequalled in their combination of dry atmosphere with a mild temperature in winter and summer, and a desirable elevation. No observations for relative humidity have been kept at any other part in the Coast Mountains, but there are doubtless many places in that range south of the Silver Gate with conditions equally favorable, as will probably appear in a few years."[1]

Atlas Peak has a mean temperature in January of 50°; in July, 74°. Dr. James Blake states that at his place, 2100 feet above the sea, near Mount St. Helena, the orange-trees suffered less one season than at Los Angeles, where a temperature of 23° was reported, while at Dr. Blake's residence the thermometer did not go below 29°. Mount St. Helena, 4300 feet high, is an extinct volcano with a flat top. The distance by travelled route from San Francisco is eighty-three miles—the last twelve being done on horseback. It commands extensive views. On the east of Napa Ridge, **Howell Mountain**, seven miles north of the town of St. Helena, five miles south of Blake's, and eight miles from Atlas Peak, has five square miles of nearly level land at an elevation of 1800 feet. There are streams and timber, and it is a favorite place for camping. Frosts are severe in the spring, but rare in the autumn.

The resort of **Napa Soda Springs** is on the west side of the same ridge, 1200 feet above the sea (see page 347). It is five miles from Atlas Peak. There is a hotel at the springs, but the Atlas Peak visitors must camp out. This last-named location was selected by the State Board of Health in 1880 as the best in the State for a hospital for consumptives. The annual rainfall at Napa City is 22 inches: at Calistoga, 30 inches. The east Napa Ridge has less frost than the fertile Napa Valley below it.

The meteorological record for seventeen years, kept at the Napa State Asylum for the Insane, shows the mean temperature and rainfall for that place to be by seasons as follows:

	Winter.	Spring.	Summer.	Autumn.	Year.
Monthly mean temperature,	45°	55°	63°	57°	55°
Rainfall (seasonal),	14.42 in.	7.66 in.	0.32 in.	3.92 in.	26.32 in.

Monthly mean temperature, January, 44°; mean minimum, 36°
Monthly mean temperature, July, 64 ; mean maximum, 76

Clear Lake is an attractive body of water twenty-five miles by six, and 1200 feet above the sea. Chief town, Lakeport, thirty-two miles from Cloverdale.

---

[1] Handbook of Pacific Coast Travel. J. S. Hittell, M.D., San Francisco, 1887.

Cobb Valley (3000 feet elevation), twenty-five miles north of Calistoga, is a favorite retreat for hunters and campers.

In Lake County are numerous springs, between 2000 and 3000 feet above the sea, having virtues of mineral waters and climate not as yet known with exactness.

Lake Tahoe is one of the most delightful summer-resorts of California. It is situated on the State-line between California and Nevada, fourteen miles by stage south of Truckee, on the Central Pacific Railroad. The elevation of the lake is 6200 feet. It is twenty-one miles long by twelve miles wide and 1500 feet deep. The water is very clear, pure, and cold, and never freezes. It is surrounded by mountains. Steamers ply on the lake. There are hotels at Tahoe City, on the west shore, and at Glenbrook, on the east shore, fifteen miles from Carson City. (For reference to the mineral springs, see Climate of California.)

SPRINGS.

There is one thing which California possesses in abundance, and that is mineral springs. In a volume prepared by Dr. Winslow Anderson, of San Francisco, in 1890,[1] the names are given of over 200 California springs, with about 100 analyses. This extensive list, however, does not include all the medicinal springs known to-day. As the work is by far the most complete one on the subject, however, the writer desires to express his sense of obligation to it for a large part of the information given in the following brief and superficial review.

Within the limits assigned it is possible to refer to only a few of the better known of California's hundreds of springs. For full analyses reference must be made to Dr. Anderson's valuable book.

Three miles from Ukiah, the county-seat of Mendocino County, are the Doolan Vichy Springs. The waters belong to the alkalocarbonated class, and are clear and sparkling and of an agreeably pungent taste. They are heavily charged with carbonic acid gas and carbonates, and contain some iron and potassium salts. They contain 268 grains of total solids in one gallon, and resemble in chemical composition the water of Vichy, France, and Ems and Fachingen, Germany. The springs flow over 20,000 gallons per hour.

---

[1] Mineral Springs and Health-resorts of California, etc. Winslow Anderson, M.D., San Francisco.

The famous Geysers of California are situated 100 miles north of San Francisco, twenty-six miles from Calistoga, or sixteen miles from Cloverdale. The geysers are in a cañon 1700 feet above the sea. There are dozens of springs, hot and warm, ranging from acid, alum, and iron to sulphurous wells boiling at a temperature of 212° F.

Twenty miles from Calistoga, at the base of a spur of the Coast Range of mountains, are the Harbin Hot Sulphurous and Saline Springs. The principal sulphur spring flows 1500 gallons per hour at a temperature of 122° F. There are also smaller chalybeate, "magnesia" and "arsenic" springs. At Harbin Springs are good accommodations, with fine natural surroundings.

Nineteen miles from Calistoga and ten miles from the great geysers are the Anderson Mineral Springs, comprising nine hot and cold springs, sulphurous, saline, chalybeate, acid, and salino-sulphurous or antacid. The resort has facilities for hot sulphur baths, and is in the midst of attractive surroundings.

Twelve miles from Cloverdale are the alkaline springs called the California Seltzer. The waters are sparkling and carbonated, and contain 187 grains of total solids in one gallon. There is a comfortable resort at the springs.

The Napa Soda Springs are charmingly located on the southwestern slope of the Coast Range, at an elevation of 1000 feet. They are six miles from Napa City. There are twenty-seven springs in all, with an average daily flow of 4000 gallons. The temperature of the water varies from 65° to 68° F. Most of the commercial Napa soda is obtained from the Pagoda Spring, an alkalo-chalybeate water, strongly charged with carbonic anhydride, delightful, clear, and sparkling, with an agreeably pungent taste. By analysis it gives 67 grains of total solids to one gallon. The resort has a fine hotel, with beautiful views of Napa Valley. It is open all the year.

In the city of Calistoga, and also just outside the town, are twenty or more mineral springs, ranging in temperature from 75° to 186° F. They are valuable light sulphurated waters, and are much used for drinking- and bathing-purposes.

Near the foot-hills, in a spur of the Coast Range of mountains in Contra Costa County, are the Byron Springs, sixty-eight miles almost due east of San Francisco and one and one-half miles distant from Byron station. There are over fifty springs, ranging from 52° to 140° F., from cold carbonated to hot sulphurated water, and located 100 feet above the sea in a small valley or basin which has the appearance of being an extinct volcanic crater. Only seven or eight of the springs are in active use. They have marked individual characteristics, and form a group of great range and value. Among them are the "Black Sulphur," containing 461 grains of total solids in one gallon (temperature, 90°); the "Iron Spring" (temperature, 79°), which contains 765 grains of total solids in one gallon—of which 670 grains are sodium chloride, with nearly half a grain of the peroxide of iron; the "Hot Salt" alkalo-chlorinated water, having a temperature of 122° F.—used for bathing; and the well-named "Surprise" Spring, containing 18,773 grains of total solids in one gallon, of which 15,417 grains are sodium chloride. This total of over 18,700 grains of mineral ingredients is about forty ounces in a gallon, or over 33 per cent. held in solution, and is more than eight times as dense as sea-water. The Great Salt Lake (Utah) contains four or five times as much salt as the ocean. Byron has a hotel with connecting cottages, and two large bath-houses.

Twelve miles west of San José, at an elevation of 1000 feet, are the **Azule** or **Blue Springs.** The water is carbonated and pungent, and is similar to the Seltzer of Nassau, Germany. It contains 261 grains of total solids in one gallon. Its action is antacid, aperient, diuretic, and tonic. It flows at a temperature of 60° F.

On the eastern slope of the Coast Range, one hundred and fifty miles south from San Francisco and seven miles south of Soledad, on the Southern Pacific Railroad, are the **Paraiso Hot Springs,** at an elevation of 1400 feet above the sea. There are several soda, iron, and sulphur springs, varying in temperature from 100° to 114° F. The hotel and cottages are of recent construction, and command fine views of the fertile valley of the Salinas River.

In the northern part of San Luis Obispo are the **Springs of El Paso Robles,** a detailed description of which will be found on page 342.

In Santa Barbara County the hot springs north of the town of Santa Barbara are described on page 340, and those situated in the Ojai Valley, in Ventura County, on page 336.

In Los Angeles County, thirteen miles southeast from the city of Los Angeles and three miles north of Norwalk station, are the

Fulton Artesian Wells, of which the two principal wells are 350 feet deep and flow copiously. The waters contain a large percentage of iron salts. An analysis made by Dr. Anderson follows :

	Grains in one gallon.
Sodium chloride	9.60
Sodium bicarbonate	2.90
Sodium sulphate	0.95
Magnesium bicarbonate	17.45
Ferrous carbonate	11.75
Calcium carbonate	12.62
Calcium sulphate	23.41
Silica	2.45
Organic matter	trace
Total solids	81.13

(Temperature, 64°.)

In San Bernardino County are innumerable thermal springs, the best known probably being the Arrowhead Hot Springs, which are described on page 330. There are a number of warm pools fed by springs, between the town of San Bernardino and the mountains, which are used for bathing-purposes.

Palm Springs. For description of the hot mineral spring and the valley, see page 332.

Near Elsinore, in Riverside County, is a valley containing a great number of hot and cold springs, ranging from 57° to 212° F. The cold springs are carbonated, containing soda, magnesia, and iron. The hot waters are sulphurous, with lime, magnesia, and borax. There are also hot mud-springs. Unfortunately this fine resort is as yet undeveloped.

In front of the village of Elsinore is the lake of the same name, six miles in length and over two miles in width. On its shores are orchards and ranches, behind which are the mountains. The elevation of Lake Elsinore is nearly 1300 feet.

In San Diego County are a number of hot springs, few of which are at present available. Near Oceanside is a valley eight miles from the ocean, containing several springs, highly charged with sulphur and sulphurous acid, and sulphates of magnesium, sodium, and calcium. The temperature of the water ranges from 85° to 135° F. They are known as the **Corral de Luz Hot Springs.**

In the Coahuila or Cabezon Valley, ten miles south of White River and fifty miles from the city of San Diego, are some hot springs known by the common Spanish name of **Aguas Calientes.** These hot sulphurous waters boil up from a granite ledge on a ridge at the easterly end of Warner's ranch. They range in temperature from 70° to 142° F. Incrustations of crystallized sulphur are deposited on the surrounding rock. The waters have been much used for drinking- and bathing-purposes.

The **Coronado Spring** water, found on Coronado Beach, is referred to on page 320, and its analysis given.

The water of Lake Tahoe, in Placer County, south of Truckee, on the Central Pacific Railroad, is remarkably soft and pure, as it contains but three grains of total solids in one gallon. The lake is situated at an altitude of 6200 feet. It is twenty-one miles long and twelve miles wide, and has an average depth of 1500 feet. From its shores rise the snow-clad peaks of the Sierras. On Carnelian Bay, at the northern end of Lake Tahoe, are the **Carnelian Hot and Cold Mineral Springs.** The waters are sulphurous and saline, and a few are carbonated. There are bathing-facilities, and tub-, plunge-, and steam-baths can be obtained.

Near the crest of the Sierra Nevada, twelve miles from Summit Station, are the **Summit Soda Springs,** which are picturesquely located in a cañon, through which winds one of the forks of the American River. The elevation is 6000 feet. The water belongs to the alkalo-chalybeate class and contains 92 grains of total solids to the gallon. It is strongly charged with carbonic acid gas.

Channel Islands of California.

Lying from twenty to fifty miles west and south of the mainland these islands possess some well-defined differences in climatic features.

"A very noteworthy fact in their climatology, and one illustrating the effect which the cold current of the Kurosiwo has upon the northern coast, and which the southern coast escapes by its deflection eastward, and through the shelter afforded by this chain of islands, is that the climate of the outer tier of islands is much harsher than those nearer the mainland. The other islands are nearer the current—possibly within the edge of that cold northern stream—while the inner chain is surrounded by the flow of the return warm current from the south."[1]

The outermost islands are dry and barren. The three largest

[1] California of the South.   J. P. Widney, M.D.

islands are nearest the coast, and are more fertile. They are resorted
to by transient summer-campers who go to enjoy the fishing and
the benefits of the ocean-air, which is much the same as if one were
on board a ship at sea. Dr. Widney says, if they "follow the general law of a diminishing rainfall as the distance from the Sierra is
greater, they must have a much less annual precipitation than the
mainland. The appearance and type of the vegetation and the
comparative scarcity of springs and running streams indicate the same
fact. They are also much freer from the strong sea-breeze which
reaches its maximum intensity near the shore of the mainland, and
also from the fog which forms along the immediate line of the coast.

"They are bathed in sunshine when the mainland opposite is
enveloped in fog. In temperature they are more equable."

The principal island is **Santa Catalina**, opposite the port of San
Pedro and about nineteen miles distant. The settlement on the east
side of the island is called Avelon. There are also safe anchorage
and landing on the west side. Santa Catalina is about twenty miles
long, with an average width of three miles. "It rises to a height
of from 2000 to 3000 feet, and is remarkable for the great trans-
verse break or depression, five miles from the northern end, running
through it and forming a cove or anchorage on each side. The land
connecting these is very low, say not over thirty feet, but the hills
rise up on each side 2000 or 3000 feet, and when sighted from the
north or south the whole appears like two very high islands. . . .
There are a number of pretty elevated valleys, several mineral springs,
and wells of good water."[1]

Dr. T. J. MacCarthy, Professor of Chemistry in the Medical College of the University of Southern California, gives the following
analysis of water from a spring the most highly charged with saline
matters of several examined. It is found at an elevation of several
hundred feet :

	Grains in one pint.
Sodium chloride . .	. 79.5
Magnesium chloride .	21.0
Magnesium sulphate .	32.5
Sodium sulphate	20.5
Calcium sulphate .	. 6.0
Magnesium carbonate .	. 2.0
Iron and aluminum .	. traces
Total solids	. 161.5

[1] California of the South. J. P. Widney, M.D.

This water he classes among commendable purgative mineral-waters.

There are on the island a hotel and a number of cottages and boarding-houses. There are many delightful spots for camping in the shady cañon. The east slope of the highest land, protected from the trade-wind, should be as fine a marine air as could be found. There is good bathing, both surf and stillwater. A steamer plies between San Pedro and Avelon.

Mr. Charles Frederick Holder, formerly editor of the *Calfornian*, who spent an entire summer on Santa Catalina Island, says it is " a mountain-range twenty-two miles long and from one to eight miles wide, rising from the ocean with grim, precipitous walls, abounding in deep cañons, and scenery grand and impressive beyond description. . . . The shore, apparently of rock, rises abruptly from the sea, facing it with a bold front, while high above ridges and peaks rise one behind the other—a maze of mountain-ranges. It would puzzle the mariner, were he not familiar with the coast, to find the harbor; but suddenly, as we near the island, a deep cañon is seen to reach down to the sea, ending in a white beach; then another, and, finally, a lofty sugar-loaf rock is passed, and the little half-moon-shaped bay comes in sight, with its sandy beach, its wide cañon reaching away to distant mountains, its scores of picturesque cottages and homes, its white tents and hotels. The town of Avelon is built in the mouth of the cañon. . . . The general trend of the island is northwest and southeast. The prevailing winds beat against the south shore, while the north is a land of calms. . . . It was my good fortune on a recent visit to accompany the survey which made a week's trip over the various ranges, during which many new and interesting features of island-life were observed. . . . The upper portion of the island is a revelation. Instead of the sharp points of mountain-peaks, here is a broad plateau, extending over to the west shore, and wide valleys, suggestive of agricultural possibilities. From a lofty point on the west I sat in the saddle and tossed a pebble that must have fallen into the ocean 1500 feet below. The afternoon breeze was blowing in the mist, which, shattered against the wall of rock, drifted up the cañon, illumined by the sun, like masses of molten silver. From far below came the roar of the sea as it broke upon the rocks, the weird cry of the sea-lion, and occasionally, out from the flying fog, dashed a white-winged gull that seemed to separate itself from the cloud-mass and become an animate being, to eye me in wonderment and soar away. The entire south coast faces the sea, with forbidding walls of rock rising from 500 to 1500 feet, breasting the sea with a bold-front, hurling the masses of foam high in the air, and in the occasional winter-storms forming a grand and impressive spectacle. Where the various cañons reach the sea are little inlets with abrupt, sandy beaches, against which the waves beat, and approachable only on calm days. . . . In the centre of the island the ocean was not to be seen. We might have been a thousand miles from it, so far as any evidence of its presence was concerned. . . . We have seen Catalina in summer, with its perfect climate, always cool, with that lack of change so desirable to the invalid ; yet the winter, if possible, is even more delightful. Then it is that the true beauties of this isle of summer are seen. The rains, which, curiously enough, are less than on the mainland, change the brown hills to a vivid green, and we have an emerald in an azure setting. Myriads of flowers spring up, and the face of the island is changed as if by magic. They grow to the very ocean-edge; their delicate forms overhanging the water, and are reflected in it. On the south coast, where high seas rage during the winter-storms, the beds of wild flowers are deluged by the spray that, hurled high in the air, is borne away over the fields to cover the delicate forms with gleaming spangles of salt. The island winter exists but in name. In February and through the winter months Catalina is still an island of summer."[1]

The Island of Santa Cruz is about twenty-one miles by four, and lies twenty miles out, opposite Santa Barbara. There is a roadstead on the north side, at the opening of a valley, where wood and water can be obtained. It is said almost all kinds of grain and fruit can be raised on the island. It rises to a height of 1700 feet. There are sea-lions on the rocks. In the interior are large flocks of sheep.

Santa Rosa Island is five miles west of Santa Cruz. It is about fifteen miles long and ten miles wide, rising to an elevation of nearly 1200 feet. Parts of the island are quite picturesque. There are numerous springs of water. It is practically uninhabited, but is used as a pasture for sheep.

---

[1] An Isle of Summer. C. F. Holder, in Californian Magazine, December, 1892.

The Climate of Oregon and Washington.

The coasts of these States are tempered, both summer and winter, by the Pacific winds. The northwest winds bring gentle rains and mists which moderate the summer heat. Hailstorms never occur, and thunderstorms rarely. The winter is the season of most rain, but the weather is mild and there is very little snow and ice. The rainfall for the year ranges from 50 inches on the southern portion of the Oregon coast to 130 inches on the northern part of the Washington coast.

At the Coast Range, which is composed of the Cascade Mountains, the precipitation lessens to about 13 inches for the plains and 22 inches on the more elevated ground. Along the coast the summers are pleasant, the temperature seldom rising above 80° F., with cool nights. The mean annual temperature is 52°. The climate is equable, the mean annual range being only 26°.

To the east of the Cascade Mountains, however, the summers are both hot and dry, and the winters are much more severe, resembling those of Pennsylvania.

These States can scarcely be considered to be of great value to invalids; but offering, as they do in the coast-districts, the advantages of a pleasant, temperate climate, like that of England, though milder, they are well suited for those who need what may be termed a negative climate. The chief cities, **Portland**, **Tacoma**, and **Seattle**, are handsome, attractive, busy places with excellent accommodations and resources of all kinds, while the **Puget Sound** district is an agreeable country for the convalescent or the tired worker to visit. The scenery is beautiful and the opportunities for sport are many.

CHAPTER XV.

MEXICO.

City of Mexico (elevation, 7400 feet; population, 350,000; distance from El Paso, 1224 miles). From the excellent meteorological record of the Mexican Government the following particulars relating to the climate of the City of Mexico have been obtained :

The dry season is from November to April, inclusive, during which period the normal rainfall (based on records for fifteen years) is 2½ inches on 24 rainy days. The other six months have 22 inches, falling on 115 rainy days. The rains begin in April or May, and are greatest in June, July, August, and September. The mornings as a rule are clear, but heavy rains fall during the afternoon and at night almost daily during the rainy season. The normal annual precipitation is a little over 24 inches on 139 rainy days.

The annual mean temperature is 60° F. The mercury in the shade rarely goes above 86° or below 35°. The daily range is, however, considerable, frequently 30° or 40°. The mean temperature for winter is 54°, with a maximum of 76° and a minimum of 30° ; for spring it is 63°, with a maximum of 86° and a minimum of 38°. The hottest months are April and May, which have a mean of 63° and 64° respectively. The coldest months are December and January, which each average 53°. February is variable, causing a proverb :

> " Febrero loco
> Porque de todo
> Tiene un poco,"

which is to the effect that February is a fool, because it has a little of everything.

The mean annual relative humidity is 60 per cent. Winter is 56 per cent.; spring, 49 per cent.; and the remaining six months 68 per cent. The wind-movement is very low, averaging for the year less than two miles per hour.

Humboldt estimated the Plain of Anahuac or Valley of Mexico

as fifty-five miles in length by thirty-seven miles in breadth.   Later writers have slightly increased these dimensions.   It is surrounded on all sides by a wall of mountains which end the view looking down nearly every street in the city.   On this plateau are five lakes, of which the lowest, Texcoco, has no outlet and is salt.   The near presence of these large bodies of water increases the dampness of the city, which is built but a few feet above the level of Lake Texcoco, and has added greatly to the difficulty of proper drainage. After several centuries of temporizing this great problem has been taken up with more energy, and relief for most of the valley was promised by the autumn of 1895.   To the invalid this question of drainage is of the greatest importance.   " Beneath the pavements of Mexico," says F. A. Ober, "is the accumulated filth of five hundred years."[1]   The foundations of the buildings are laid in marshy soil, and it is but a few feet down to stagnant water.   During the rainy season an unusually protracted downpour turns the streets into rivers of mud, and the town for a time is almost uninhabitable.   Malarial and typhus fevers are more prevalent than they ever should be in such a healthy climate.   The new system—now about completed— consists of forty miles of canal and six miles of tunnel, cut through the mountains, and flowing north into the river Pánuco.   Although the completion of this canal and tunnel, by draining the southerly portion of the valley through its lowest lake, will be of general benefit and preserve the City of Mexico from future inundations, it will still be necessary for the sewers of the city to be newly con- structed and connected with this system of drainage—a work of great labor and expense, requiring many years.   Until this has been done the drainage will remain inefficient.

The Mexican capital has a delightful climate—not quite so dry, so far as that factor is concerned, as some of the cities near the plains further north.   The moderate height of the mercury in summer during the rains is surprising.   On the few occasions when a cold wind sweeps down from the north, the cold is trying to delicate persons, on account of the great thickness of the stone- walls and the lack of fireplaces or stoves.   It is more chilly indoors than outdoors at such times.

Mr. Ober quotes the saying of a French traveller that " Mexico is a grand city in the Spanish style, with an air more inspiring,

---

[1] Travels in Mexico.   F. A. Ober.

more majestic, more metropolitan than any city of Spain except Madrid. Crowned by numerous towers, and surrounded by a vast plain bounded by mountains, Mexico reminds one somewhat of Rome. Its long streets, broad, straight, and regular, give it an appearance like Berlin. It has some resemblance to Naples and Turin, yet with a character of its own. It makes one think of various cities of Europe, while it differs from all of them. It recalls all, repeats none."

The city is well laid out, with wide streets and avenues which are adequately lighted at night. The principal streets are paved with asphalt, and kept clean. There is an unusually large *plaza*, on which the band plays, a beautiful *alameda*,[1] and several public gardens. The fashionable drive is on the noble avenue called the Paseo de la Reforma, leading out to Chepultapec. The view from the towers of the grand cathedral is very fine, although it embraces only a portion of the southern half of the great valley. There are a number of good hotels and restaurants, four theatres besides the circus, and many fine public and business buildings. There are a great many magnificent residences, always built around a *patio*, containing flowers and palms and fountains. There is an extensive system of tram-cars drawn by mules. French and English, as well as Spanish, are spoken in the hotels and shops.

Guadalajara (elevation, 5100 feet; population, 100,000), one hundred and sixty-one miles west of the main line of the Mexican Central Railroad at Irapuato, is a beautiful city, the capital of the State of Jalisco, situated one hundred and thirty miles from the Pacific Ocean. It is well laid out and possesses some of the finest public buildings, parks, and gardens in Mexico.

Guadalajara is well supplied with restaurants and hotels. As in the capital, the service most satisfactory to Americans is found in establishments managed by a French or German proprietor. The cathedral contains an Assumption ascribed to Murillo. The Governor's palace and the Degoldado Theatre are large and costly buildings.

The drainage and general sanitary condition are superior to those of most Mexican towns.

The climate is fine, although warmer than that of the City of Mexico, the average for the year being about 8° higher for

[1] Alameda (ar-lar-may'-da). A shady street or walk planted with poplars or cottonwoods (from *alamo*, poplar).

Guadalajara. The mean annual temperature is 67° F. During five years the maximum was 96° and the minimum 24°.

The seasonal record for the year 1885, taken at 12 M., was as follows :

		Temperature.		Relative humidity. 12 M.	Total rainfall.
	Mean.	Max.	Min.		
Winter,	60°	82°	40°	49 per ct.	1.40 in.
Spring,	71	91	51	39 "	2.12 "
Summer,	72	88	54	61 "	23.25 "
Autumn,	66	83	51	60 "	10.64 "

The rainfall from November to April was about 2 inches, leaving about 35½ inches for the months from May to October. The heaviest precipitation was in June, July, August, and September. The normal annual rainfall for ten years is 34 inches, and the annual mean of the relative humidity at noon 53 per cent. The wind-movement is usually low. For 1885 it barely reached the rate of two and one-half miles per hour.

As these observations in Guadalajara were taken but once a day, which is unusual, it will be instructive to compare the temperature and relative humidity with the noon-record of Colorado Springs for one year, taken at the same hour of the day :

Colorado Springs, Col. Latitude 38° 50', elevation 6000 feet. Observations for the year 1893.	Mean temperature. 12 M.	Mean relative humidity. 12 M.
Winter	40°	33 per ct.
Spring . .	51	31 "
Summer . .	75	33 "
Autumn .	58	27 "
Year . .	56	31 "

To continue the comparison, the normal annual rainfall for Colorado Springs is about 14½ inches, of which the five months from April to August, inclusive, are entitled to 11 inches. The winter rainfall (December, January, and February) at Colorado Springs is 0.7 of an inch, and at Guadalajara 1.4 inches. There are usually 57 cloudy days in the year at Colorado Springs and 126 cloudy days at Guadalajara. In winter, cloudy days at Colorado Springs (four years), 13 days ; at Guadalajara (five years), 25 days. Except in the summer, it is seen to be 8° to 20° F. cooler at noon at the springs, and the much lower record of relative humidity shows a drier climate. This humidity-record for the year at Colorado

Springs, taken at noon, is 20 per cent. lower than the yearly mean, based, as is customary, on daily morning and evening observations.

If this rule is applied to the record of Guadalajara, the actual annual normal relative humidity will then appear to be over 70 per cent., instead of 53 per cent., as shown by the above noon-observations.

Guadalajara is partly protected on the west from the Pacific winds by broken ranges of mountains. There are occasional early morning mists, and on Lake Chapala heavy mists are frequently noticed.[1]

Lake Chapala is a beautiful sheet of water forty miles south of Guadalajara. It is 6000 feet above the sea. Its length is fifty miles and its average width about eighteen miles. The northern shore is well timbered and backed by picturesque hills. A steamer plies on the lake.

The Falls of Juanacatlan, said to be the next in volume in North America to Niagara, which they resemble in some ways, are fifteen miles east of Guadalajara, and can be reached by tram-car. The river Lerma, or Rio Grande de Santiago, is here 560 feet wide and falls sixty-five feet in a single leap.

The pleasant suburb of San Pedro, on a ridge higher than the city, is resorted to in summer by the wealthy residents.

A sanitarium in Guadalajara is being carried on under American management.

The cities of Mexico and Guadalajara were the only ones for which detailed weather-observations could be obtained. Table XIII. contains the annual mean for twelve Mexican cities.

A brief description of many of the more important towns of Mexico follows. The elevations of most of the principal cities were furnished by the Government Weather Bureau.

Monterey (latitude, 25° 40′ north; elevation, 1600 feet; population, 30,000), on the Mexican National Railroad, one hundred and sixty-eight miles from Laredo, is situated in a beautiful valley, and is much frequented as a winter-resort, as it has a mild, although quite moist, climate. There are two hotels. The annual rainfall is excessive, amounting in some years to 124 inches. The mean annual temperature is 70° F.; maximum, 92°; minimum, 53°. Mean winter-temperature, 55°. The prevailing direction of

---

[1] Summerland Sketches. Felix L. Oswald, M.D.

the wind is southeast. Monterey is the most Americanized of the Mexican towns. It is distant about two hundred miles in a direct line from the Gulf of Mexico.

The **Topo Chico Hot Springs** are situated three miles east of Monterey. There are two springs, one a tepid arsenic spring, the other very hot, said to be 208° F. There is a hotel at the springs.

Saltillo (latitude, 25° 25′ north; elevation, 5350 feet; population, 20,000). Saltillo is about sixty-five miles southwest of Monterey, and 3700 feet higher, on the northeastern edge of the great Central Mexican plateau. There is a fertile valley between the two towns. The principal interests are manufacturing; there are several cotton and woollen mills. There is but one hotel beside the railroad restaurant. The town is regularly laid out, with several *plazas* and an *alameda*. The houses are built usually of sun-dried bricks. Water is brought from the mountain in an aqueduct. It is not so modern a town as Monterey, but is drier.

The baths of San Lorenzo are three miles from the town.

Saltillo has a mean annual temperature (from a record for four years) of 62° F.; maximum, 93°; minimum, 27°. Mean annual relative humidity, 61 per cent. Mean annual absolute humidity, 3.75 grains of vapor to the cubic foot. Mean annual precipitation, 21 inches. Mean annual velocity of wind not given; prevailing direction north.

Chihuahua (elevation, 4700 feet; population, 20,000) is on the Mexican Central Railroad, two hundred and twenty-five miles south of El Paso. The town is situated on a level plain, surrounded by hills, on the small Rio Chubisca. The railroad station, as is customary in Mexico, is one mile from the town. Horse-cars run to the *plaza*. The flat, Moorish-looking buildings are usually of one story—sometimes two, and on some streets they are built with arcades over the sidewalks. The fine cathedral of San Francisco, formerly known as the church of La Parroquia, faces the small *plaza*, which has some trees and a fountain. There are also two *alamedas* or tree-shaded streets—cottonwoods predominating. In some cases these trees are five feet in diameter. There are two newspapers, large markets, a mint, an ice-factory, and a flouring-mill. The band plays on the *plaza* three evenings a week. Not much can be said for the hotels. The hottest months are from May to August. The rainy season is from June to October. No weather-record for Chihuahua was obtainable.

There is a town-supply of water, brought in an old aqueduct. The country immediately around the town is fertile and cultivated, there being a number of large ranches or *haciendas*. Fine grapes are grown here. Chihuahua is sufficiently near to the United States to have a few "American" methods and ideas.

Durango (elevation, 6200 feet; population, 30,000).[1] At Torreon, five hundred and eighteen miles from El Paso, passengers on the Mexican Central Railroad can take the train on the Mexican International Railroad for Durango, one hundred and fifty-seven miles distant to the southwest. The journey requires about six hours. Durango is situated in the plain of San Antonio, on the southeast slope of the mountains, the main range of the Sierra Madre being about thirty miles west. The foot-hills extend to the city, which is sheltered by the mountains from the influence of the prevailing Pacific winds. The distance from the ocean is about one hundred and forty miles in an air-line.

The streets are regularly laid out and well shaded. The *plazas* are attractive, with flowers all the year. The town relies on surface-drainage; but the streets slope to the south, and, as the subsoil is sandy, the drainage is not particularly objectionable. The water is soft and pure and the supply plentiful. There are thermal baths supplied from springs flowing in the town. The usual rainy season of the Cordilleras prevails from June to October, the heavy rains falling usually during the afternoon. The rainfall for the year, June 1, 1894, to May 31, 1895, was 21½ inches, of which 1.85 inches fell during the seven months from November to May, inclusive, and 19.75 inches during the five months from June to October, inclusive. The temperature by seasons for the same period was as follows: monthly mean, winter, 50° F.; spring, 67°; summer, 71°; autumn, 61°; annual, 62°. The maximum temperature was 88° in May, and the minimum 20° in February.

Zacatécas (elevation, 8180 feet; population, 50,000). This mining-town is picturesquely crowded into a ravine, about half a mile above the level of the surrounding plain. A tramway leads from the station to the business-centre—a ride of nearly half an hour. There are two *plazas*, a market, cathedral, mint, State and municipal palaces, etc. An inadequate supply of water is brought in

---

[1] The elevation given by the engineers of the Mexican International Railroad is 6207 feet. Humboldt estimated the elevation of Durango at 6840 feet, and its environs as averaging 6560 feet. The railroad officials estimate the population (1895) at possibly 35,000.

by means of an old aqueduct and then distributed from the fountains or basins, into which it trickles, by *aquadores* or water-carriers. The streets are poorly paved and badly drained. The buildings are usually built of stone, and are two or three stories in height. One of the hotels (the "Zacatecano") occupies a portion of what was formerly the Convent de Agustinos. Zacatécas is situated in an arid region, which, with its elevation of a mile and a half above the sea, accounts for its cool and dry climate. The silver mines are within convenient distance for visitors.

The mean monthly temperature at Zacatécas for 1894, by seasons, was as follows : winter, 55° F.; spring, 64°; summer, 64°; autumn, 59°; annual, 61°; mean for January, 54°; for May, 70°; for July, 62°.

Guadalupe, a suburb six miles distant from Zacatécas and at a lower elevation, is reached by tram-cars. The town, although small, has a grand old church, with chapel, cloisters, and garden, an orphan asylum, a picturesque market, and some fine private gardens.

Aguas Calientes (elevation, 6100 feet ; population, 35,000). An attractive but "slow" and quiet city, built on a fertile plain, which is highly cultivated. The usual tramway runs from the railroad station. The hot springs are two miles east of the town and one mile east of the station, at the end of a beautiful *alameda*. The bath-houses are of stone, and contain large pools of warm water (about 96° F.). There are three groups of bath-houses, Los Banos Chicos having the prettiest surroundings. The natives bathe publicly in the tree-shaded canal that flows by the side of the Paseo.

Aguas Calientes has several *plazas* and many beautiful gardens. The band plays in the principal *plaza* two or three times a week. The hotels are good. There is a market, and fruit is abundant and cheap.

Aguas Calientes is half a day's railroad journey from Zacatécas, and being 2000 feet lower has a more balmy and less chilly air than the mining-town.

The mean monthly temperature for 1894, by seasons, was as follows : winter, 57° F. ; spring, 67°; summer, 71°; autumn, 64°; annual, 65°; mean for January, 55°; for May, 74°; for July, 71°.

San Luis Potosi' (elevation, 6200 feet ; population, 60,000) is situated on a branch of the Mexican Central Railway running to

Tampico, and is one hundred and thirty-nine miles from Aguas Calientes. It is also on the main line of the Mexican National Railroad. It is a busy, progressive city, owing much of its importance to the silver-mines in the vicinity. It is surrounded by a broad and fertile plain, and, being about two hundred miles from the Gulf of Mexico, is protected from moist winds by ranges of mountains to the east and south. San Luis Potosí has clean streets that are regularly laid out and well paved. The drainage is poor, but not dangerous. There is an imposing cathedral, several handsome *plazas*, and a spacious *alameda*. The garrison military band plays three evenings a week. The churches and public edifices are built of stone, adorned with carving. Most of the buildings are two stories in height. There are electric lights and tram-cars. The railroad station is near the centre of the city.

San Luis Potosí has a fine climate, with a moderate yearly rainfall. The mean monthly temperature for 1894, by seasons, was as follows: winter, 57° F.; spring, 68°; summer, 69°; autumn, 62°; annual, 63°; mean for January, 57°; for May, 72°; for July, 68°.

Leon (elevation, 5900 feet; population, 100,000). Leon is a manufacturing city in the centre of a rich valley, where there are farms and grazing-lands. The streets are narrow, with workshops on every block. The principal objects of interest are the main *plaza*, the *paseo*—which is a part of the highway to Silao—the cathedral, and a fine theatre. The water-supply is abundant. A tramway runs from the railroad station to the centre of the city. In 1894 the mean monthly temperature was for winter, 58° F.; for spring, 69°; the hottest month (May) having an average of 74°.

Silao (elevation, 5900 feet; population, 15,000), situated twenty miles south of Leon, is an attractive town with several fine churches and many handsome gardens. Twenty miles south of Silao is Irapuato, a town of 14,000 inhabitants, noted for its strawberries. The railroad station is, as usual, a mile away from the town. Mean annual temperature for Silao for 1894 was 66° F.; for the winter months, 59°; spring, 69°. May averaged 75°.

Guanajuato (elevation, 6750 feet; population, 50,000), fifteen miles northeast from Silao, is reached by a branch road, the last three or four miles being tramway. Guanajuato is a mining-town, situated in a narrow ravine, which has been terraced on each side to afford room for building. The streets are narrow and steep and are frequently only stairways from one terrace to another. It is a quaint and interesting old place, with costly churches, a mint, and a fine theatre. There are also, in spite of the steep hillsides, a pretty little *plaza* and an *alameda*. Guanajuato is picturesque, but unsanitary. Its open drains are unhealthy, and it is a draughty place for invalids.

The mean annual temperature is 63° F., with a range during the year from 34° to 87°. In 1894 the mean temperature for January was 57°; for winter, 59°; spring, 67°. The mean annual relative humidity is 58 per cent. and the mean annual rainfall 38 inches.

Near Guanajuato is the famous Valenciana silver-mine, with its village and splendid church.

Queretaro (elevation, 6060 feet; population, 50,000) is a beautiful city on a smiling plain, encircled by distant mountains. The climate of Querétaro is considered very fine. The streets are narrow and winding, but usually clean. There are several *plazas* with palms, shrubs, roses, and luxuriant vegetation, and a beautiful *alameda* with great trees. Water is conveyed from a stream five miles distant by means of a stone aqueduct architecturally impressive. There are a number of public fountains. In addition to the cathedral of San Francisco, recently "restored," there are several fine old churches, and the constant ringing of bells is noticeable, even for Mexico. Near the city is the melancholy Hill of the Bells, where Maximilian and two of his generals were shot, June 19, 1867. In 1894 the seasonal temperature in Querétaro was as follows: winter, 59° F.; spring, 69°; summer, 68°; autumn, 62°. The monthly mean for January was 60°; and for May (the warmest month) 72°; for July, 67°; annual mean, 65°.

Pachuca (elevation, 8070 feet; population, 30,000) is forty-four miles east of Tula and is reached by a branch from the main line of the Mexican Central Railway. Pachuca has about the same elevation as Zacatécas and a similar climate, although situated two and one-half degrees further south. The town lies in a basin surrounded by mountains, and is in the centre of a rich silver-mining district. The streets are steep and narrow.

As in the country east of San Luis Potosi, a sharp change of climate can be effected by going to the eastern slopes of the mountains, which are kept fresh and green by the moist air coming from the Gulf of Mexico. In 1894 the mean monthly temperature for Pachuca for January was 54° F.; for July, 67°; for the year, 59°. By seasons: winter, 56°; spring, 60°; summer, 63°; autumn, 58°.

Puebla (elevation, 7100 feet; population, 80,000) is distant from City of Mexico one hundred and fifteen miles by rail. Puebla is one of the cleanest and best-drained cities in Mexico. It is a city of churches, with broad streets, a well-kept main *plaza*, and no offensive smells. There are other small *plazas* and two *paseos*. The grand cathedral is second only to that of the City of Mexico in size, and its interior is more richly decorated. From the hill of Guadalupe a fine view is obtained of the city and its rich and fertile plain.

Puebla has cooler evenings and nights than the capital, owing to the closer proximity of the snow-covered mountain-peaks.

The Pyramid of Cholula is seven miles from Puebla to the west, and can be reached by tram-cars.

Toluca (elevation, 8650 feet; population, 20,000). A charming town forty-six miles west of the City of Mexico, in a fertile valley surrounded by lofty mountains. Toluca has clean streets, an *alameda*, and market containing all the fruits of the tropical country below. The hotels have *patios* filled with flowers. Baths are usually attached to the hotels. The town is provided with two theatres and a line of horse-cars. The climate of Toluca is cooler than that of the City of Mexico. About forty miles from the capital, where the railroad crosses the Sierra de las Cruces, an elevation of 10,550 feet is reached on the Continental Divide.

Morelia (elevation, 6200 feet; population, 25,000), lying nearly two hundred miles west of Mexico, on the Mexican National Railroad, is one of the prettiest cities in the republic, with an impressive cathedral, a beautiful *plaza*, where the Eighth Regiment band plays three times a week, a shady *paseo*, and the inevitable *alameda*. The many fine residences contain *patios*, or inner courts, with fountains and a luxuriant wealth of flowers. The water furnished for domestic use is muddy. The town lies in a basin on the western slope, surrounded by mountains, and has a softer, damper climate than the cities on the great inland plateau, owing to the Pacific-influence.

No detailed meteorological data were obtainable.

Pátzcuaro (elevation, 7000 feet; population, 10,000) is thirty-seven miles west of Morelia, situated on Lake Pátzcuaro, which is about thirteen miles long and thirty miles in circumference. This Moorish-looking town is two miles from the station, and is perched on the hillsides, whence extensive views may be obtained of the lovely lake, with its forest-clad islands, and of the surrounding mountains. It is a quaint town, with narrow, crooked streets and many shrines, and a sleepy, tree-shaded *plaza.*

Across the lake, which Humboldt compared to Lake Geneva in beauty, is the Indian village of Tzintzúntzan, where there is a venerable church containing a famous picture by Titian, called "The Entombment." A steamboat plies on the lake.

There are hot springs in the neighborhood of Pátzcuaro, flowing at a temperature of 100° F.

The annual rainfall is more than 40 inches.

Orizaba (elevation, 4090 feet; population, 15,000). Orizaba lies on the eastern slope, within the Gulf-influence. It is on the verge of the *Tierra Caliente*, or "hot lands," and is supplied with an abundance of fresh fruit. The red-tiled town is finely situated in a valley surrounded by mountains. It is a favorite winter- and summer-resort. The name is said to mean "joy in the water," the presence of which is indicated by numerous cascades within view. The climate is very good, except that it is damp and somewhat warm. During February, March, and April Orizaba is visited almost nightly by violent wind-storms from the mountains.[1]

Jalapa (elevation, 4335 feet; population, 18,000). A quaint and very old town in a green valley, swept by mists from the Gulf. Jalapa is in the coffee-raising district. It has a delightful but exceedingly humid climate, as it lies within the belt of 100 inches annual rainfall.

---

[1] Face to Face with the Mexicans. Mrs. F. C. Gooch.

RELIEF MAP OF SOUTH AMERICA

CHAPTER XVI.

SOUTH AMERICA.

THE general climate of South America is chiefly determined by two facts: first, the latitude, three-quarters of the continent lying in the tropical zone; and, secondly, the position and height of the Andes Mountains, which run parallel to the western coast. Within the tropics, as has been said before, the prevailing winds, the "trades," blow from the east, and bring with them a great amount of moisture. The eastern elevations are not high enough to obstruct entirely the passage of these winds, and, although some of their moisture is precipitated here, they carry the remaining portion with them, distributing it over the interior until, being finally brought into contact with the high peaks of the Andes range, it is condensed and falls in showers on their eastern slopes. The western coast, being thus shut off from the moisture-bearing winds by the extreme height of this range, receives very little rain and remains dry until a latitude of thirty degrees south is reached. Here, however, the conditions change. The prevailing winds, being westerly, carry their moisture but a short distance into the interior, when they are intercepted by the Andes, through whose cooling influence the aqueous vapor is condensed and precipitated, this time on the western declivities. The country to the eastward of these mountains, below the thirtieth parallel south, is, therefore, dependent for its rainfall mainly upon the occasional Atlantic winds.

The Amazon, the monster river flowing through the northern part of this continent from west to east, discharges, according to the *Encyclopædia Britannica*, more water than the eight principal rivers of Asia, and the La Plata more than all the rivers of Africa; and these rivers, the Amazon especially, form practicable waterways far into the interior of the continent.

Near the equator the humidity of the atmosphere is so great that, although 150 inches of rain may have fallen on the east coast, the air still retains sufficient moisture to keep the interior, as far as the Andes, well watered. At Rio Janeiro the mean yearly temperature is about 74° F., and in the district where the Paraguay takes its rise it is about 65°. Lima (540 feet elevation) has a mean of 72°, and at Buenos Ayres the mean is 68°.

The northern part of the continent is covered with enormous forests.

The tropical character of the seacoast and lowlands is so marked that they are not available for invalids. On the slopes of certain portions of the Andes, however, are climates which commend themselves, and have under certain conditions been successfully used by the foreign health-seeker. It was on this continent that the altitude-cure was first employed.

All these Andean health-stations, except Arequipa, have almost insuperable objections for a great invalid. The long sea-voyage, the heat and discomfort of the coast-towns, the inaccessibility of the mountain-resorts, necessitating long and fatiguing journeys of perhaps two weeks on muleback, with the danger of exertion at great altitudes; the trying heats and long and severe rains; the annoyance and trials of bugs, beetles, and all the abounding insect-life of tropical regions; the unsanitary conditions and utter lack of conveniences and accommodations and of necessary supplies; the difficulty of always finding an abundant supply of pure, soft water; together with the experience of Spanish food and cooking, are certainly sufficient cause for Dr. Williams's conclusion that "the Andes can only be considered a fit resort for energetic young men with limited tubercular lesions, capable of enduring fatigue and able to accommodate themselves to conditions of life unlike those to which they are accustomed."

The Andes.

Portions of these magnificent mountains in the countries of Ecuador and Peru have long been noted for their peculiar climatic advantages of warmth and equability, with rarefaction, caused by great elevation being obtained in the tropical zone. These advantages are, however, more apparent than real, as is shown by careful investigation.

In Ecuador the Pacific watershed receives rain copiously during the winter and spring season, and in many valleys and plains the country is fertile. North of latitude 40° south the rains become more frequent and profuse, and in most of the United States of Colombia the vegetation is scarcely inferior to that of the Amazon.

Further south, along the coast of Peru, the moist Atlantic currents deposit their moisture on the eastern side of the Cordilleras.[1] On the Peruvian coast while the prevailing cool south winds are blowing rain never falls. The houses are built with flat roofs. Except for scant verdure within the fog-regions on the hills, animal- and plant-life exists only where an occasional river, fed by the snow-mountains, reaches the coast. During the colder months, between April and October, the winds are variable, partly blowing from the north; and from the sea upward to an elevation of 3000 or 3500 feet dense fogs prevail, which pass into misty rains.[2] The mean annual rainfall of Lima is 9 inches, while over a large part of the arid plains of interior Peru it is practically *nil*.

The traveller, Frank Vincent, refers to the climate of Lima as bad, with five months of rain and snow and the rest of the year hot and dry.[3] Another traveller[4] says Lima is very unhealthy, with fog and dampness for weeks in winter. The mean temperature during the cold season, from June into November, is 56° F. Between November and May the maximum is 82°. Back from the coast, within a limited range, a variety of climates can be found, and fruits of the tropical and temperate regions are to be seen in the same market. As Williams puts it: extensive plains exist at high elevations, on which populous cities are built, and in some of these the climate is temperate and genial.[5]

Bogota, the capital of the United States of Colombia, is 8665 feet above the sea, in latitude 4° 6′ north. It is said to have a climate like Malaga and an annual mean temperature of 59° F., with but little variation for the different seasons. One writer gives it a range of 55° to 70°. The city is situated in a finely fertile country. It lies on the eastern edge of a great plain or valley below two hills, called Guadaloupe and Monseratte, which rise 1800 feet and 1500 feet, respectively, above the plain. The capital has a population of about 100,000, but is isolated and difficult of access.[6] It is 700 or 800 miles inland from the coast, and ten or twelve days or more are required for the journey—partly by boat up the Mag-

[1] The Inter-continental Railway. W. D. Kelley, in Cosmopolitan Magazine, August, 1893.
[2] American Resorts, with Notes upon their Climate. Bushrod W. James. Chapter transplanted from Dr. A. Woeikof's Die Klimate de Erde.
[3] Round and About South America. Frank Vincent.
[4] Peru. E. G. Squier.
[5] Aërotherapeutics. Charles Theodore Williams.
[6] The Capitals of Spanish America. W. E. Curtis.

dalena River and partly by muleback from Honda over wretched roads.[1] The city is well built in the Spanish fashion, with streets at right-angles. Most of them are poorly paved. In the business section the houses are usually two stories in height. The dwelling-houses are built of sun-dried bricks, whitewashed, and are usually one story only, on account of earthquakes. There are said to be three months of fair weather, alternating with three months of intense tropical rains—the rainiest season being in the winter. The records of rainfall and humidity of Bogotá were not obtainable. The town has a line of telegraph to the outer world.

Quito (elevation, 9350 feet; population, 40,000), one hundred and fifty miles (air-line) north northeast from the port of Guayaquil, is the capital of Ecuador; it is situated in a basin fourteen miles south of the equator, on the eastern skirts of the volcano of Pichincha (15,924 feet), yet not in full view of the crater on account of an intervening hill.

The climate is said by Mr. F. Hassaurek, formerly United States Minister to Ecuador, to resemble perpetual autumn,[2] being fairly cool, with a mean annual and seasonal temperature of 59° F. or less. Crops do not await the succession of the seasons; flowers and leaves fall while fresh ones bud into life. Roses and wildflowers bloom all the year. It is a pilgrimage of 270 miles from Guayaquil, of which seventy miles can be done by steamer and rail and the remaining 200 miles on mules, requiring ten days for the journey for able-bodied men. The mule-trail is called "el camino real," and has been the highway of commerce for three hundred years. There is a telegraph-line to the coast. Quito is much higher at the north end of the town. It has no proper water-supply. The streets and sidewalks are narrow and indescribably filthy. Few streets are paved, and they are cleaned only by the violent rains which prevail in the afternoons and nights during the rainy season. The dry season is from June to December, although it may begin to rain in October or November. In January the rains are frequent and descend with great force. There are no hotels. Furnished houses of nine or ten rooms can be hired for from $20 to $40 per month. There are no fireplaces, or even chimneys, and few carts or wheeled vehicles. The buildings are flat and heavy in appearance, with projecting roofs to afford protection against the severe rains. Travellers complain of feeling

[1] Round and About South America. Frank Vincent.
[2] Four Years Among Spanish Americans. F. Hassaurek.

chilly in Quito, and of being troubled with cold feet. The changes from midday to night are sharply felt, and throat- and lung-troubles are said to be prevalent. This evidently does not refer to consumption. Overcoats are worn in the evenings, and woollen socks are advised. The temperature is said rarely to go above 90° F. or below 40°. It is usually 70° or 75° during the day and 60° or 65° at night.

Mr. Whymper quotes from a bulletin of the observatory at Quito a table of monthly mean maxima and minima for a complete year, as follows:[1]

		Max.	Min.
December,	1879	69°	41°
January,	1880	74	45
February,	1880	75	45
March,	1880	73	42
April,	1880	72	45
May,	1880	72	46
June,	1880	73	43
July,	1880	73	44
August,	1880	73	38
September,	1880	74	39
October,	1880	68	42
November,	1880	68	41

The annual mean of the maxima was 72° F. and of the minima 42°, with a monthly mean temperature for the year of 57°. The highest maximum was in February and the lowest minimum in August, with a range between them of 37°. The mean annual rainfall is 70 inches. There does not appear to be a record of the humidity, but from the equability of temperature it is safe to say that it is moist. At greater mountain-heights in the interior of Ecuador the sun is usually obscured by clouds or fog, except for a short time at midday. December and January are usually warmer than June and July.

Mr. Whymper's experience in the Andes with "mountain-sickness" differed somewhat from that of the engineers in charge of the recent survey for the Trans-continental Railway, who found that the fatigue of exertion in the rarefied atmosphere of these high altitudes began to show itself at 8000 or 9000 feet above sea-level, and at 14,000 feet was apt to take a form of sickness called there

[1] Travels Among the Great Andes of the Equator. E. Whymper.

" sorojchi,"[1] with dizziness, bleeding at the nose and ears, and perhaps bowel-troubles.   Mr. Whymper was not affected by the sickness at the greatest heights, but at 16,600 feet suffered from intense headache and weakness, with labored respiration and a general feeling of illness.   He did not suffer from nausea, but had no appetite.

Arequipa (elevation, 7650 feet) is in latitude 16° 22' south, about eighty miles direct, or one hundred and seven miles by rail, inland from its port, Mollendo, on the Pacific.   On account of the steep grades the time required is nine hours.   It is the third city in importance in Peru, with a population of 35,000, largely Indians, and is on the line of railway from the coast to Lake Titicaca.

The city—which is known as the "Gem of the Andes"—is well situated on a green and irrigated sandy plain, through which flows the river Chile.   There are few trees except eucalypti.   The houses are usually of one story, built of a white stone of volcanic origin, called "sillar."   The streets are narrow, and partly paved with cobblestones.   There are tram-cars, fair hotels, and a club.   Many of the churches are several centuries old and covered with quaint carvings, but the greater part of the town is rebuilt on the ruins of the great earthquake of 1868.   The Harvard College observatory, which has an establishment near Arequipa, on a hill 400 feet above the town, has also erected a meteorological station on the summit of the neighboring volcano, El Misti, at the great height of 19,300 feet above the sea.

Although noted for its delightful climate, the heat in Arequipa is said by Mr. Vincent to be intense.[2]   Professor Pickering, of the Harvard Astronomical Observatory, on the contrary, in a brief meteorological summary for the year 1891 (published in *Astronomy and Astro-Physics*, May, 1892[3]), states that the hottest day was June 3d, when the thermometer rose to 79° F., and the coldest was eight days later, when it fell to 38°.   From April 1st to November 1st it was rainless and absolutely clear.   During January and February 2 or 3 inches of rain fell—invariably during the afternoon.

The temperature never fell below the freezing-point, but there

---

[1] "This abused word is pure Aymará. The j has the sound of German *ch;* the *ch* its invariable Spanish value, as in *church*." Charles F. Lummis.

[2] Around and About South America. Frank Vincent.

[3] Fuller results of the observations taken at the Observatory in Arequipa will probably appear later in the Annals of Harvard College Observatory. See article by A. Lawrence Rotch in American Meteorological Journal, October, 1893, referring to the first station on Charchani. The higher station on El Misti was building about the time of that publication in October, 1893.

were occasional frosts. The wind-movement was moderate, the greatest velocity having been 17 miles per hour in December. The air was usually clear and dry, although the lowest relative humidity mentioned, 35 per cent., would not indicate a condition of dryness quite as great at the same temperature as that of the Rocky Mountain plateau in Colorado or New Mexico.

Jauja and Tarma, situated in the well-sheltered valley of Jauja, at about 10,000 feet elevation, are health-resorts of considerable repute, the former being used by the Peruvian Government as a sanatorium for military consumptives. The mean annual temperature of Jauja is stated by Archibald Smith to be between 50° and 60° F., and that of Huencayo, which is slightly less elevated, as between 51° and 63°.

Argentine Republic. Scrivener recommends the mountain-districts of this region, particularly for consumptives.

CHAPTER XVII.

EUROPE.

EUROPE is the smallest of the great continental divisions of the globe, and is but a third larger than the United States exclusive of Alaska. It projects so far from the vast body of the Asiatic continent as to have been termed the Western Peninsula of Asia. From its eastern frontier stretches for five thousand miles the land-surface of Asia, giving a continental character to the climate of Eastern Europe, as shown in the cold winters and hot summers of the great plain of Russia, which is too far removed to receive more than a slight and intermittent influence from the Atlantic.

Western Europe, on the other hand, the shores of which, with their numerous promontories and off-lying islands, project into this great ocean, has an equable climate, owing to the currents which touch the coasts and to the moist, warm winds which blow off this immense body of salt water. This sea-influence is increased by the Mediterranean, which washes the southern shores. The climate of southern Europe is also affected by its proximity to the continent of Africa, and, as its most southerly points lie within 9° of the Tropic of Cancer, its temperature is higher than that of western and middle Europe. These are chiefly situated within the cool belt, though the northern limits are actually within the frigid regions of the Arctic Circle. The bulk of the European continent, therefore, has a cool, temperate climate, its equability being insured by the modifying influence of the seas which partially surround it. Owing to the Atlantic winds and to that great branch of the Gulf-stream which flows toward its shores, the climate of western Europe is much milder than that of the Atlantic and Arctic coasts of North America, which lie at an equal distance from the equator.

The moisture which is absorbed by the westerly winds as they cross the Atlantic brings a heavy rainfall to western Europe, the precipitation lessening, but still remaining high, until it reaches the eastern plains, where it becomes very moderate.

"Europe," writes Davis, "may be divided into three regions—

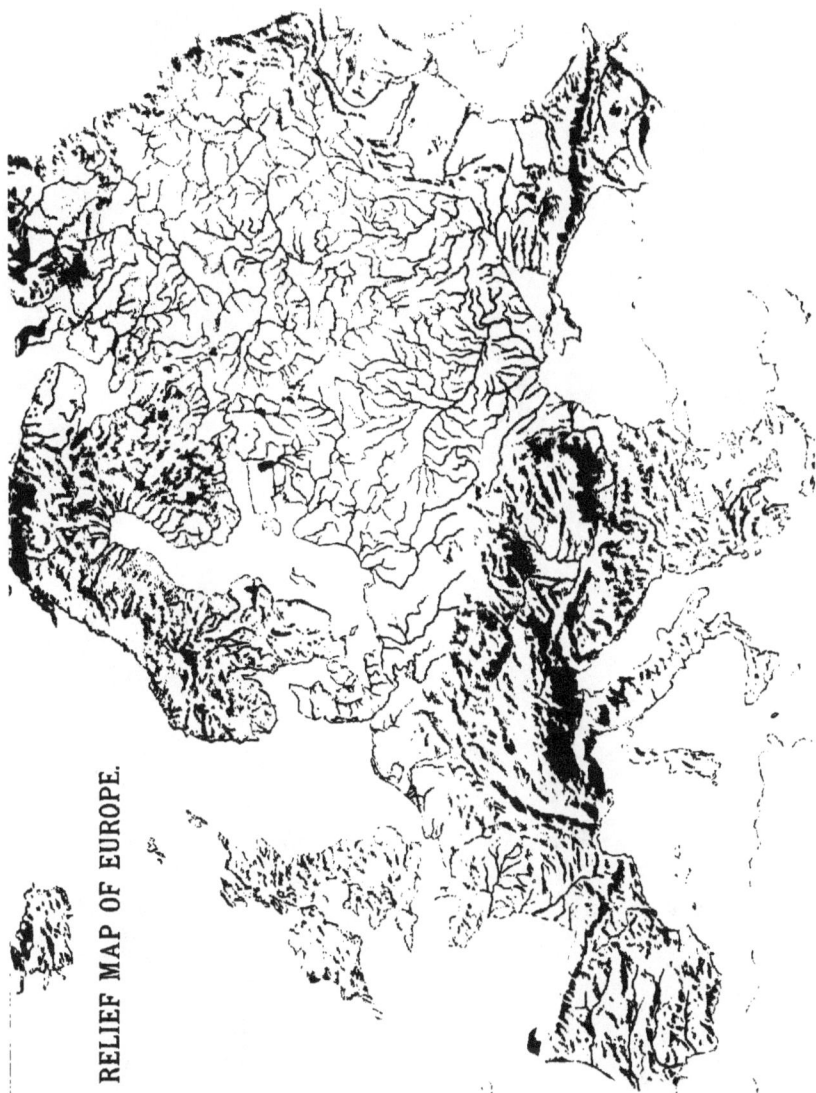

RELIEF MAP OF EUROPE.

From BUTLER'S GEOGRAPHIES, by permission of E. H. BUTLER & CO. Copyright, 1888.

mountainous highlands in the southwest, lower highlands in the
northwest, with lowlands between the highland regions, and also
spreading far to the northeast. Many peninsulas and seas make
the coast of Europe more irregular than that of any other continent."[1]

With respect to its rainfall, the greater part of Europe belongs to
the zone of irregular seasonal distribution. While the southern
portion, with its dry summers, belongs to the subtropical zone,
"the line of demarcation runs at a little distance to the north of
the Spanish coast of the Bay of Biscay, continues along the northern
slope of the Pyrenees, turns northeastward to the neighborhood of
Valence on the Rhone, curves southward to Genoa, follows the line
of the northern Apennines, strikes across the Adriatic from Rimini
to the neighborhood of Zara, and proceeds by way of Seraievo, Novi-
Bazar, and Sofia to the coast of the Black Sea, south of Zozopoli.
Within the subtropical zone the maximum rainfall occurs during
winter in the south of Spain and Italy; during autumn and winter
in central and northern Spain, the south of France, and northern
and central Italy. In the zone of irregular distribution Scotland,
Ireland, and western England have their maximum in winter;
western France, eastern England, the coast regions of the Low
Countries and Denmark, and the greater proportion of Norway
have theirs in the autumn; while in eastern France, the German
Empire, Austria, Hungary, Russia, and Sweden it falls in summer."[2]

In northwestern England, at Stye Head Pass, the greatest precipitation is recorded (189.49 inches), while the least occurs in the
almost rainless region around the Caspian Sea.[3]

The prevalent winds of western Europe are southwest and west
southwest; in southeastern Europe winds most frequently blow
from the north and east, the latter being more prevalent in the
winter and autumn.[4] Hurricanes are rare, but westerly storms
cross the continent six or seven times during the winter. The chief
local winds are the dry, warm Föhn of the Alps, which resembles
the "chinook" of the Rocky Mountain region of North America,
the violent Boro wind of the upper Adriatic region, the Etesian
winds of the Mediterranean, the mistral of southern France, and
the hot, desiccating sirocco from the African deserts.

[1] Frye's Geography, p. 75. [2] The Encyclopædia Britannica.
[3] Dr. Otto Krümel: Ztschr. für Erdkunde zu Berlin, 1878.
[4] Wesselovski: Encyclopædia Britannica.

The snow-line varies greatly in different localities, ranging from 8860 feet in the western and central Alps to 1400 feet on the eastern slopes of the Caucasus Mountains. The average height is lower in the Alps than in the Rockies.

The British Isles.

This dominion consists of two large islands and many small ones. Great Britain, the largest of these, although it is the most important island in the world, is only one-fortieth the size of the United States of America.

While they lie in the same latitude as Labrador, which has a cold and severe climate, that of these islands is mild, with even seasons, owing to the ameliorating influence upon the west winds of that great branch of the North Atlantic eddy which flows along the British coast.

The rainfall is, generally speaking, heavy, especially on the western shores, and the air is humid. To these prevailing conditions there are, however, many remarkable local exceptions.

The land in the northern and western parts of Great Britain is moderately high, and in northern Scotland it is very rugged. In these regions the rocky shores are much indented by lochs and bays which are studded with small islands. The southern and eastern portions are chiefly lowlands, and the shores are of sand or clay.

Ireland, lying furthest to the west and being first exposed to the western gales, receives more rain, and the climate is, for the same reason, damper. The surface of Ireland is chiefly low and flat, and there are large areas of bog. Near the coast, however, there are some moderate elevations.

Ireland.

The coasts of Ireland and Scotland are dotted with beautiful resorts, many of which afford good sea-bathing. The climate of Ireland is unsuited to the needs of most health-seekers, especially consumptives; the death-rate from consumption is extremely high, being 14.9 per cent. of all deaths, which is higher than the percentage for any other part of the British Isles. But the coasts afford some pleasant summer-resorts, and the sedative character of the air renders them advantageous to certain classes of invalids. On the Atlantic coast the resorts, while affording good bathing and possessing a healthy and bracing climate, are, of course, more windy and

more subject to fogs and mists. Among them may be mentioned
Bandarem, on Donegal Bay; Kilkee, in County Clare; and Kilrush, at the mouth of the river Shannon.

Glengarriff, near Bantry Bay, has, like Queenstown, a southern
aspect, and is much sheltered from cold winds. Its situation is
beautiful and interesting, and the climate lacks severity and is
equable.

Queenstown, situated in the Cove of Cork, is built in terraces
on a hillside, which protects it to the north, but leaves an open
aspect to the south. It has a mild climate, not subject to sudden
changes of temperature. The mean annual temperature is 51.9° F.;
the mean temperature for spring is 50.17°, and for winter 44.2°;
and the annual rainfall is 34 inches.

Among popular summer-resorts with good bathing may be mentioned Bray, Dundrum, Kingstown, on the eastern coast; Howth,
on Dublin Bay; Rosstrevor, New Castle, Holywood, and Dona-
ghadee. The climate for all these places may be characterized as
mild and damp.

Portrush and Port Stewart, lying on the north coast, have a
climate more tonic in its effects than the resorts just mentioned, and
Portrush especially is quite popular.

County Clare has also sulphur and iron springs, located at the
town of Lisdunvarna, and their local fame is considerable.

Scotland.

Scotland is, on the whole, a harsher but more invigorating climate
than Ireland, and the consumptive death-rate is not so high, being
13.8 per cent. of all deaths.

The west coast is very beautiful and attractive, and its summer
climate is mild but very humid; the number of rainy days is great
and fogs are frequent. The east coast is drier and brighter, but not
so equable.

The air of the moorlands of the interior is keen and bracing, but
except for a few weeks in the early summer rains and mists are
common. For invalids who are strong enough to exercise, especially if they are able to take advantage of the opportunities for
sport, and to endure the humidity, the moorlands often prove a good
tonic during the summer and early autumn.

During the same season the climate of some of the islands off the coast is very pleasant because of its mildness and equability, and the primitiveness of the surroundings often affords a healthy change for the invalid accustomed to luxury.

Scotland has many resorts which would deserve a longer notice could more definite statistical information in regard to them be obtained. We may, however, mention the following :

Oban is prettily situated on the shores of a bay in Argyleshire, and has about 2000 inhabitants. The scenery has been highly praised, and the writings of Mr. William Black have made the place widely known. The neighborhood is interesting and beautiful. It is coming more and more into favor as a summer health-resort.

Helensburgh, which is not far from Dumbarton, is situated on the river Clyde. Like many of the Scotch summer-resorts, its climate possesses both mildness and equability, and it is recommended for throat and bronchial affections.

Rothesay, on the west coast and to westward of the Firth of Clyde, is used both in winter and in summer. Situated amidst beautiful scenery and affording good bathing, it has in its climate an additional advantage, the temperature being characterized by both mildness and equability. It is seldom that the heat in summer is greater than 70° F., and the thermometer rarely falls below freezing-point.

Ardrossan, on the coast of Ayrshire, affords good sea-bathing. The summer climate is mild, damp, and equable.

North Berwick, situated about twenty miles from Edinburgh, possesses a good beach and large golf-links. Its climate is bracing.

St. Andrews is located on the coast of Fifeshire. The climate is healthy, but the place is much exposed to the northeast winds which prevail during the spring season. Here are the oldest and most celebrated golf-links in the world, and the town and its situation are interesting and delightful.

Stonehaven, a few miles south of Aberdeen, is charmingly situated and affords good bathing. The atmosphere is tonic and invigorating.

Strathpeffer, situated in a valley near Dingwall, in the northern part of Rossshire, has strong sulphur springs, a bathing and drinking establishment, and a fine hotel. The climate is fairly good, but the advent of a rainy season may render the place undesirable.

Nairn, on the bank of the river Nairn, is about eighteen miles to the northeast of Inverness. It is a much-patronized resort, has a good beach, and is easy of access. It has been recommended as a winter-resort, but detailed information as to its merits is not forth-coming.

At **Crieff,** a finely situated resort in Perthshire, is a well-arranged hydropathic establishment. The climate is bracing.

At the **Bridge of Allan,** a beautiful spot near Stirling, is a salt spring. The situation of the place is protected and the climate mild.

Moffatt, in Dumfriesshire, lying at an elevation of 400 feet above the sea-level, has a cold sulphur spring, the waters of which contain sulphuretted hydrogen, sulphate of sodium, and common salt. The climate is healthy. There is a very well and favorably known hydropathic establishment at Moffatt.

England.

Different parts of England show considerable variation in their climates. Speaking broadly it may be said, however, that it is not so mild and damp as Ireland, nor so bracing and harsh as Scotland. The death-rate from consumption is lower than in either, being 12.3 per cent. of all deaths.

The resorts on the west coast of England are both warmer and moister than those on the east coast, for they border upon a warmer sea, and warmer, damper winds prevail. In winter the difference in temperature is from 3° to 6° in favor of the west coast, but this is lessened upon the approach of spring, and, in summer, warm winds blowing off the continent may cause a positive reversal of the conditions mentioned above. Resorts on the east coast, being cooler and drier, are consequently more bracing.

The County of Cornwall, a promontory which constitutes the southwest corner of England and terminates in Land's End, has a climate differing from the rest of the island and resembling that of the Riviera in winter. It is 5° cooler, the relative humidity being about 10 per cent. higher. The rainfall and the number of rainy days are decidedly greater. The temperature-range is, however, much less, and the climate is milder and more equable, but less stimulating; while fairly sunny, it is not so bright as that of the Mediterranean shores. Here will be found flowering in winter many exotic and subtropical plants ; the early flowers and vegetables of the Scilly Isles are especially celebrated, and the climate of these attractive islands has been recommended for invalids.

This promontory is eighty miles in length, or thereabout, and forty at its base, narrowing down to an average breadth of twenty miles. On its southern shore is the English Channel and on its northern the Atlantic Ocean. Being about one hundred miles from the coast of France, it is so situated that it enjoys the advantages of an island climate, which gives to it its equability, while the influence of the Gulf-stream increases its temperature above that normal to its latitude and adds to its humidity. The land rises from the shore to a height of 600 feet, this being the average height of the range of hills which runs through the length of Cornwall. This range gives shelter to its shores, and especially to the southern coast, which is thus protected from Atlantic gales. Throughout this district, therefore, but particularly on the southeast coast, is found a climate well adapted for English invalids who seek an easy and pleasant escape from the fickle chilliness of their homes in other parts of the island.[1]

Only some of the most frequented and characteristic of the numerous English resorts can be described.

Dr. Weber, in comparing the resorts of England with those of the European continent, writes as follows :

"On the whole, we might say of English resorts that the climate is healthy and invigorating, though not agreeable, and that to a certain extent it requires powers of resistance.

"The climatic characteristics of English seaside-places may thus be summarized : greater warmth than is due to latitude ; equability of temperature as regards different seasons and times of day ; a comparatively high amount of humidity ; a dull atmosphere, with little sunshine ; and very favorable hygienic and dietetic conditions."

COAST RESORTS.

Dr. Braun says : "England is remarkably well provided with seaside-places, with excellent air and good opportunity for sea-bathing."

Llandudno, a pleasant but somewhat rainy resort in Wales, also comes under the head of equable and moist climates, and is sometimes used in the winter season. The mean winter temperature is 43.7° F. The mean humidity is about the same as that of Hastings.

Tenby, lying on the south coast of Wales, has a moderately dry, mild marine climate. It is nine hours' journey by fast train from London. It stands upon a rocky peninsula which attains an eleva-

---

[1] Sir Joseph Fayrer writes from personal experience very enthusiastically of this district in an article upon Falmouth, in the Journal of the British Medical Association of August 29, 1896, from which much of this information is taken.

tion of 100 feet above sea-level. There are admirable sanitary
arrangements, the water-supply is good, and the houses are, for the
most part, well built and substantial. There is a magnificent beach
which affords good bathing-facilities, and there are all the diversions
usually to be found at seaside-resorts. During the year 1882 the
lowest temperature was reached on December 10th, when the thermometer stood at 32° F., and the highest on August 9th, when it
registered 90°.

Ilfracombe, on the shore of the Bristol Channel, is finely situated
on a beautiful rocky coast, has a bracing climate, somewhat like that
of Brighton, and is very pleasant for summer residence. The arrangements for taking sea-baths are unusually good, and they include a
swimming-bath of considerable size, which is filled by every incoming tide. There is not so much rain as at Torquay. Many enjoyable
trips may be taken from Ilfracombe through a picturesque country.

Falmouth, which lies on the southern coast of Cornwall, is an
example of the climate of the southwestern part of England. It is
situated on the shores of a fine harbor, and as the ground rises from
the shore the town is built in terraces upon a hillside. The coast
consists of strips of beach alternating with stretches of high cliff,
from which the views are very fine. Owing to the mildness and
dampness of the climate the vegetation is luxuriant and beautiful,
and the neighborhood has many charming country-seats. The
accommodations at Falmouth are very fair, but far less has been
done to attract visitors than at other seaside-resorts possessing such
natural advantages. Although the amount of rainfall is large, the
storms are rarely of long continuance, and the soil, being gravelly,
is porous and dries quickly. Annexed is a table taken from Sir
Joseph Fayrer's article, already referred to, which shows the temperature of Falmouth as compared with that of some islands and of
several towns belonging to the Riviera and to the Pyrenean regions:

	November.	December.	January.	February.	March.
Falmouth,	47.8°	44.3°	44.1°	45.1°	44.7°
Penzance,	47.26	45.17	45.21	45.20	45.32
Scilly,	49.8	46.7	46.3	46.9	46.4
Cannes,	52.6	46.3	48.0	48.8	57.0
Montpellier,	50.7	45.7	42.1	44.8	48.9
Mentone,	54.0	49.1	48.7	49.1	52.8
Nice,	53.8	48.5	47.1	46.2	51.8
Pau,	47.0	42.8	41.2	43.6	48.8
Madeira,	64.96	62.58	61.89	62.7	64.0

Torquay, beautifully situated on the north shore of Tor Bay, has a particularly well-sheltered location and a warm, moist, very relaxing climate. The annual rainfall is over 39 inches, distributed through 200 rainy days. The town is well regulated; there are an adequate supply of pure water and a good sewerage system. There are ample and excellent accommodations. It is warmer and more relaxing than Bournemouth.

Bournemouth. This health-resort has a sheltered situation on sand-hills wooded with pine. The bay lies open to the southwest, but is protected on the northwest, north, and northeast. It has a moist, equable temperature, drier than Torquay. In January the average of the minima is 35° F, while in July the average of the maxima is 71°. By seasons the mean temperature is: for winter, 42°; spring, 49°; summer, 60°; autumn, 52°. The yearly rainfall is about 30 inches, with from 120 to 160 rainy days. The relative humidity is from 75 to 86 per cent.—usually in the vicinity of 80 per cent. It is a very much-used resort in winter for chest-cases. There are first-class hotels and lodgings.

Ventnor is charmingly placed on the south coast of the Isle of Wight, the town being built in terraces on the side of the Undercliff. It is used both in winter and summer as a health-resort. The situation is sheltered and the climate mild and free from sudden changes of temperature. There is a National Hospital for Consumptives at Ventnor, the arrangement of which is exceptionally good.

Brighton is on the south coast of England, and within an hour's journey by rail from London. The season is during the winter. There are good hotels and all the attractions of a city by the sea. It is regarded as having a climate generally warm but stimulating, and there are much more sunlight and less fog than in London. The air on the chalk-downs behind the town is considered very fine. The mean temperature for January is 39° F.; for April, 49°; for July, 62°; for October, 52°; for the year, 50°. The soil is dry and the air less moist than at Bournemouth. The annual rainfall is 26½ inches. It is somewhat windy.

Eastbourne, situated between St. Leonards and Brighton, on the coast of Sussex, is a well-laid out town, extending some three-quarters of a mile back from the seashore. Thus, patients who come here may be directly under the influence of the sea-air or they may enjoy the more bracing atmosphere of the extensive downs which lie inland. The water-supply is good and the sanitary arrangements are excellent. There are many amusements for the visitor at Eastbourne—riding, driving, swimming, and varieties of sea-baths, lawn-tennis, etc., and other pleasant diversions are to be found in the agreeable Devonshire Gardens. There is a magnificent view over the ocean from the fine promontory known as Beachy Head. The promenades are furnished with seats and sheltered lounges, and are lit by electricity. Eastbourne is a somewhat more fashionable resort than those previously mentioned, and has a larger permanent population.

St. Leonards, also known as a winter-resort, may, in all its main characteristics, be classed with Hastings, to which it is joined. St. Leonards is more open and rather more bracing than Hastings, and is a very attractive place, with handsome hotels and residences.

Hastings, situated about thirty miles from Folkestone, looks toward the south, and is sheltered by high cliffs from winds blowing from the north and northeast. It is known chiefly as a winter-resort. The autumn here is pleasant, but the spring is cold and windy. The soil is sandy and porous, and the water-supply and the drainage-system are good. Fogs are rare, and the temperature, both for winter and summer, is very even. There is a pier at Hastings, and a pavilion in which daily concerts are given, and the neighborhood abounds in points of interest, affording an object for excursions.

Folkestone, built upon a cliff of green sand, fifty feet above sea-level, which gives it a pleasant elevation and openness, is a very popular health-resort. There is an ample supply of pure water, and the town has a good drainage-system. The rainfall for the year is said to be about 25 inches. The air is stimulating. The roads are good, there are plenty of attractive walks and drives, and the resort is an eminently cheerful place.

Dover, situated on the south coast, has, during most of the year, a climate dry, tonic, and invigorating; but the locality is subject to the disadvantages of great cold in January, high winds in March, and extreme heat in July. It is, to some extent, protected from winds blowing from the north, northeast, and northwest by the chalk-hills which lie at the rear of the town. The subsoil is chalk, but most of the houses are built upon the beach, where the surface-soil is, of course, porous and drains quickly. The water-supply, derived from wells sunk in the chalk to a depth of 226 feet, is pure and plentiful, and the sanitary arrangements in general are good.

Ramsgate is reached over the Chatham and Dover Railroad in two hours, or by steamer from London Bridge in six hours, and, during the season, small steamers also run between this point and Sandwich, Dover, and Deal. The climate is warmer than that of Margate, because the situation is more sheltered from the sweep of the northerly and northeasterly winds. Ramsgate has a fine bathing-beach, and there are attractive drives through the neighborhood to Canterbury and Folkestone, but it is not so lively a place as Margate. The town itself lies between two cliffs, the east and the west, where most of the visitors congregate. The Eastcliff Hotel, with its abundant and complete facilities for all sorts of sea-baths, deserves especial mention.

Broadstairs, lying between Margate and Ramsgate, is a much quieter, less pretentious resort than either of them. It has a good beach, which affords safe bathing, but the land in the immediate neighborhood rises to a height of from 120 feet to 150 feet above sea-level. Like Margate, Broadstairs stands upon ground having a chalky subsoil. The water-supply is pure and abundant, and sanitary regulations are strictly enforced.

Margate, reached in two hours from London by rail and in five hours by river steamer, is one of the most popular of English resorts. The subsoil of this locality is chalky, so that after a rain the ground dries quickly. Although it is subject to high winds, Margate has a bracing and comparatively dry climate, and the extensive beach affords good bathing, particularly for those who do not care to try deep water. The place is considered especially good for scrofulous troubles, and is the seat of the Royal National Hospital for Scrofula. While situated near Ramsgate on the east coast, Margate, from the peculiarity of its location on the peninsular jut of land just below the Thames's mouth, has a northerly or northeasterly exposure, and in the spring, when northeasterly winds prevail, it is, in consequence, one of the few resorts in this vicinity which have a sea-breeze. There are good accommodations to be had, both as regards hotels and lodging-houses. The Cliftonville Hotel is situated on the higher ground known as the Cliff, but the less expensive hotels and boarding-houses are on lower ground close to the harbor and pier. During the summer Margate is a very popular point for cheap excursions.

Felixstowe is situated on the Suffolk coast, near Ipswich, and has a southerly outlook. The air is bracing and comparatively dry, and the weather is fairly good up to the latter part of February. The water is pure and contains a little iron. There is a first-class hotel, "The Bath", and every facility for good and safe bathing. There are also very well-known golf-links here.

Bridlington and **Filey** are two pleasant seaside-places a little to the south of Scarborough.

Scarborough, in Yorkshire, five and one-half hours by rail from London, is a favorite seashore-resort. It is reached over the Great Northern Railroad. It has an elevation of 100 feet above the sea, being built upon precipitous cliffs. Castle Hill, to the northeast of the town, and Oliver's Mountain, more to the south, reach a height of 285 feet and 500 feet, respectively. Besides its healthful situation and its tonic climate, Scarborough has claims to favorable consideration as a health-resort in that it is a well-regulated town, but its many diversions and gaieties render it more particularly agreeable to those who are dissatisfied with the quiet, monotonous life which is, nevertheless, most suitable for a large class of invalids. The mean annual temperature is 46.7° F., and the annual and daily range is small. Scarborough possesses a fine beach, and at the Spa are chalybeate and saline springs.

Whitby, a town whose chief industries are fishing and shipbuilding, is also well known as a summer-resort of no mean attractions. It lies on the coast of Yorkshire. There is a good water-supply and adequate and complete sanitary arrangements. Boating, bathing, and fishing are among the amusements, and there is a building for concerts, balls, etc., and a good public library. There are two excellent hotels, the "Royal" and the "Crown," and many good lodging-houses.

INLAND RESORTS.

Ilkley, in Yorkshire, not far distant from Matlock, is a health-resort possessing a fine air and a bracing climate. Here are located establishments for the purpose of enabling patients to undergo hydropathic treatment.

Buxton has an elevation of 1000 feet above the ocean, and the air is pure and stimulating, but the rainfall is great. It is more shut in than Ilkley. Buxton has simple thermal springs which Yeo speaks of as resembling Ragatz and Plombières. The waters, used chiefly for bathing, are, he says, efficacious in gout, rheumatism, joint-affections, etc.

Matlock, in Derbyshire, situated amidst beautiful scenery, possesses a fine, bracing climate and several springs, of which carbonate of lime is the chief solid element. Here is located a well-known establishment for the hydropathic treatment of such disorders as dyspepsia, chronic rheumatism, and glandular affections.

Woodhall Spa, in Lincolnshire, some eighteen miles from the coast, is situated in a district which is said to have the smallest rainfall of any in England, very little over 20 inches. The subsoil is dry sand, and the climate is invigorating. The "bromo-iodine spring" at Woodhall contains, as stated by Dr. Frankland, 4.396 grains of bromine, 0.616 grain of iodine, and a little arsenic to a gallon of water. There are good accommodations to be found in comfortable boarding-houses or at the excellent Victoria Hotel.

Droitwich, situated in Worcestershire, three and one-half hours' journey by rail from London, has strong salt springs. There is a large swimming-bath, the water of which has a temperature of 80° F., and the water, used in this way, is said to be beneficial in certain forms of gout, rheumatism, etc. There is an excellent hostelry, the Royal Brine Baths Hotel, and boarding may also be procured in the town.

Malvern, also in Worcestershire, lies on the eastern slope of hills, which render its situation very sheltered. The climate is both bracing and equable, the water-supply is pure, and the accommodations are good. Dr. Charteris says of this resort: "I do not know of any health-resort, at home or abroad, more effective than Malvern for restoring to its original vigor the unstrung nervous system or for bringing back the appetite, jaded by late hours or the luxuries of town-life." The hotels are excellent and the views fine.

Bath, once so popular a resort and bidding fair to rise again into notice, has indifferent, earthy, thermal springs. The chief solid ingredient of these waters is sulphate of lime. There are four hot springs, the temperature of which ranges from 104° to 120° F., and the bathing-facilities are good. The situation of the town is somewhat sheltered from the north and east winds, and the winter-temperature is said to be from 3° to 5° warmer than that of London. The climate is mild, moist, and sedative. Bath is a handsome city, with excellent hotels and lodgings.

Harrogate possesses a number of sulphur springs, containing sodium sulphide, free sulphuretted hydrogen, and different amounts of common salt. There are also chalybeate springs of varying strength. They are used much as are the waters at Homburg and Kissingen, and it is claimed that they are useful in cases of dyspepsia, constipation, congestion of the liver, and in some forms of skin disease. Harrogate has an elevation of 430 feet, and the air is pure and bracing, but the place is rather windy.

Norway and Sweden

possess very pleasant summer climates with beautiful scenery. A trip around their coasts in one of the well-fitted, comfortable tourist-steamers sailing from England during the summer months affords the advantages of an agreeable and short sea-trip for convalescents, for jaded citizens, or for the martyrs to hay-fever. Journeys through the interior are easily made, and to the fisherman Norway is especially attractive.

The North Sea.

On the coasts of the North Sea are many more or less well-known resorts which have a bracing, tonic atmosphere, and there are also some islands, lying near the mainland, which are somewhat used as summer health-resorts.

Heligoland, owned by England, is the best known of these island-resorts. It lies six miles from the mouths of the rivers Weser and Elbe, and steamers run between it and Hamburg and Bremen. The island is formed of sandstone rock, one portion being considerably elevated over the other. Owing to its situation and to its small extent, Heligoland has as pure a sea-climate as if, to quote Dr. Yeo, it were a "ship anchored out at sea."

Norderney, an island lying to the southeast of Heligoland and much nearer the mainland, has the same climatic characteristics.

Holland.

In the summertime, for those who need a mild, sedative climate and do not fear the dampness, Holland is an attractive country where sight-seeing is interesting if fatiguing. There are several pleasant seaside-places, the most notable being Scheviningen.

The resorts on the coast lying nearly opposite the English shores have, to speak generally, an air which possesses more dryness and quality of stimulation than the resorts further to westward, and less equability.

Scheviningen, situated on the coast of Holland, about three miles from the Hague, offers all the advantages of a seaside-resort, having a quaint and pleasing character quite its own. It possesses a good beach, every facility for sea-bathing, fair hotel-accommodations, and a soft, healthy, though not bracing climate. The wind is sometimes high, but rarely sharp. Many pleasant and interesting excursions may be taken to the Hague, to Delft, to Haarlem, and to Amsterdam.

Belgium

is somewhat drier and warmer than Holland, but has otherwise much the same peaceful characteristics, and a few weeks' loitering among the quaint old Flemish towns, such as Bruges and Ghent, with their beautiful old buildings, or in the cheerful, handsome capital, Brussels, is pleasant and profitable to many delicate travellers, while on its coasts the resorts of Ostend and Blankenberghe are attractive to those who seek a sea-climate amidst gay surroundings.

Ostend is a seaside-resort with a magnificent beach, where good sea-bathing may be enjoyed. The climate is stimulating. The popularity and gaiety of Ostend, however, make it rather expensive and less valuable to most health-seekers.

Blankenberghe, to the east of Ostend, has an even finer beach and a better seaside promenade; but it has become, of late years, almost as gay and expensive a resort.

Spa, situated in Belgium, but lying close to the German frontier, is a very favorably known resort, with a population of 7000 or more. It possesses iron springs to the number of sixteen, the principal one being called " Pouhon". An analysis of sixteen fluid ounces of the water of this spring shows, according to Dr. Charteris, four-fifths of a grain of iron, salts of sodium, calcium, and magnesium, and eight cubic inches of carbonic acid gas. Spa has an elevation of over 1000 feet, but the climate is somewhat enervating.

The Channel Islands,

which still belong to England and are the last relics of the Norman possessions, are situated in the Great Bay of St. Michael. They have a warm winter climate, with a fine but humid air, and are somewhat subject to winds. The accommodations in Jersey and Guernsey are good and the rates of living are low.

Jersey, the largest of these islands, lies thirteen miles northwest of France and thirty-five miles south of England. Its two chief towns, St. Aubin and St. Heliers, are much frequented by invalids.

France.

It may be said of the French seaside-resorts that, generally speaking, the climate is drier and sunnier than that of English resorts of the same character, and much time and money have been spent in developing their natural attractions and adding artificial ones.

Boulogne-sur-Mer, a favorite watering-place, with all the attractions that a good beach, a fine casino and bath-houses, a good theatre, and the diversions of a resort of the gayer sort can offer, is, as might be expected, rather more expensive than other watering-places on the coast, though very good accommodations may be had at a moderate price. The climate is comparatively dry and bracing.

Dieppe is very similar in climate and resources to Boulogne. It has in its castle and cathedral interesting Norman relics, and in its neighborhood is a very attractive country-side, with many picturesque churches and ruins to be found among its pretty lanes and orchards.

Etretât, lying between Fécamp and Hâvre, has a different aspect from the other resorts. It is situated between two fine cliffs which attain a height of 270 feet, and the scenery is very picturesque. It is a favorite resort for artists and authors, whose pictures and writings first brought it into notice. It has now excellent accommodations and a small casino, and many pleasant excursions may be made through the beautiful country surrounding the village.

Trouville, originally a little fishing-village, has become one of the best known of watering-places, the resort of all the fashion and wealth of Paris. The season is during the summer months. The fine, sandy beach is divided into three parts, the left being assigned to the use of ladies only, the centre to that of ladies and gentlemen, and the right for the use of men exclusively. The accommodations are, of course, all that can be desired, and, pre-eminent in its line, stands forth the magnificent Hôtel des Roches-Noires. Trouville has also a splendid casino and a good hydropathic establishment.

Granville. The coast-line turns southward at Cap de la Hague, and near here lies this seaport-town, which has a fine, smooth beach and good bathing.

St. Malo, situated on a peninsula, is connected with the main land by a causeway. There is a good beach, fine sea-bathing, and the location is adjacent to a beautiful and interesting section of country. There is a casino at St. Malo, and, during the summer months, diversions of various kinds are provided for the amusement of visitors.

Dinard lies on the opposite side of the Rance from St. Malo, and a steam-ferry running between them leaves and arrives hourly. There is good sea-bathing here. It is a distance of only ten or eleven miles to the town of Dinan, which lies amidst the finest scenic surroundings in Brittany.

On the coast of Brittany, surrounded by fine scenery, are **Paimpol, Tréquier,** and **Roscoff.** They lie, however, too far out of the regular tourists' route to be much frequented. Roscoff has a climate noted for equability.

Douarnenez, Audierne, Concarneau, Le Croisic, Pornic, and **Royan** are towns, some of which are engaged in the sardine industry and all of which afford good sea-bathing. The coast near Audierne is more exposed to the Atlantic gales; consequently it is more frequently visited by storms than the majority of resorts here. Concarneau has a large aquarium and marine laboratory.

Arcachon (latitude, 44° 7' north), thirty miles southeast of Bordeaux and ten miles from the coast, is situated on the shores of a salt-water lake, which is open only to the north, being sheltered on the other three sides by sand-hills covered with pine-woods. It is considered to resemble Bournemouth. The Bassin d'Arcachon, a land-locked inlet from the Bay of Biscay, communicates with the sea by a channel two miles long. It is thus perfectly protected from the Atlantic rollers, and, as it has a gently sloping beach, the bathing is much safer than at Biarritz.

Biarritz is located on high cliffs overlooking the Bay of Biscay, five miles southwest of Bayonne and in the same latitude as Pau, 43° north. The situation is much exposed to gales blowing off the Atlantic. Biarritz has a climate resembling that of places on the English Channel, but warmer. It has a fine beach and surf-bathing. There are excellent hotels and a good casino, where many amusements are provided for visitors, and the neighborhood, owing to the proximity of the Pyrenees, abounds in interesting and pleasant walks and excursions. The winter season extends from November to March, when the warm weather commences. The coast is rocky and picturesque. Bennet did not regard the climate as equal to that of the Riviera, but it is bright and exhilarating, both bracing and sedative in its effects. Yeo, however, thinks it too blustering and humid for the majority of chest disorders. The mean winter-temperature is 41° to 46°; spring, 51° to 53°; summer, about 64°. The relative humidity for the year is 80 per cent. Annual rainfall about 49 inches. A resident physician, quoted by Dr. Bennet, stated that for three years the average rainfall from October to April was 25 inches on 76 days. The soil is absorbent and dries quickly.

St. Jean de Luz, located on the shore of a bay a little to the southward of Biarritz and near the westernmost spurs of the Pyrenees, has a situation sheltered by the surrounding hills from winds blowing from the northeast and southwest; but the social atmosphere is duller than that of other Pyrenean resorts, nor are the accommodations so good.

INLAND RESORTS.

It would be entirely unnecessary, and within the limits of this book manifestly impossible, to give full descriptions of all the inland resorts and watering-places in which France abounds; but it will be well to mention a certain number of them to serve as an indication in making a choice.

Vichy, reached from Paris in eight and a half hours and possessing excellent hotels, is the most popular of the French spas. The situation is amidst rather uninteresting country, but everything has been done to render the place attractive to visitors. There is a magnificent casino, where are given balls, concerts, and good dramatic performances, and the arrangements for taking the waters are unusually perfect. The springs at Vichy may be divided into two groups, plain alkaline waters and alkaline iron waters. Bicarbonate of soda predominates in all the Vichy springs, and is by far the chief ingredient in the plain alkaline waters. Some of the springs are cold, but the greater number are hot.

The climate is said to be much like that of Paris. Fogs and thunderstorms are, however, rather frequent.

ANALYSES OF THE THREE CHIEF SPRINGS AT VICHY.[1]

	La Grande Grille. Grammes.	L'Hôpital. Grammes.	Les Célestins. Grammes.
Bicarbonate of soda,	4.883	5.029	5.103
Bicarbonate of potash,	0.352	0.440	0.315
Bicarbonate of magnesia,	0.303	0.200	0.328
Bicarbonate of strontia,	0.003	0.005	0.005
Bicarbonate of lime,	0 434	0.570	0.462
Bicarbonate of oxide of iron,	0.004	0.004	0.004
Sulphate of soda,	0.291	0.291	0.291
Phosphate of soda,	0.130	0.046	0.091
Arseniate of soda,	0.002	0.002	0.002
Chloride of sodium,	0.534	0.518	0.534
Silicic acid,	0.070	0.050	0.060
	7.914	8.222	8.244
	Litre.	Litre.	Litre.
Free carbonic acid,	0.908	1.067	1.049

The amount of common salt (chloride of sodium) gives the waters their pungent taste.

Royat-les-Bains, situated a little to the southwest of central France, in the department of Puy de Dôme, has an elevation of 1380 feet above sea-level. The waters contain some lithia and a little arsenic. There is, as is usual at French spas, every possible arrangement for using the waters, and Royat has a hydropathic establishment, so that patients may go through a regular water-cure. The use of the water is prescribed for gouty and rheumatic affections and glandular enlargements. The soil on which Royat stands is volcanic and porous.

As to the climate, sudden storms of wind and rain are not rare, but the atmosphere is dry, and there is much sunshine.

Mont Doré-les-Bains lies in central France, at an elevation of 3400 feet above the sea, and is also within the boundaries of the Puy de Dôme. The situation is good, the surroundings are beautiful and picturesque, and many interesting excursions may be made through the neighborhood, but the accommodations at Mont Doré by no means equal those at Royat. There is a large établissement des bains and an établissement des vapeurs, also extensive. The springs at this spa are, to quote Dr. Yeo, but "feebly mineralized"; the waters are alkaline, and the main ingredients are bicarbonates of soda and potash and chloride of sodium. They have a tempera-

---

[1] Yeo's Climate and Health-resorts.

ture of 107° to 115° F., are used for both drinking and bathing, and are taken in cases of lumbago, sciatica, intercostal neuralgia, rheumatism, and throat affections.

La Bourboule lies at a distance of five miles from Mont Doré, and has an elevation of 2850 feet. The waters, which have a temperature of 140° F., are saline, effervescent, and arsenical, and are efficacious in cases of chronic skin-diseases and in rheumatism. They are used for drinking and bathing. There are the usual establishments for the application of the waters, a large swimming-bath and a fine casino, and many delightful excursions may be taken from here through the beautiful surrounding country.

Aix-les-Bains, in the province of Savoy, has an elevation of 850 feet, and is prettily situated. The climate is mild and equable, though the summer heat is sometimes great. The springs are sulphurous, the chief element being sulphuretted hydrogen, and their temperature ranges from 110° to 115° F. The waters are used for both drinking and bathing, and for the latter mode of application the arrangements are very perfect. There are also rooms for inhaling the sulphurous vapors and the spray of the waters. The season is during the summer, but the thermal établissement is open during the entire year. The place is resorted to by sufferers from rheumatism and gout and from some forms of chronic catarrh. The hotels are excellent, and there is a fine casino. It enjoys a great vogue with the fashionable world, and may perhaps be regarded to-day as the most important of the French spas.

ANALYSES OF TWO SPRINGS AT AIX-LES-BAINS ACCORDING TO VINTRAS AND BRACHET.[1]

	Source de Soufre.		Source d'Alun ou de St. Paul.	
	Vintras.	Brachet.	Vintras.	Brachet.
Calcic carbonate,	0.1485	0.1894	0.18100	0.1623
Magnesic carbonate,	0.02587	0.0105	0.01980	0.0196
Ferric carbonate,	0.00886	0.0010	0.00936	0.0008
Calcic sulphate,	0.01600	0.0928	0.01500	0.0810
Magnesic sulphate,	0.03527	0.0735	0.03100	0.0493
Sodic sulphate,	0.09602	0.0327	0.09240	0.0545
Aluminic sulphate,	0.05480	0.0081	0.06200	0.0003
Sodic chloride,	0.00792	0.0300	0.01400	0.0274

Silica, calcic phosphate in minute quantities, and traces of lithium and iodine; also some organic matter (barégine).

[1] Yeo's Climate and Health-resorts.

Pau (latitude, 43° 17' north; elevation, 620 feet), an inland town, noted for the stillness of the air, where "a plentiful amount of rain falls perpendicularly." It is finely situated on a high ridge overlooking the chain of the Pyrenees—twenty-five miles distant. The mean winter-temperature is 42° F. The annual amount of rainfall is 43 inches. There are 119 rainy days. The stormy season is in September, October, and November. The climate is cold and sedative. There are beautiful promenades and good accommodations for visitors.

Barèges, lying at an elevation of 4000 feet above sea-level, is not well situated, but is rich in mineral springs. There are twelve of these, ranging in temperature from 88° to 113° F., and the waters are used for both drinking and bathing. Its sulphur waters, among the strongest existing in the Pyrenees, abound in a nitrogenous substance to which has been given the name of barégine. There is a hospital for soldiers, and one for nuns and priests at Barèges. The climate is very variable, and the place is subject to cold mists and strong winds.

Cauterets is not fashionable like Luchon, but is much patronized by invalids who intend to make a business of getting well. It has an elevation of 3000 feet, but the climate is rainy, subject to sudden changes, and, on account of the very sheltered situation of the place, not particularly bracing. Entirely surrounded by lofty mountains, it is apt to be hot and sultry in the middle of the day, though the mornings and evenings are usually fresh and cool. There are many mineral springs at Cauterets, used for drinking and bathing, and the bathing-establishments have every appliance for the use of the waters, which are of the sulphurous saline variety and have a temperature ranging from 55° to 145° F.

Bagnères-de-Luchon, lying in the Pyrenees, near the Spanish frontier, has an elevation of 1900 feet. There are about fifty sulphur springs here and some chalybeate waters. It is a very much patronized resort, and has most excellent bathing-establishments. In fact, the features of Luchon are beauty and luxury. The waters are taken for skin-diseases, scrofula, and rheumatism. It has a mild climate, but during the months of May, June, and September the air is quite fresh and bracing.

Spain

is generally too warm for summer, and the accommodations and methods of travel do not make it attractive to the delicate or refined tourist. There are, however, some resorts in the Pyrenees and some on the Mediterranean coast which possess good climates and certain advantages. Of the latter, Malaga is the most frequented.

San Sebastian and **Gijon** lie on the northeastern coast. San Sebastian is used as a winter-resort, and the situation is both beautiful and protected. Because of the latter characteristic and of its mild climate it might be a suitable resort for patients who have lung-troubles, but in the absence of exact information it is impossible to make positive assertions.

Malaga is one of the mildest climates on the Spanish shores of the Mediterranean. It is well sheltered, warm, moderately moist, and slightly stimulating. The mean winter-temperature is 56° F. The heat becomes intense early in the spring. Rainfall for the year, 16½ inches on 40 rainy days. Dr. C. T. Williams refers to this climate as being warmer and "drier" than the Riviera, and it has apparently a smaller daily and seasonal range.[1] The great drawback has always been the lack of suitable accommodations for invalids, especially in the shape of suburban villas and hotels away from the densely populated and poorly drained town. It is picturesque and bright.

Panticosa, a Spanish spa lying on the southern slopes of the Pyrenean range, at an elevation of about 5000 feet, is said to be very suitable to phthisical cases. It possesses mineral waters of no great strength but containing a considerable amount of nitrogen and of sulphuretted hydrogen, and having a temperature of 77° to 91.4° F. It is customary to drink from twenty to thirty glasses of this water daily. Little exact information in regard to this resort can be obtained.

Seville, interesting as a city, likewise offers to the health-seeker the advantages to be derived from a warm, sunny, usually dry climate. Ice and snow are said never to exist here. The hotel-accommodation is good.

Portugal.

Among the cities of Portugal the two large commercial ports, Oporto and Lisbon, on the shores of the Atlantic, are somewhat used as health-resorts by those whose chief desideratum in a climate

[1] Aërotherapeutics.

is mildness, but the country generally, like that of Spain, is lacking in resources for the comfort of foreigners.

The Riviera.

Certain parts of France and Italy have usually been treated under the comprehensive name of the Riviera, and both countries possess resorts characterized by the climatic conditions peculiar to this district. It has, therefore, seemed better to the author to follow the established custom of treating such resorts under this head than to describe each place with others in the country in which it belongs.

The temperature of the Riviera during the winter is not usually more than 8° or 10° F. warmer than the south coast of England, but it is less moist and more stimulating, and compared with the English coast the Riviera has half the number of rainy days and four or five times the number of bright ones.

Dr. Williams[1] has well and accurately described this region. He refers to the entire coast as having a mean winter-temperature of 50° or 51° F., the minimum of 42° to 46° being reached in December. Occasionally it sinks to the freezing-point, and snow sometimes falls. A feature of the climate is the rapid fall of temperature after sunset, especially at any distance from the sea. The relative humidity varies from 61 to 74 per cent. The rainfall is generally 31 inches. A rainy season occurs in October and November and again during the latter half of March. The rainfall increases eastward on the Riviera and diminishes westward. The principal winds are the northwest or "mistral," a steady, cold, and dry wind, prevailing chiefly in March and much dreaded, although it usually brings fine weather ; the northeast or "bisé," a cold wind ; and the southeast or "sirocco," a warm, enervating one. The westerly winds are dry and the easterly moist, the opposite of what prevails in Great Britain and Ireland. The amount of lime in the soil renders the dust in the more windy places, like Hyères and Nice, quite objectionable. The winter climate of the Riviera is clear and bright, with a good deal of wind, but devoid of fog or mist. The influence of the sea-air in rendering the air moist is not usually strongly marked, on account of the prevalence of land-winds.

Dr. J. Henry Bennet warns the visitor to the Riviera not to expect an absence of winter. It is only a retreat affording a certain

---

[1] Aërotherapeutics; being the Lumleian Lectures for 1893. C. T. Williams, 1894.

local protection in the warmer temperate zone. He says: "The descriptions of the winter climate of Nice, Cannes, Hyères, and Italy in general contained in most books of travel, works on climate, and guide-books, are mere poetical delusions. The perpetual spring, the eternal summer, the warm southern, balmy atmosphere, described to the reader in such glowing terms, only exist in the imagination of the writers. Although there is so much sunshine, so much fine weather, such immunity from fog and drizzling rain, it is still the continent of Europe, with ice and snow behind for more than a thousand miles, to the North Pole. It is still winter. Wind, rain, a chilly atmosphere, and occasional cold weather, with snow on the mountains and flakes of ice in exposed situations, have to be encountered. . . . The existence of orange and lemon trees, of geraniums, heliotropes, verbenas, and roses, flowering throughout the winter, does not necessarily imply the absence of cold weather, but merely the absence of absolute frost."[1]

The French Riviera.

Hyères is situated three miles from the sea, which makes it somewhat less stimulating to nervous people. It has a climate said to be warmer than that of Nice. The minimum (or night) temperature for Hyères is 29° and for Nice 27° F.[2] The mean midday temperature for Hyères for winter is 52° and for spring 63°. A record for twenty-six years, quoted by Dr. C. E. Cormack, of Hyères, shows the average number of rainy days for the eight months—October to May—to be 36 days as against Nice 51, Cannes 54, and Mentone 64 days. The average number of rainy days for Hyères for winter is 14 and for spring 11. The usual number of stormy days for each month may be considered as 4. The wettest months are October and December and the finest month March. Much of the rainfall occurs at night. The greatest record of rainfall in one year is 52 inches and the smallest 15 inches—a mean of 33½ inches. Snow falls perhaps once in three years, but seldom remains more than an hour or two. There are occasionally light frosts at night. The prevailing winds at Hyères in the winter are from the north, northeast, and northwest (the mistral), while in the spring they are from the east, southeast, or northeast.

[1] For comparison, see Santa Barbara; Table XIV.; and Tables V.-X.
[2] The French Riviera. Charles Cormack, in the Climatologist, January, 1892.

The winter-winds are sometimes quite violent. Hyères is favored with good hotels.

Cannes is prettily situated in a recess of the Gulf of Napoule. It is sheltered to the north and northeast, and to some extent to the northwest. On the east and south it is exposed. It has an annual mean temperature of 60° F., with an average for the winter months of 50° and for the spring of 62°. January is the coldest month. The prevailing wind is from the east and southeast. The annual rainfall is 25 inches. The annual number of rainy days in Cannes is 54. It is a fashionable resort, with superior accommodations.

Nice (population, 90,000) has an annual mean temperature of 59° F. The mean for winter is 47° and for spring 56°. The relative humidity during the winter is 66 per cent., and in the spring 68 per cent. The annual rainfall is 32 inches on 60 rainy days. Nice is a southern capital, with its Italian opera and French theatre, its daily fashionable promenade and drive, its military band, and its swarm of gaily dressed people.[1] It is comfortably convenient to Monte Carlo. The city is situated on the verge of a valley, surrounded on three sides by the Maritime Alps and open on the other, or south side, to the Mediterranean. Dr. Madden, of Dublin, refers to the climate as warm, dry, tonic, and exciting.[2]

The average mean temperature is a little higher than at Cannes.

Mentone, twenty-three miles from Nice, can be reached by express from Paris in twenty-two hours. It has a warm, equable climate, and is exempt from harsh, cold winds. It is one of the least dry resorts on the Riviera, because so much of the dry land-wind is kept away by the neighboring elevations. It has an annual mean temperature of 60° F.; a temperature for winter of 49° and for spring of 60°. The relative humidity for the year averages 72 per cent. The Mentone amphitheatre is better protected than other portions of the Riviera by mountains, and to this is due the locally warm climate. Its lemon-groves are only equalled in localities much further south. The nights are usually chilly, the thermometer reading 46° to 54° F. with south winds, and with winds from the north, which are drier, ranging from 40° to 45°, or even lower than 40° between December and April. In the daytime the shade-temperature is 50° to 56° when the sun shines. For invalids north

---

[1] Winter in the South of Europe. J. Henry Bennet, London, 1865.

For comparison with the coast of Southern California, see Santa Barbara.

rooms are to be avoided, as they are 4° to 8° colder than south rooms. The rainy days come between October and May. Dr. Bennet notes 29 rainy days during the four months December to March. But little rain falls during the summer. The amount of annual rainfall is not obtainable, but it is undoubtedly greater than at Nice, as there are at Mentone during the year an average of 80 rainy days. The severest rains usually come during the months of October, November, and March.

The water-supply for domestic use should be boiled as a necessary safeguard.

Hotel charges and the cost of housekeeping in Mentone are high.

The Italian Riviera.

Bordighera, the first resort beyond the Italian boundary, is sunny, but more exposed than the eastern part of Mentone. It has a fine growth of palm-trees.

San Remo is fifteen miles further from Nice than Mentone, and is less picturesque, but has a similar climate. The town is larger than Mentone and thoroughly Italian. It faces southwest on a beautiful bay. It is well protected by encircling mountains from the most severe north and east winds. The season is from November 1st to April 30th. The mean shade-temperature for that period for eleven years is reported by Dr. Hassall to have been 53° F. For the three months of winter it was 48°. Freezing-point was usually reached once during each season. The mean seasonal relative humidity was 70 per cent. The average seasonal rainfall was 16 inches, distributed over 34 rainy days. Since the completion of the new aqueduct the drinking-water of San Remo is considered excellent.

Along the entire Riviera, from Hyères to Spezia, are a number of small towns with various natural advantages, in some cases as great as the more famous resorts, although the shelter of the mountains is less complete at the eastern end.[1]

Genoa, with a climate neither warm nor equable and subject to frequent rains and winds, can in no sense be characterized as a health-resort, but it is in many ways an interesting town, and is the terminus of a voyage taken by means of the North German Lloyd steamers from New York across the Atlantic and through Mediter-

[1] Dr. Rohden on Climatic Health-resorts, in Braun's work on Balneology.

ranean waters, which is most delightful and which may be extremely beneficial. The ships are comfortable and well-found, so that the daily life of the passengers is under favorable auspices, and the conditions are very suitable to those who are suffering from overwork or to invalids whose general condition is good, especially if sea-sickness is not greatly to be feared.

Pegli and Nervi, near Genoa, are quiet, well-sheltered places, and have good hotels; but too little attention has as yet been given to the requirements of invalid life to recommend them highly for the more delicate class of patients. Nervi has, during the months from November to April, a rainfall of 25½ inches, distributed through fifty-four or more rainy days.

Spezia, a town of about 20,000 inhabitants, has a mild climate, rather moist and fairly equable, and the air is free from dust. There is every facility for sailing, boating, and riding. The town is situated upon a gulf, the west coast of which is rugged; but the northern and eastern borders are lower and allow of walking and driving. The hotels are spacious and comfortable.

Islands of the Mediterranean.

Corsica is about 115 miles in length and 54 miles wide at its broadest point. It is 90 miles from the coast of France and about 54 miles from Italy. The island is a mass of Alpine mountain-ridges rising out of the sea. In the centre the principal range rises to a height of 8000 or 9000 feet. The eastern coast is marshy and unhealthy, but the climate of the western coast is healthy. The season extends from the last of October to June.

Ajaccio (latitude, 41° 55′ north) has some repute as a winter-resort, and is a "clean and smiling little French town"; it is situated on the northwest side of a picturesque bay. Twenty miles inland is a semicircle of mountains rising to 6000 and 9000 feet, the highest peaks of which are covered with snow even in summer. There is a marked absence of the strong winds that usually prevail on the Mediterranean in winter, although there is a regular sea-breeze during the day and a land-breeze at night.

Dr. J. Henry Bennet says: "The vegetation of Ajaccio and the neighborhood indicates a climate at least as warm as that of Nice, perhaps even a shade warmer. The olive, the orange, the prickly pear, thrive with great luxuriance. The lemon grows also, and bears fruit outdoors, but only, as in Nice, in sheltered and protected spots."[1]

Dr. Bennet considers Ajaccio as the best town on the island for a winter-residence.

There are mineral baths at Guagno, twenty miles northeast of Ajaccio, and the water of a fine sparkling chalybeate spring at Orezza, thirty miles south of Bastia, is bottled for the market.

Dr. Hermann Weber, of London, spent the winter of 1895 in Ajaccio, and, in a private letter, states that he considers it a good place for invalids requiring a moderately warm winter, as it possesses a climate less dry and exciting than that of the Riviera. There are pleasant walks among the olive-trees and pines. There is a good hotel. Steamers go from Nice and Marseilles in from fourteen to twenty hours. They are small boats, and far from comfortable in rough weather.

Corfu. This island is reached by steamer from Brindisi or Trieste and lies near the coast of Albania, from which it is separated by the strait known as the Channel of Corfu. It is picturesque and interesting, is traversed by very good roads, and affords many pleasant excursions, and the hotel-accommodations are good. The winter climate is not equable, nor is the place free from fogs and cold winds, though the wind from the southeast is said to be unpleasantly enervating, but there is little dust. The mean temperature for January and February is 50° F. or thereabout, and for April it is 60°, and the mean relative humidity ranges between 70 and 80 per cent. The rainfall is large, and there are from November to April about 72 rainy days.

Malta (latitude, 35° 54' north) cannot be said to present many advantages to the health-seeker. Its appearance is barren, its accommodations (outside the two hotels in Valetta and its suburb, Sliema) are poor, the possibilities for the daily life of an invalid limited and unsuitable, and the climate, while equable, sunny, and moderately dry, is relaxing. It is true that, as it has nowhere a greater elevation than 600 feet, the great exposure of Malta to winds may somewhat counteract this relaxing tendency; but the remedy, like the disease, has its disadvantages. Fever appears to be rather prevalent. Of course, here as elsewhere, there is a great difference

---

[1] Winter in the South of Europe. J. Henry Bennet, M.D.

between seasons, and during a fine autumn the author found Malta very pleasant.

The climate is remarkably equable, the night and day temperatures varying little more than 5° in winter and only 8° in summer.

The mean annual temperature is 65.95° F. Winter-temperature, 55.99°; spring, 61.18°; summer, 76.17°; autumn, 70.44°.

The annual rainfall is 23.93 inches, of which two-thirds fall in the winter. The influence of the sirocco is sometimes felt, and it is apt to be windy.

A short stay at Malta during the winter may be beneficial to an invalid of the robuster sort, and the place affords all the gaiety and amusement consequent upon its being an important garrison-town.

Sicily. The chief town of Sicily, Palermo, possesses a climate not subject to sudden change. The mean winter-temperature is about 54° F., but the winter climate is moist as well as warm. The city is handsomely built, and its situation, opposite the bay and surrounded by mountains, is incomparably fine. According to Dr. Yeo, Palermo, as compared with Nice, has 131 rainy days to 60; but the rainfall, on the contrary, is only 21 inches to 25 inches at Nice. The African sirocco occasionally blows over the island, and, while its warmth probably serves to raise the temperature, its effects are bad, rendering the air dry and irritating.

Catania and **Aci Reale**, which have also been recommended as health-resorts in Sicily, need but a short notice. The climates are rather similar. The mean winter-temperature of Catania is 53.8° F. and the daily range in temperature is about 14.5°. The temperature at night is much lowered, owing, it is said, to the snow-fields of Mount Etna, the town being situated at the foot of the volcano. Aci Reale has mineral springs and a fine bathing-establishment, and its elevation is greater than that of Catania, affording a fresher, purer air. The hotels in both towns are said to be good.

Italy.

COAST RESORTS.

Sorrento is delightfully situated opposite the beautiful island of Capri. In the spring it is resorted to by pulmonary invalids, and in May and June there is sea-bathing.

Castellamare di Stabia, situated on the south shore of the Bay of Naples, is used as a health-resort during the fall and winter seasons. It possesses saline waters, containing also a small quantity of iron. On account of its climate it is resorted to by phthisical invalids, and its mineral waters are said to be useful in cases of calculi, disorders of the liver accompanied by dropsy, and so forth.

Ischia, an island in the Bay of Naples, has alkali-saline waters, whose temperature is 145° F. There were formerly extensive bathing-arrangements, but many of the establishments were destroyed by the earthquake of 1883.

Naples has a dry, mild winter climate, and is still a little used as a resort at that season. It possesses many different kinds of mineral springs. The city, though dirty, is attractive, and the surrounding scenery is most beautiful. The hotels are good and excursions through the environs are very interesting.

Inland Resorts.

Rome is too well known to need more than the briefest mention. As a health-resort its reputation, which was at one time great, is on the wane. The mean winter-temperature is from 45° to 50° F., but the air is not so dry as that of the Riviera. Dr. Charteris very sensibly says that medical advice should always be taken before deciding on winter-quarters in Rome because of the prevalence of fevers in certain localities.

Lucca, in Tuscany, is situated on a hillside with good shade-trees about it. There are sulphate of lime waters with a temperature ranging from 70° to 125° F., which are said to be efficacious in cases of rheumatism and gout and in skin-diseases. There are adequate arrangements for receiving and caring for invalids, and the climate is agreeable.

The Italian Lakes. On the shores of these lakes one finds charming, cool, moist summer climates, with beautiful scenery and good accommodations. The climate is much influenced by the proximity of the Alps, which render them too hot in summer and too windy in winter; but there are several places whose chief recommendation to invalids is that they make good resting-places in the spring and autumn for those who are journeying to or from the winter- or summer-resorts of Italy, the Riviera, and Switzerland. Among them may be mentioned :

Lugano, on Lake Lugano (elevation, 900 feet). Pretty situation and good accommodations.

Pallanza, on Lake Maggiore (elevation, 600 feet). Good climate and capital arrangements for the reception and care of invalids. The hotel is open during the winter months.

Cadenabbia. Good place to avoid the ill-effects of the seasonal changes of spring and fall.

Bellagio, on Lake Como, one of the most noted of these resorts, is, during the summer and autumn, a favorite, but not a well-chosen spot for those who suffer from pulmonary troubles. It is beautifully situated and has a delightful climate.

Santa Caterina, lying 5700 feet above sea-level, has a most charming situation. It is on the southern slopes of the Alps, and its climate is, therefore, less stimulating than that of the Engadine resorts. It possesses, however, a strong iron spring, and this, combined with the elevation of the resort, makes it particularly suitable for anæmic patients.

Austria.

Trieste, on the shore of the Adriatic, beside being a large seaport-town, is used during the summer months by those who wish to enjoy good bathing on a fine beach. The hotels are good and the prices of living by no means exorbitant.

Teplitz, one of the oldest of spas, has been resorted to for over 1100 years. It is situated in a good-sized valley, surrounded by granite hills, and may be reached from Carlsbad by rail in three hours. It has a thermal spring, the waters having a temperature of from 98° to 120° F. They are used chiefly for bathing, and are supposed to be good for all troubles resulting from gout.

Carlsbad, although it lies in the valley on both sides of the river Tepl, has an elevation of 1200 feet above the sea. The river is small and the valley narrow, and the lodgings on the hillsides are therefore most desirable as being cooler and more breezy than those lower down. The hotels and lodging-houses have no *table d'hote*, and it is customary to take one's meals at the restaurants, where the food is under strict medical supervision. There is a fine Kurhaus, with daily concerts and theatrical performances, but, as is the case at most of the gayer resorts, the rates of living during the season are rather expensive. The waters have a temperature of 167° F.; the chief ingredient is soda, and they are highly charged with carbonic acid. They are peculiarly efficacious in cases of constipation, intestinal catarrh, hepatic and splenic enlargement, diabetes, etc. A rigorous diet is imposed upon patients using the waters. Below is given a table, with analyses of three of the principal springs. As Carlsbad is one of the most famous of European spas, it goes without saying that the bathing-arrangements are complete and the accommodations good.

ANALYSES OF THREE OF THE PRINCIPAL SPRINGS AT CARLSBAD.[1]

	Sprudel. In 16 oz.	Mühlbrunnen. In 16 oz.	Schlossbrnn. In 16 oz.
Sulphate of soda .	18.21 gr.	17.96 gr.	17.24 gr.
Sulphate of potash .	1.26	1.71	1.46
Chloride of sodium	7.91	7.89	7.52
Carbonate of soda .	10.45	10.86	9.66
Carbonate of lime .	2.28	2.02	3.06
Carbonate of magnesia .	0.95	0.26	0.38
Carbonate of protoxide } of iron . }	0.02	0.02	0.01
Carbonic acid .	11.80 cub. in.	14.80 cub. in.	20.60 cub. in.
Temperature .	. 164.2°	125.6°	124.7°

Marienbad, lying at an elevation of 2000 feet, is in Bohemia, and is reached over the railroad running from Eger to Pilsen. Its waters closely resemble those of Carlsbad, but they are cold. They are prescribed for obesity, lack of digestive power, constipation, chronic dysentery, etc. The principal well is located at the upper end of the town, and is called the Kreuzbrunnen. Here, as at Carlsbad, there is a regular diet for those partaking of the waters.

Luhatschowitz, in Moravia, at an altitude of 600 feet, in the midst of interesting and attractive surroundings, possesses alkaline, iodine, and bromine waters, which are the strongest of their kind known. They are considered good in uterine diseases and scrofula. The milk-cure is also administered here.

At **Baden,** which may be reached from Vienna in an hour, are found saline sulphurous springs having temperatures ranging from 82° to 95° F. There are arrangements for various kinds of baths. Baden is beautifully situated on a mountain-slope and is surrounded by pine- and beach-forests, and many delightful and interesting excursions may be taken in its vicinity. The resort is a fashionable one and the accommodations are good.

[1] Yeo's Climate and Health-resorts.

Voslau, near Vienna, has an altitude of 700 feet, and is delightfully placed among hills whose sides are covered with vineyards. The climate is healthy and the air fresh and pure. There are waters at Voslau whose temperature is 75° F., and good facilities for brine- and pine-baths. The whey- and grape-cures are given at Voslau.

Ischl, situated in Upper Austria, at an altitude of 1600 feet, is reached by rail to Gmunden, and has a population of 9000. Its position at the meeting-point of three valleys, with an extensive outlook over cultivated country, is beautiful, and its climate, refreshing and agreeable, is another attraction. The thermometer rarely rises above 86° F. as a midsummer temperature. The season begins in May and lasts until October. The waters, which are saline and sulphurous, are reputed to be efficacious in nervous affections, chronic bronchitis, etc.; there are good bathing-appliances, including a large swimming-basin.

Wildbad-Gastein, situated near Salzburg, has springs whose waters range from 75° to 170° F. in temperature. They are said to be good in low nervous troubles and in gout and rheumatism. Gastein has an altitude of about 3000 feet. Braun states that Gastein resembles Teplitz in everything but its climatic conditions, which make it particularly suitable to cases complicated by a much weakened or lowered constitution and vitality, requiring, therefore, great cautiousness in treatment.

The highest temperature which has been observed here is 86° F., and the climate is not subject to sudden or extreme changes of temperature.

Bregenz, on the east shore of Lake Constance in the Tyrol, has an elevation of 1900 feet. It is a popular summer-resort, and its advocates claim that it is especially good for patients suffering from chest-troubles. There are pine- and beech-woods near Bregenz.

Meran, situated in the Austrian Tyrol, on the southern slopes of the Alps, has an elevation of 1050 feet. Except to the southward, it is surrounded by mountains which rise as high as 10,000 feet, and its situation is consequently both sheltered and picturesque. It may be reached by diligence from Tarasp or by rail from Innsbruck *via* Botzen.

Having an open southern aspect, Meran is exposed to occasional fierce winds which blow from that direction, but the climate is usually sunny, still, dry, and not too warm. There is an average of 7 snowy days during the winter and of 52 rainy days during the entire year, only 13 of which belong to the winter season.[1] Mists rarely settle in the valley. The medium winter-temperature is 41.67° F., showing a climate somewhat warmer than that of Montreux, but decidedly colder than that of Torquay, whose winter-records show a mean of 44.6°. Monthly means: September, 62.6°; October, 55°; November, 42°; December, 35.4°; January, 32.6°; February, 38.1°; March, 46°; April, 54.7°. This cold is not, however, depressing to the vitality because of the dryness of the atmosphere. The humidity-records are as follows: January, 80 per cent.; April, medium, 67 per cent.; minimum, 41 per cent.; yearly mean, 67 per cent.; rainfall from September to December, 11.77 inches, and from January to April, 4.33 inches.

Every attention is given to making Meran as attractive as possible to visitors. There is a fine Kurhaus, located on one bank of the river Passer, and containing reading- and smoking-rooms, reception- and dining-rooms.

Beside the grape-cure, which is the especial feature of Meran, there is the whey-cure, administered during the spring, and medicated baths and treatment by variations of barometric pressure in a pneumatic chamber may be taken at all seasons. Speaking of Meran as a winter health-resort, Dr. Yeo says: "The class of individuals to whom the climate of Meran seems best suited is those suffering from pulmonary disease, who find by experience that a dry and bracing climate suits them better than a warm, moist one, and who can bear a certain amount of cold in winter without being made uncomfortable by it."

Dr. St. Clair Thompson, who was at Meran in the year 1885, observes: "For a short stay it is of use to patients to break their journey in travelling either north or south, to avoid too sudden transitions of climate. To the large number of invalids with whom its climate agrees it offers the advantage of allowing a more prolonged stay than most health-resorts—a boon to many to whom travelling is unpleasant or impossible. With a summer visit to the neighboring mountain-villages, many spend the whole or greater part of the year here."

Hungary.

The resorts in Hungary are comparatively little known, and it is always hard to get statistical information from such a distance;

---

[1] See Yeo's Climate and Health-resorts.

but among the places which have come under the notice of medical men of this and other countries may be mentioned the following :

Bartfeld, in the northern part of the Theissan district, near the Galician border, has a good establishment for the application of the waters of the alkaline chalybeate springs found there.

Szobrancz, also in the Theissan district, but lying to the south-west of Bartfeld, is surrounded by beautiful and picturesque scenery, and possesses saline sulphurous waters.   This spa is very popular.

Herculesbad, called also Mehadia, lies in the extreme south-eastern part of Hungary, in a wild and mountainous region.   Here are found alkali-saline sulphurous waters, having a temperature of from 80° to 130° F., and a large bathing-establishment.

Baassen possesses saline-bromo-ioduretted waters, with a temperature of 50° F.   They are said to be good in cases of chronic uterine enlargement and in secondary syphilis.

Harkanyï is picturesquely situated, and has alkaline sulphurous springs whose waters have a temperature of 130° F.

Russia.

The resorts of this extensive portion of Europe are few in number and are rarely used by foreigners.

The greater part of the country is flat.   It is very cold in winter and hot in summer.   There is no district of Russia where the summer heat is not comparatively great, the thermometer rising to 86° F. or thereabout and descending in winter as low as 13° and even 22° below zero.   These temperatures have been noted both at Astrakhan, in the far south, and at Archangel, in the extreme north.   The range for the country generally averages 140°, but on the shores of the Black Sea it is reduced to 108°.   The summer climate of the Crimea is reported as pleasant and healthy, and there are said to be fairly cool summer-resorts on the coast of the Black Sea.   The rainfall throughout the country is small, from 16 to 23 inches for the year, and by far the greater precipitation is during summer.   The spring is warmer than the fall.   The effect of the Atlantic winds is, however, sufficiently felt to qualify the Continental character of the Russian climate.   These winds lessen the cold of the unusually long winters and add to the dampness in summer, at which season they also sometimes bring thunderstorms.

The steppes of Tartary, at least the barren portions of them, are characterized by a fine, dry, desert-air, and have an established reputation for the relief of consumption—the value of the climate being increased by the fact that these districts in many parts have considerable and varying elevation, from the sinks around the Caspian Sea to the higher benches approaching the mountain-ranges. The benefit derived from residence on these portions of the steppes is doubtless enhanced by the open tent-life and the drinking of koumiss, a fermented preparation of mare's milk made and used by the Tartars.

Germany.

Pyrmont, in the northern part of Germany, near Hanover, is delightfully situated in the Emmer valley, and has claims to considerable importance as a spa. It has two salt springs, the weaker one of which contains a good deal of carbonic acid. There is also an iron spring of some strength, more than is possessed by those at Schwalbach and St. Moritz; but it is by no means so agreeable to the taste, owing to the presence of a small amount of sulphate of magnesia. Pyrmont has a mild, healthy climate, and is resorted to by those who have chlorosis or anæmia or suffer from obesity or atonic dyspepsia.

ANALYSIS OF STEEL-SPRING AT PYRMONT.[1]

Bicarbonate of protoxide of iron	. 0.57
Bicarbonate of lime .	. 10.47
Sulphate of magnesia	. 3.88
Sulphate of lime .	. 9.05
Free carbonic acid .	. 29 cub. in.
Temperature .	. 55.8° F.

Aachen or **Aix-la-Chapelle** lies one hour's journey from Cologne. It has a rather damp climate. Here are found hot sulphur springs, the principal one being the Kaiserquelle, and the waters are thought to be specially good in cases of syphilis and useful in a less degree for gout and rheumatism. In sixteen ounces of the water from the Kaiserquelle were found 31.9 grains of solid and 26½ cubic inches of gaseous constituents.

Near Aachen is Burtscheid, where are warm sulphur springs, which are used both for drinking and bathing. The Mühlenbad Quelle, the waters of which are the hottest in Germany, has a temperature of 171.5° F.

---

[1] Braun's Curative Effects of Baths and Waters.

Johannisberg, lying on the river Rhine, is one of the places known as a station for the grape-cure. It has a hydropathic establishment, of which one of the features is a system of electro-therapeutics. The accommodations are good.

Neuenahr, in the valley of the Ahr, a small tributary of the Rhine, lies between the cities of Bonn and Coblenz. Between Remagen and Neuenahr is one hill, 900 feet in height, called the Landskron, and near this is located the Apollinaris Brunnen, the source from which comes that mineral table-water now so famous all the world over. The Landskron is, however, the one notable point in the scenery thus far, the surroundings being otherwise rather uninteresting. Above Neuenahr we come to the ancient and interesting town of Ahrweiler and the village of Altenahr, and through this district the views are most picturesque. The banks of the river are covered with terraced vineyards, the grapes from which are made into a red wine which is highly esteemed. Neuenahr has some not very powerful alkaline springs, charged to saturation with carbonic-acid gas. The waters have a temperature ranging from $72\frac{1}{2}°$ to 104° F. There is a creditable Kurhaus pleasantly situated, the mode of life is simple and wholesome, and the food is good. It is stated that the climate is dry, healthy, and equable, but that heavy mists are not infrequent in the morning. The town is somewhat warm, although there is usually a breeze through the valley.

Görbersdorf, in Silesia, situated at a height of 1700 feet above the sea, is well known as a resort for phthisical invalids and as the seat of Dr. Brehmer's sanitarium. This fine building is located near a hill covered with pine-woods, and has covered walks and a winter garden which is adequately heated. The grounds are peculiarly attractive. There is nothing especially beneficial in the climate for consumptives except its purity and the opportunities afforded for living out of doors.

Laubbach, not far from Coblenz, is a summer- and autumn-resort lying amidst beautiful scenery and said to be especially healthful for those who suffer from kidney-troubles. There is treatment by inhalations, and the milk- and grape-cures are also administered here. The climate is invigorating, the place amusing, and the charges moderate.

Falkenstein, another Rhenish health-resort, is important chiefly because of the establishment there by Dr. Dettweiler of a sanitarium for the treatment of consumptives, anæmic patients, and convalescents. The building, standing at an altitude of 1700 feet, is in a protected situation, being sheltered to the north by the Taunus Mountains and to the eastward and westward by lesser elevations. The most careful and firm supervision is maintained over every detail of an invalid's daily life, and the arrangements of the sanitarium are very complete. It is chiefly to these facts that Dr. Dettweiler attributes the undoubted excellence of his results. Hydrotherapy and the application of electricity are included in his treatment.

Wiesbaden, not far from Schwalbach, is one of the oldest watering-places in Germany, and is a town of considerable size, having over 50,000 inhabitants. The waters here have for their main solid ingredient common salt, and the temperature of the chief spring, the Koch Brunnen or boiling well, is 153° F. There are very good bathing-arrangements at Wiesbaden, and the milk- and whey-cures, and, in the autumn, the grape-cure, are also administered. The waters are particularly good for chronic rheumatism. Wiesbaden has first-rate hotels and numerous and good lodging-houses, and the rates of living are moderate. The magnificent Kursaal contains library and reading-rooms and restaurant, and here are given concerts, weekly balls, etc., and beautiful walks and drives may be taken through the surrounding country. Wiesbaden is used as a resort both in winter and summer, and its winter climate, although cold, is dry and sunny.

Schwalbach, situated at an elevation of almost 1000 feet, amidst a picturesque district in Nassau, is a small village lying at a distance of nine miles from Eltville, a Rhine station. The new part, where are the visitors' quarters, is built upon the sides of two valleys, which meet just where the new Kursaal is built. This is *the* iron-cure of Germany, and is especially prescribed in cases of anæmia and bloodlessness resulting from hemorrhage. The waters are used for both drinking and bathing, and contain, beside iron, some carbonic acid gas. The climate is stimulating. Many amusements are provided at Schwalbach for the diversion of visitors, and enjoyable excursions may be taken through the surrounding country.

Nauheim, lying at the foot of the Johannisberg Mountain, about twenty miles from Frankfurt and twelve from Homburg, has many claims to popularity. The springs give forth muriated saline waters of some strength, which are charged with carbonic acid gas. The waters, which have a temperature of 96.4° F., are useful in cases of enlargement of the liver and spleen and in scrofula. There are good bathing appliances and a salle d'inhalation.

ANALYSIS OF SCHWALBACH WATERS.[1]

	Stahlbrunn.	Weinbr.	Paulinenbr.
Bicarbonate of protoxide of iron	0.64 gr.	0.44 gr.	0.51 gr.
Bicarbonate of magnesium .	0.14	0.07	0.09
Bicarbonate of magnesia . .	1.63	4.46	1.23
Bicarbonate of lime . .	1.67	4 39	1.65
Bicarbonate of soda . . .	0.15	1.88	0.13
Carbonic acid . . .	50 cub. in.	45 cub. in.	40 cub. in.
Temperature, 47.8° to 50°.			

Kreuznach is located on the river Rhine, not far from Bingen, at an elevation of 280 feet. Neither the town itself nor its situation is so interesting as many of the Rhenish resorts. The springs which are used for bathing are at some distance from Kreuznach, but the one spring which is used for drinking—the Elizabeth—rises very near the Kursaal. Although the waters contain a very little of certain compounds of bromine and iodine, they must be regarded practically as strong salt springs. The Elizabeth spring contains in one pint of water 94 grains of solid ingredients, and of these 73 grains are of chloride of sodium, about ¼ of a grain of bromide of magnesium, 0.03 of a grain of iodide of magnesium, and 13 grains of chloride of calcium. The dose is at first small and is gradually raised to about a pint a day. The bath is, however, the most important method of application.

Baden-Baden lies at the head of the Black Forest, in the midst of the most beautiful scenery. The season here is from May to October, during which time people come to take the waters, which are of the simple saline variety and have a temperature ranging from 110° to 150° F. Baths are given for paralysis and rheumatism.

At **Reiboldsgrun**, in Silesia, which has an elevation of 2200 feet, is a sanitarium in charge of Drs. Koeppe and Wolf, physicians well known through their experiments as to the influence of altitude on blood-changes.

Rippoldsau, situated below the Kniebis, the highest peak in the Black Forest, and surrounded by pine-woods, is a village with a population of 800, lying amidst romantic and beautiful scenery and

---

[1] Braun's Curative Effects of Baths and Waters.

resorted to both for its climate and its mineral waters. The waters may be classed as saline chalybeate. The whey- and milk-cures are also features of this resort, which is much frequented by phthisical invalids during the summer. The beautiful surroundings have been rendered still more attractive by every resource which art can bring to the aid of nature.

Kissingen, in Bavaria, has an elevation of over 600 feet, and though not so gay as Homburg, it is celebrated for its waters, which are efficacious in cases of dyspepsia, abdominal plethora, and disorders of the liver. There are three springs—the Rakoczy, the Pandur, and the Maxbrunnen—the latter being weaker than the others; these have muriated saline waters. Two springs are used for bathing—the Sool-sprudel and the Schön-sprudel. Prince Bismarck lives at Kissingen for two months out of every year. It is a place of 10,000 inhabitants, and everything is done to make the town charming for visitors. Between the Kurhaus and the Kursaal is a fine promenade or Kurgarten, and Kissingen has also a small theatre. There are two smaller spas near by, Brocklet and Brückenau.

ANALYSES OF RAKOCZY AND MAXBRUNNEN AT KISSINGEN.[1]

	Rakoczy. In 16 oz.	Maxbrunnen. In 16 oz.
Chloride of sodium	44.71 gr.	17.520 gr.
Chloride of potassium .	2.20	1.140
Chloride of lithium .	0.15	0.004
Chloride of magnesium	2.33	0.510
Sulphate of magnesium	4.50	1.820
Sulphate of lime . .	2.99	1.060
Carbonate of lime. . .	8.14	4.620
Carbonate of protoxide of iron .	0.24	....

Free carbonic acid, 41 cubic inches.

Switzerland.

Seelisberg, in the Canton of Uri, is a summer health-resort possessing a mild climate, a humid air, and attractive surroundings. The scenery is fine and the district is also historically interesting. The situation is sheltered. There are good baths, and the resort is much patronized. No account of Seelisberg would be complete without mention of the really splendid Hotel Sonnenberg, with its luxurious arrangements for visitors.

1 Yeo's Climate and Health-resorts.

Montreux is situated upon the northeastern shore of Lake Geneva, on a small indentation known as the Bay of Montreux. Its situation is more sheltered than that of other villages in the district.

The climate is fairly open, dry, and sunny, there being about 60 rainy days during the year; the rainfall is, however, large. The mean yearly rainfall, as shown by the records for seven years, was 50 inches, while that of Geneva was only 32 inches. Winds are rare.

There are numerous good hotels and pensions.

The *grape-cure*, given, among other stations, at Meran and in Montreux and its vicinity, and said to be efficacious for various disorders, but especially in phthisis and general debility, claims, as chief among its advantages, the prevention of waste of the system and the properties of a safe and mild aperient. It is given in the autumn, during the months of September and October. Patients take the grapes before a light breakfast, going out to purchase them, and eating them while they wander about. The ordinary dose is from two to four pounds daily, but it may, in exceptional cases, rise as high as six pounds.

Arosa is situated in a narrow valley which joins with the Schonfigg valley at Langwies. It lies at an elevation of more than 6000 feet. The climatic characteristics are said to be similar to those of Davos, but Arosa has from half an hour to an hour more sunshine, even in the midst of winter, than Davos. There are three hotels beside the Pension Brunald, which is the only one open for the reception of visitors during the winter. The village is picturesque, but it is, or was as late as 1890, reached only by a bridle-path, which, of course, would render it difficult of access for many invalids. At Arosa were conducted the elaborate, exact, and successful experiments of Dr. Egger to investigate the changes in blood caused by altitude. The place is resorted to in winter by phthisical invalids whose general health is not too much impaired, and in summer by patients who are suffering from nervous troubles.

Davos. The district of Davos (an elevated valley of the Canton des Grisons) is some fourteen miles in length. The portion of the valley containing the well-known health-resorts—Davos-Platz and Davos-Dörfli—lies at an altitude of about 5200 feet, and is surrounded by Alps ranging to 9000 and 10,000 feet above sea-level. It is sheltered from the north and west winds. The principal village, Davos-Platz, is situated on the northwestern side of the valley.

and contains nearly all the hotels, pensions, and shops. Davos-Dörfli in winter receives the sun's rays some two hours earlier, although it loses them earlier also.

There has been much complaint from the older residents of the rapid overcrowding of Davos and consequent danger of poor drainage and smoke-nuisance; but Dr. C. T. Williams stated in his Lumleian Lectures for 1893 that there were no grounds for this feeling yet, as the valley is wide and the hotels and houses scattered.

The climate here in winter is very cold, with a few hours of sufficient sunshine to allow of sitting out in the sun's rays; but it is usually freezing in the shade, and at night the mercury falls far below the freezing-point. The night-air, while cold, is dry, and invalids usually sleep with their windows open.

TABLE III.—DAVOS-PLATZ.[1]

Monthly mean.	Temperature. Mean.	Relative humidity. Per cent.	Absolute humidity. Grains per cubic foot.	Dewpoint.	Rainfall (total). Inches.	Clouding. 0 cloudless. 10 sky-covered.	Wind. Miles per hour.
Winter . .	23°	82	1.21	18°	6.1	4.4	1.6
Spring . .	35	74	1.74	27	6.4	5.2	2.5
Summer	52	74	3.29	44	11.9	5.1	3.4
Autumn	34	79	1.81	28	9.2	4.9	2.4
Annual . .	38	77	2.03	31	33.6	4.9	

NOTE.—Temperature, humidity, and rain and snowfall are based on observations for twenty-one years.
Wind. January, February, March, April, November, and December from records for four years; August from records of two years. In above seasonal report for the wind-movement *Winter* includes December, January, and February; *Spring* includes March and April; *Summer* includes August; *Autumn* includes September, October, and November. Other months are missing.

The mean annual temperature in Davos, as given by Mr. Waters from the records for twenty-one years,[2] was 38° F. The mean temperature for winter was 23°, for spring 35°, and for summer 52°. This is slightly higher for the winter-temperature than the record of the meteorological station for eight years, which was kindly furnished by Dr. Carl Ruedi. That was 20° F., with a maximum of 42° and a minimum of —7.6°.

There is freedom from wind and fog. The wind-record in

[1] Some Meteorological Conditions of Davos. Arthur William Waters, Esq., 1890.
[2] Loc. cit.

Davos during the winter, taken by Mr. Waters, showed from 9 to 1 o'clock a movement of 0.6 mile per hour, and from 1 to 3 o'clock a movement of 1.3 mile per hour. The mean velocity was for winter 1.6 mile and for spring 2.5 miles per hour.

(For comparison with Davos-Platz, see Denver, Estes Park, Colorado Springs; also Table II. and Tables V. to XII.)

The relative humidity of Davos is 82 per cent. for winter and 74 per cent. for spring and summer; but the absolute humidity is low during the cold months on account of the extreme low temperature.

The annual rainfall and melted snowfall at Davos (from the records for twenty-one years) are 33.6 inches, of which the seasonal fall for winter is 6.1 inches; for spring, 6.4 inches; for summer, 11.9 inches; and for autumn, 9.2 inches.

The soil of the valley is sandy and dry, except near the stream that flows down the centre. During the winter season, however, the ground is usually covered by two or more feet of snow.

The snow-melting generally begins before the middle of March, and the usual exodus of patients takes place about the first week in April.

Tarasp, lying in the district known as the Lower Engadine, at an altitude of between 4000 and 5000 feet, has a bracing Alpine climate, less severe than that of resorts in the Upper Engadine and less subject to sudden changes of temperature. The air also is neither so dry nor so rarefied, and is free from dust. The "season" is from June until the end of August. Tarasp has many and different mineral springs, some being saline-alkaline, others chalybeate, and still others sulphurous; the waters also contain Glauber's salt. The saline-alkaline springs are the most important, and are used for both drinking and bathing. The Tarasp water is very rich in free and half-free carbonic acid, and Dr. Yeo says "that there is scarcely a spring in Europe that is known to possess so many important qualities."

Wiesen (elevation, 4770 feet) is well situated on sloping ground, and has a dry surface. The scenic surroundings are picturesque and pretty, and the place is rather more sheltered from cold winds than Davos, near which it lies. The hours of sunshine on the shortest winter-days are from 10 o'clock in the morning until 3 in the afternoon. The place is quieter than Davos or St. Moritz, and the accommodations are spoken of as excellent.

Andermatt, reached by the Gotthard Railroad, and therefore easily accessible to invalids who would otherwise be debarred from its advantages, is used as a health-resort both in winter and summer.

In this district the snow lies from about the middle of October until the latter part of April, and the number of snowy days from October to March is, as stated by Dr. Loetscher, 52. Of dull and cloudy days he gives 67, and says that there are more misty days during the winter than at Davos.

In the Ober-Alp Lake there is fishing, and many beautiful excursions may be taken from Andermatt. The accommodations are ample and good.

Grindelwald, situated in the Bernese Oberland at an elevation of 3400 feet, and surrounded by the finest scenery, is a climatic station. The narrow valley in which it lies runs from east to west, and is sheltered from north winds and to some extent from east winds, but is exposed to the blasts of the Föhn, which blows from the south and southwest.

The climate is bracing and the winter-temperature is mild and equable, but in summer it varies greatly.

There are baths for the benefit of patients, and excellent accommodations, one hotel, the Hotel Zum Schwarzen Adler, being especially well arranged.

Leysin is situated on a lofty plateau where the Ormont valley joins that of the Rhone. The climate is rigorous and stimulating, and is recommended by Dr. Bezencent for scrofulous and spinal disorders in children.

Mürren, also in the Oberland, has an altitude of 5000 feet, and is a climatic health-resort possessing the grandest scenic surroundings and sheltered from the north and northeast winds. Although the climate is tonic, the air is milder and less bracing than that of the Engadine district. Baths and the whey- and milk-cures are administered. The accommodations are good.

Samaden, 5700 feet above sea-level, is a small town lying about three miles from St. Moritz, and has a less interesting situation than many of the Swiss resorts. It is somewhat used as a winter-resort, and possesses a Kurhaus for the housing of patients during that season. This establishment is well heated and has arrangements for baths and douches. One of the hotels has also been kept open during the winter for some time past. The view of the Bernina range from Samaden is very fine.

St. Moritz lies in the Upper Engadine, at an elevation of 6080 feet, surrounded by lofty mountains and fairly well sheltered from the wind. The winter season is longer than at Davos, and a greater amount of snow falls. From some observations made by Mr. Waters in St. Moritz[1] in 1882–'83 the temperature during the day was found to be in January, taken at 9, 1, and 3 o'clock, 18°, 27°, and 27° F., respectively, and in March, at the same hours, 20°, 26°, and 25°. The highest wind-movement was between 1 and 3 P.M., and recorded in January 3.3 miles and in March 7.7 miles per hour during those hours.

The baths of St. Moritz lie in the plain, 300 feet below the upper village. St. Moritz has a summer season from June 15th to September 1st. Rain may be expected in summer one day out of three.[2] The mean temperature for the four months June to September, for eight years, based upon three daily observations, was 51° F. Maximum, 78°; minimum, 32°. In October the weather begins to be quite cold. Heavy snows begin in November. The hotels at St. Moritz are excellent.

Pontresina, situated in the Upper Engadine, at an elevation of nearly 6000 feet, is a small, but much-visited hamlet, having a population of 400 and containing good hotels. There is a magnificent view to eastward and southward.

From the peculiarity of its situation, at the open meeting-point of two valleys, Pontresina has a great many hours of sunshine; but the same peculiarity makes it subject to cold winds and mists. Its climate is very bracing, and is rendered more so by the proximity of glaciers.

It is a fashionable summer-resort from which consumptives are, as far as possible, excluded, and it is much used as a starting-point for climbing the adjacent mountains and glaciers.

Maloja (6000 feet) is a plateau situated at the higher or southwest end of the Upper Engadine, where the Föhn or warm southern wind is seldom felt. Dr. Tucker Wise took meteorological observations at Maloja for several winters. The mean day-temperature of 1883–'84 was 25° F. for November, December, January, and February. The maximum temperature during these four months was 45° (December 26th) and the minimum —7° (February 19th). The maximum solar radiation was 143° (on February 13th). The

---

[1] Observations made in St. Moritz, in the winter of 1882–'83. Arthur William Waters, Esq.

[2] The Climate of St. Moritz, Upper Engadine. Walter B. Platt, M.D., Baltimore, 1887.

mean day-temperature of 1884–'85 (November to March) was 26°, and of the winter of 1885–'86 (November to March) 23°. The average relative humidity for the last-named period was 76 per cent. Dr. Wise says the temperature is more equable than at Davos, but there is more valley-wind. The winter season at Maloja lasts a little longer. There is a fine hotel at Maloja, open throughout the winter.

CHAPTER XVIII.

AFRICA—ASIA—AUSTRALASIA.

Egypt.

THIS country has long been used as a winter sanitarium. It has the advantages of the desert in its dry air and the absence of rain.

Cairo (latitude, 30° 3′ north; population, 400,000) has an annual mean temperature of 72° F., with extremes during the year from 35° to 111°. The mean temperature for winter is 58°. There is a great difference between the day- and night-temperatures, due to radiation, which in winter amounts to 23° or even 38°. Freezing-point is not reached in Cairo, but absolute minima of 35° and 36° have been recorded.

From *Egypt as a Winter-resort*, by Dr. F. M. Sandwith, of Cairo, the following temperature-records have been taken:

	Mean temperature.		
	Max.	Min.	Monthly.
October	84°	65°	74°
November	74	56	64
December	68	50	58
January	61	47	54
February	65	49	57
March	73	53	63
April	81	60	70

The annual rainfall is 1¼ inches, with from 12 to 15 rainy days. The annual mean relative humidity is 56 per cent.; for the winter it is 66 per cent. The difference between the dry and wet bulbs sometimes amounts to 24° F.

The air of the desert is exceedingly pure. The climate suffers from hot southeast winds, which are very distressing. A visitor[1] speaks of the weather from November to March as for the most part bright and sunny, and not too warm. There are occasionally cloudy and comparatively cold days, with rain, between the middle of December and the middle of February.

[1] Winters Abroad. R. H. Otter, M.A., London, 1882.

PLATE X.

RELIEF MAP OF AFRICA.

From BUTLER'S GEOGRAPHIES, by permission of E. H BUTLER & CO. Copyright, 1888.

Residence in Cairo should not be too prolonged. The strongest objection to a long stay on the part of an invalid lies in the fact that the greater part of the European quarter of the city is built on what was once a swamp and is still in parts very swampy ground. After sunset a rising fog may be seen, which is most dangerous to invalids. Moreover, the country around Cairo is most thoroughly irrigated, and where the rich crops of grass and clover are cultivated the exhalations from the moist land after sunset are extraordinary. The sand and dust constantly inhaled during the day are also objectionable.

Visitors do not usually go to Cairo before October, and leave before the end of April.

The sewers of Cairo are bad and the death-rate in the town population is surprisingly large.[1] Carefully filtered water should always be used. In Lower Egypt and near the Nile dew forms every night. At Luxor and near the first cataract it is hardly noticeable.

The town has good hotels, but the means for heating during the rare periods of cold and damp weather are insufficient.

Among the sources of discomfort and danger to delicate invalids in making the otherwise delightful voyage up the Nile in a dahabiyeh are the fogs on the river for one hundred miles south of Cairo, the cold and frosty nights in the desert, and the strong north wind which is frequently annoying to boats coming down against it. But even for delicate persons the Nile voyage is a pleasant mode of spending a portion of the winter. It affords plenty of air without the labor of exercise. The nights are cool and the days warm and clear.

Gizeh, across the river, seven miles from Cairo, with one good hotel, and **Helouan,** fifteen miles from Cairo, where there are two hotels, and villas that may be hired for the season, are said to possess advantages as health-resort stations.

Luxor, 450 miles south of Cairo, has a still drier air. The population of Luxor is 4000, and there are two hotels. It seldom rains here and the wind-movement is very small. The only objectionable wind is the " khamsin," from the south or southwest, which usually blows for three days at a time. It is seldom troublesome during the winter months, but is more prevalent in the spring.

---

[1] Wintering in Egypt. Frederick Peterson, M.D., in New York Medical Record, August, 1892.

Algeria.

This French province on the north coast of Africa has been greatly used as a winter-resort, although it has not won unreserved praise as a suitable climate for sufferers from pulmonary disease.

The Atlas Mountains traverse Algiers in three chains, running east and west, and rising in places to a height of 7000 feet. South of the last chain stretches the Great Sahara Desert, which, according to Dr. Bennet, is the key to the Algerian climate and converts what would otherwise be a dry climate into a moist one. This is explained by the fact that the atmosphere overlying this rainless tract of desert, becoming heated both winter and summer, rises into the higher strata, and thus a vacuum is formed, which the cooler air from the Mediterranean basin rushes in to fill, being sucked in over the summits of the Atlas ranges. Consequently the regular winds are and must be northeast and northwest, and south winds only blow exceptionally, though the sirocco, when it blows, is a terrible blast from the desert. These northerly winds coming from the Mediterranean or the Atlantic are laden with moisture, and striking the Atlas ranges are at once cooled and condensed and deposit their moisture in the form of frequent and abundant rain over the entire Algerian region, reaching into the desert two hundred and fifty miles from the sea. The rainfall is consequently heavy, the annual average for Algiers being 40 inches, distributed over about 87 days, principally in winter; the greatest number of rainy days usually occurs during November, December, and January, which is a decided drawback for visitors. The rainfall increases to the eastward, the westerly province of Oran, where there are few or no forests, having the least rain.

Algiers (latitude, 36° 47′ north), a town with a population of 75,000, has a mean annual temperature of 64° F.; the mean for winter being 55° and for spring 66°. The annual mean of relative humidity is 73 per cent., which varies but little during the year. The dews of summer-nights are very heavy. Williams considers the climate as milder and moister than that of the Riviera and occupying an intermediate place between it and that of Tangier (Morocco), where the equalizing influence of the Atlantic is more felt in moderating the extremes. Bennet is authority for the statement that "the monthly temperature is one-third higher than the Riviera, on account of the warmer nights." Except for a month or thereabout during each year the wind is from the north. Algiers has a choice of suburbs and hill stations, the principal suburb being *Mustapha Supérieur*, where furnished villas may be hired. The drawbacks of residence in Algiers, aside from the dirt, odors, and bad drainage, are the amount and season of the rainfall and the periods of dust and extreme heat, although the winters differ greatly in these respects.

Oran (latitude, 35° 44' north ; population, 70,000). This portion of Algiers is reported to have a good winter climate, with less rainfall than Algiers. Oran is an old town, lying at the head of a bay. There are several beautiful promenades. The water-supply is said to be good. Regular steamers go to Marseilles, Barcelona, Valencia, Maloja, and Gibraltar. It is 261 miles by rail to Algiers.

Biskra or **Biskara** (latitude, 35° 27' north ; population, 7200). An important French military post, reached by rail from Algiers or Phillippeville *via* Constantine. Biskra is situated on an elevated tableland on the south side of the Aures Mountains, an extension of the Atlas range. From this oasis on the northern edge of the Sahara caravans start for their journeys across the great desert. The streets of the town are broad, with one-story houses, built largely of brick. The accommodations have been well spoken of. The winter season is prolonged well into spring. The summers are unbearably hot.

Tangier (Morocco) (latitude, 35° 47' north ; population, 15,000). This Arab town, which contains usually but 300 or 400 resident Europeans, is finely situated on a bay on the Strait of Gibraltar, fourteen miles east of Cape Spartal. The distance across to the town of Gibraltar is thirty-eight miles northeast.

Tangier is a wretchedly built place, with few comfortable houses. The streets are narrow, crooked, and very dirty. The whole appearance is that of an old, decaying town. The water-supply is poor and particularly bad in summer. As the streets are poorly paved and there are no wheeled vehicles, visitors find it necessary to go about on horseback.

There is a modern hotel offering good accommodations.

Dr. Williams says: "Tangier has long been noted for its remarkable climate, which apparently combines the warmth of the Mediterranean with the equability of the Atlantic, and, being separated by a series of mountain-ranges from the Sahara Desert, does not share all the features of the Algerian climate, though its rainfall is large—30 inches—occurring chiefly in October and November. The mean winter-temperature is about 60° F. and the diurnal variations are slight. The climate is mild, with a bracing element, owing to Atlantic winds."

South Africa.

The **South African Highlands** have a number of stations which have been used by English invalids, situated in Cape Colony, the Orange Free State, and the Transvaal or South African Republic.

There are two lines of mail steamers plying between London and the Cape of Good Hope, touching at Madeira and sometimes at Ascension and St. Helena. The voyage on the larger ships has been reduced to about seventeen days.

Since the importance of the diamond and gold mines has been realized railway communication throughout South Africa has been greatly extended, and nearly all the towns are now fairly accessible. The low belt of country fringing the coast rises into plateaus and mountains of considerable elevation on proceeding inland. A large portion of this veldt or elevated rolling prairie is from 2000 to 5000 feet above the sea. The highest mountain in Cape Colony, the Compassberg, has a height of 7800 feet.

Invalids who go to the Cape are advised not to remain for any length of time within one hundred miles of the coast nor below an altitude of 1500 feet.

The rainy season is in the winter in the west and southwest, and in the summer in the eastern part of the colony, where invalids usually go. In the high and more exposed regions the severity of the winter is the more trying, while the ordinary conveniences for heating are absolutely wanting. Referring to this, Dr. Symes Thompson says: "The houses are usually built without fireplaces, and coal and wood are almost unobtainable, dried cow-dung doing duty for peat, as well as cement for flooring and stucco for the walls."

The greatest drawbacks to the South African plains appear to be the winds, dust, heat, and absence of shade, which is much felt during the warm weather, and general scarcity of good water and the lack of suitable accommodations for invalids. Winter is the best time of the year for visitors, who are usually rare at that time, as it is then summer in Europe. The months from February to April, when it is the height of the grape season, are pleasant in portions of Cape Colony, and the rains of late autumn and early winter have not commenced.

Cape Town is not recommended as a residence for delicate persons. The water-supply is inadequate and the drainage bad, while the prevailing wind, known as the "Cape Doctor," raises dust-storms which rage furiously in the afternoons. It is a latitude subject to frequent and heavy storms, which are less felt inland, where the climate is also more equable. The annual rainfall at Cape Town is 24 inches; at the attractive suburb of Wynberg, nine miles distant, 43 inches; and at Bishop's Court, half-way between the other two, 57 inches—a difference produced in each case by the local influence of Table Mountain.

The Central Karoo district has an average altitude of 3000 feet, while the Upper Karoo plateau ranges from 2700 to 6000 feet. The climate of the great or central Karoo is characterized by dryness and prolonged droughts. The annual rainfall is from 9 to 18 inches and the number of rainy days is small. The heat of summer (December, January, and February) is intense, reaching 110° F. in the shade; but the air is said to be dry and the nights cool. Thunderstorms of great violence follow in the wake of the north-west winds, converting large tracts of country into temporary lakes in a few hours, but these storms are rare. The soil is baked clay, usually utterly denuded of trees. Snow appears in winter on the high mountains, but not in the veldt. The air is clear, bright, and bracing. The most suitable stations in Cape Colony for pulmonary invalids seem to be as follows:

Beaufort West (2792 feet); annual rainfall, 8 inches on 25 rainy days.

Lemoenfontein (3192 feet).

Aliwal North (4318 feet). Observations for four years give an annual rainfall of 23 inches, with 89 rainy days. This is a large village on the Orange River, a stream which is never dry, but runs low for nine months in the year. The mean temperature for winter is 48° F.; for summer, 67°, with an annual range from 24° to 102°. Relative humidity in winter, 77 per cent.; in summer, 55 per cent.

Burghersdorp (4552 feet), the chief town of the eastern division of the Karoo, is hot during the summer, with cold winter-nights; the rest of the year is delightful. Principal rainfall is in the summer. Fogs are unknown on the plains, but of frequent occurrence in the mountains. Annual precipitation, 11 inches on 41 rainy days.

Colesberg (4407 feet) has 12½ inches of rainfall on 33 rainy days.

Tarkastad (4280 feet) is not shut in by hills, and has constant

breezes.   It is situated in a fertile region and surrounded by large
farms.   It has an average of 9 inches annual rainfall with 55 rainy
days.

Kimberly (4000 feet elevation) is 647 miles from Cape Town by
rail and 485 miles from Port Elizabeth.   It has 21½ inches of rainfall on 64 rainy days.   The mean temperature in winter is 50° F.
and in summer 70°.   The winter is short and mild.   In summer,
although the air in December, January, and February may be heated
to 104° or 105°, it is said to be dry and not oppressive.

Cradock, in the Karoo district, is recommended by Dr. P. C.
De Wit, in the *British Medical Journal*, 1894.   The elevation is
2853 feet, and many portions of the district are 1500 feet higher.
It is easily accessible, but the accommodations for visitors are
poor.

Bloemfontein, the capital of the Orange Free State, with a population of about 3500, is the best-known health-station of the district.
It is about the same distance from Cape Town and Port Elizabeth
as Kimberly.   The town is sheltered by neighboring hills and stands
at an elevation of 4540 feet.   It is dry and not very cold in winter,
while the summers are long and hot.   The mean annual temperature is 76° F.   Mean minimum for winter, 55°.   Mean maximum
for the six hot months, 82°.   The annual mean relative humidity is
55 per cent.   The annual rainfall is 17 inches on 70 rainy days.
The town drainage is bad, but improvements are said to be contemplated.

Ladybrand (5000 feet) has 27 inches of rainfall on 87 rainy
days.   The plains of the Orange Free State are dry, but open and
dreary.

Pretoria, the capital of the Transvaal (elevation, 4000 feet; population, 5000), is said to be a well-sheltered, attractive little town.
The surrounding hills, which rise to a height of 8000 feet, are covered with mist in the summer, yet the country is said to be healthy;
the summers are hot.   Pretoria has 31 inches of rainfall and 64
rainy days.

Johannesburg (5000 feet) is a large place of over 20,000 inhabitants, about forty miles south of Pretoria, in the gold-mining district.

In Bechuanaland the elevation is from 4000 feet to 6000 feet.
More rain falls than in Cape Colony.   Above 4500 feet there is less
liability to fevers.   It is well not to live near or to leeward of newly
cultivated or irrigated land.

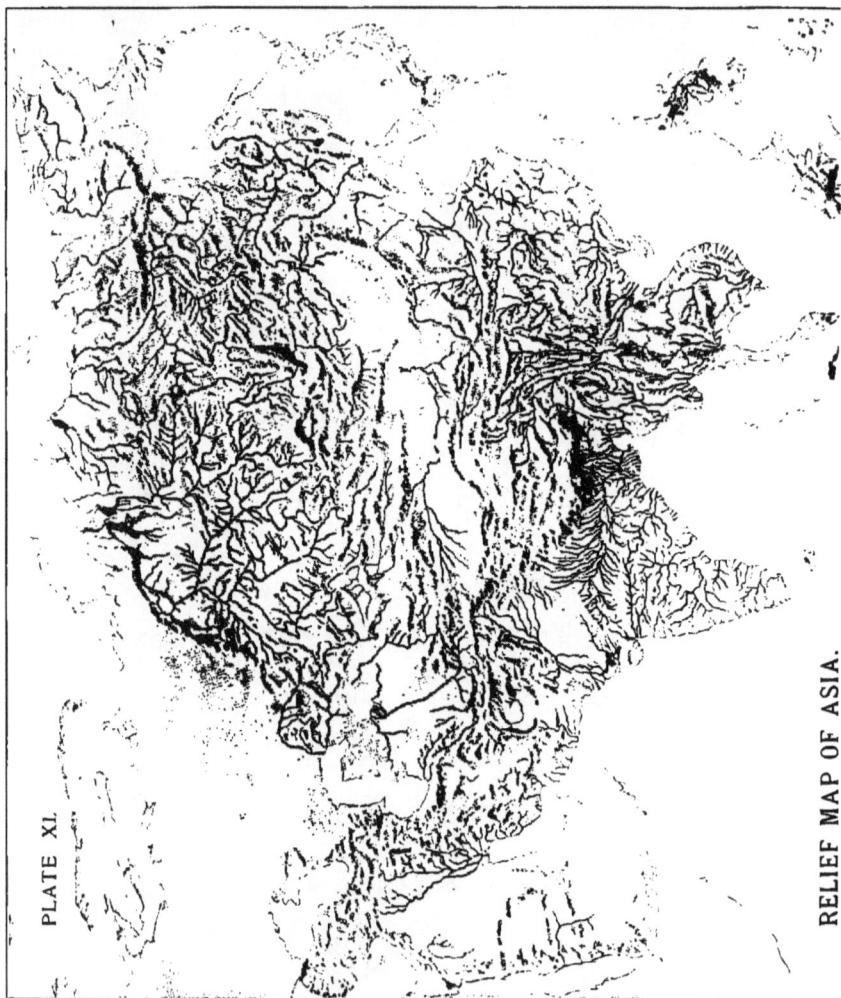

Asia.

Practically the only portions of Asia which can be regarded as health-resorts for Europeans are the mountainous districts of India.

The Himalayas. Through the northern part of this country run the numerous ranges of the immensely high Himalayan system, which in many ways characterizes and modifies the climate. The southern slopes of the Himalayas are dotted, from Darjiling to Simlah, with hill-stations ranging in elevation from 4000 to 8000 feet. While this country is much cooler than the plains, it is subject during the summer to very heavy rains, owing chiefly to the influence of the so-called southwest monsoons, which are very damp winds and bring a great amount of moisture.

It is stated that fifty miles north of Dalhousie, a British station in the North Punjab, the influence of the monsoon dies away, so that, while no climatic records are obtainable, there is little doubt that the humidity of this section is greatly lessened. Good, dry health-resorts could undoubtedly be established on the north slopes of the Himalayas; for this, being the lee side of the range, receives only a small amount of the moisture brought by the monsoon, which, reaching first the south slopes of the mountains, deposits upon them, as has been said, the chief part of its burden. The country on the northern side is, however, too remote from European settlements to be at present available.

Kassauli and Murree, and **Dagshai and Nynee Tal** are stations lying between Simlah and Umballa. Kassauli and Dagshai are used as military sanatoria. These resorts stand at elevations ranging between 6000 and 8000 feet.

Darjiling, in Bengal (latitude, 27° north; elevation, 8200 feet), is one of the best-known resorts in India. The town has a population of about 4000. It is 308 miles north of Calcutta, and is much resorted to in summer by residents of the hot plains. There are grand views of the mountains on the north and west. The mean annual temperature is 54° F.; winter, 41°; summer, 73°. The rainy season is from June to September, and the total rainfall is stated to be 132 inches. The mean annual atmospheric pressure is 24.058 inches. During the tremendous summer-rains the atmosphere is very moist.

Simlah, the hill-metropolis, is a cool mountain-resort which, during the summer, is the seat of the British Government. It is also used as an invaliding station, is said to be dry, and is one of a series of stations of similar quality. It is very gay during the season. The climate is characterized by an absence of wind and a usually cloudless sky.

Nilgiri Hills. Portions of the Madras Presidency, lying in the Nilgiri Hills, inside the Western Ghats, possess very good resorts, having an elevation of from 5000 to 7000 feet. The monsoon, striking the Western Ghats which form the coast-range, sheds most of its moisture before reaching these hills, so that the annual rainfall is only 55 inches—less than half that of most Himalayan stations. It is stated that the climate is also more equable. The mean winter-temperature is 60° F.; that of summer, 65°. Occasional fogs are reported. Utakamand, the summer capital of the Presidency, with an elevation of 7000 feet, is situated in these hills.

Australasia.

Australasia has been used by the English for health-purposes especially on account of the long sea-voyage required to reach it. A description of such a voyage and of the meteorological conditions has been well given by Dr. Williams in his Lumleian Lectures for 1893, when he presented an instructive record and chart showing the meteorology of a sailing-voyage from England to Australia, around the Cape of Good Hope, by a gentleman connected with the British Meteorological Office. Starting in October, "the temperature ranges from 53° to 58° F. for the first five days. Off the Azores it rises to 60°, and passing Madeira to 69°. In crossing the line the maximum, 82°, is attained, but breezes are present and temper the heat; afterward it gradually falls; in 30° south latitude 70° F. is the average, and this sinks to 58° on reaching the Cape. After rounding the Cape the temperature, owing to the mixture of the warm Agulhas current with the antarctic current, is uncertain, varying from 47° to 56°, the currents overlapping each other and causing great varieties of atmospheric temperature. The vessel reaches 43° south latitude and steers eastward, and, on account of the influence of the antarctic circle, the temperature ranges from 47° to 55°, and rises to 67° on approaching the continent of Australia."

The actual range of temperature on the voyage is from 47° to 82°, the number of rainy days being about 20. The relative humidity varies from 74 per cent. to 91 per cent., the average percentage being 82. The average length of such a voyage is usually from eighty to ninety days. If the voyage is commenced in May or June, the heat is greater near the equator, and Australia is reached in midwinter. The return-voyage to England *via* Cape Horn is not so favorable. "Near Cape Horn the temperature falls to 41°, and for several days is often little over 43°—a rather wretched state of things. In 40° south latitude a rise of 50° F. occurs, and off Rio de Janeiro 78°. England is reached in the early part of June."

The great objection to the route *via* Suez is the intense heat of the Red Sea, which is overpowering for many invalids.

Speaking of the voyage to New South Wales, Dr. G. L. Mullins, of Sydney, writes: "The following, as a rule, bear the long voyage and are benefited by a sojourn in New South Wales: those who are predisposed to phthisis; the scrofulous; those in the early stage of phthisis, with consolidation of the lung around the tubercle; those in whom the disease is quiescent; the subjects of frequent small hemorrhages; those with slight lung-mischief, with irritable cough but no fever; the anæmic, or those with defective appetite; those fairly free from dyspnœa and cough and able to take exercise.

"The following may be considered unsuitable and should not be allowed to travel so long a distance: weak, nervous, or excitable subjects; those who suffer severely from mal-de-mer; the subjects of severe hemorrhages; those in the acute or fever stage; the subjects of other organic diseases in addition to slight phthisis; those in whom the phthisis is advanced and the mischief still extending.

"As to the suitability of any particular district in the colony, a local practitioner should be the judge in each individual case."

The Intercolonial Medical Congress, which met at Dunedin, New Zealand, in February, 1896, adopted resolutions to the effect that professional attention in England should be directed to the importance of the nature of cases of tuberculosis sent to Australia, and that such as are sent should be directed to avail themselves of the climates of the interior.

Australia.

The climate of the Australian seacoast is very variable at all times of the year, and the air is peculiarly irritating.[1]

[1] Winters Abroad. R. H. Otter, M.A., London, 1882.

The highland regions of Australia present a choice of elevations from 2000 to 7000 feet, but there are as yet few accommodations for invalids.

Regarding the value of the climate of Australia for consumptives, an Australian correspondent of the *Lancet*[1] recently referred to the need of sanatoria in suitable locations. He stated that the deaths from phthisis in each of the colonies of Victoria and New South Wales exceeded a thousand a year, and thought that there was "quite sufficient tuberculous disease among the native-born population to keep such institutions well filled".

"As a general rule, it will be found that a suitable climate will possess the two characteristics of being inland and having a relatively high altitude; and this is the case all over the colonies.

"In regard to accessibility, the different places vary most materially. The greater number in Victoria and New South Wales are accessible by rail, but often it is a very long journey, as to Hay. To reach some places involves a rough coach-journey. These, of course, are all points which demand attention. Places deserving a trial are as follows: in Victoria, Ballarat, Geelong, Echuca, Mount Macedon. In New South Wales: Blue Mountains, Bathurst, Orange, Hay, Dubbo, Blayney. In Queensland the Darling Downs. I by no means wish it to be thought that this list is exhaustive."[2]

Melbourne, 37° 5' south latitude, lies at the mouth of the Yarrow River, and is the capital of the province of Victoria. It has a mean annual temperature of 58° F., with extreme ranges during the year from 111° to 27°. The average annual rainfall is 26 inches. The seasonal mean temperatures are for winter, 49.2°; for spring, 57°; for summer, 65.3°; for autumn, 58.6°. The mean daily range is 22.3° for summer, 18.5° for autumn, 15° in winter, and 20° in spring. The rainfall on the coast of New South Wales varies from 20 to 50 inches, with from 100 to 150 rainy days.

Mr. Otter, in a charming account of his sojourn in Australia, says that out of several invalids in various stages of disease whom he met travelling for their health the one who had made real progress had spent most of his time at an elevated station in New South Wales. He speaks very highly of the station-life for invalids.

Sydney has a mean annual temperature of 62° F.; winter, 55°; summer, 74°; the annual rainfall is 48 inches. The amount of

[1] Lancet, 1894, ii. p. 57.
[2] G. A. Van Someren. M.B.C.M., in British Medical Journal, October 10, 1896.

rainfall decreases steadily on going inland, while the temperature increases, rising to 100° in the shade, and in some places to 140°. The hot north winds during the summer are very trying, and the dust is so annoying that nearly all the residents wear dust-coats to preserve their clothing. To the north of Sydney are the resorts of **Port Maquaire** and **New Castle,** and to the south **Wollongong, Cape St. George,** and **Eden.** There are other stations lying at some distance from the sea, but on the ocean-side of the mountain-range.

Orange, a place of 5000 inhabitants, lying about one hundred and ninety miles west of Sydney, has an elevation of 2400 feet above the level of the sea. The mean annual rainfall for twenty years is 38.95 inches, distributed over 101 rainy days; but, although the rainfall is large, the atmospheric humidity is not proportionally high, and the winter atmosphere, although stimulating and tonic, allows of much outdoor life. The variation in temperature is less during the winter months than at other times of the year, but the summers are spoken of as "never oppressive". These statements are on the authority of Dr. Van Someren, of Orange.

Tasmania.

This island, known to an older generation as Van Diemen's Land, and more especially defined in their minds as a station for transported convicts, lies south of Victoria, a province of Australia. It is reached from Melbourne in twenty-four hours by the steamships of the Australian Steam Navigation Company, whose Tasmanian port is Launceston.

There is no lack of water in Tasmania, and the vegetation is consequently luxuriant. Fruits fine as to size and quality are produced, and the hop-growing industry has become very important. A very good beer is produced from these hops, lighter than the English beer and much esteemed in the colonies. Mines of gold, iron, and tin have been developed and found to be profitable. Fishing, rabbit-shooting, and kangaroo-hunting are among the attractions which this island offers to sportsmen, the kangaroos being hunted by means of dogs especially bred to the work. Quail, duck, and snipe abound.

The climate is not extreme, the mean annual temperature being 54° F. and that of the summer 63°, while in the winter season it is 46°. Inland for a short period during the winter the climate is quite rigorous. In summer the same hot wind which blows in Australia sometimes visits Tasmania also, but, on the whole, the island is a valuable and cool summer-retreat, especially for Australians. The annual rainfall is 24 inches. Martin says that at Hobart Town the number of rainy days for dry years is 100 and for wet years 120. It is said that snow does not lie on the plains and valleys of the lowlands, and on the highest peaks only during two or three of the winter months. The eastern part of the island is reputed dry and the western rather wet.

Hobart Town, the chief town of Tasmania, is a city of about 20,000 inhabitants, and is charmingly situated at the mouth of the Derwent River, near Mounts Wellington and Nelson. It has several comfortable hotels and a small club, and the social atmosphere is pleasant. There are walks, rides, and drives through the beautiful surrounding country, and many delightful and interesting excursions may be taken, among them one to Port Arthur, where live such of the convicts as have not yet served out their sentences. The drainage of Hobart Town, however, is spoken of as defective.

New Zealand.

New Zealand consists of two large islands and one smaller one, with the adjacent Auckland and Chatham groups. It lies southeast of Australia, between latitude 34° and 47° south. Both of the principal islands are mountainous, North Island having its highest peaks in its southern half, while in South Island Mount Cook, the highest point of the range which borders the entire west coast, reaches an elevation of 12,800 feet; these have lately been explored by Fitzgerald. There are few rivers, although each island has one about 200 miles long; but running streams are numerous, so that the islands are well watered. There are many lakes, and in North Island is the well-known and beautiful lake-district where are found, besides geysers and sulphur springs, placid pools, like baths, containing warm, clear water, azure in hue. All of these waters have a reputation for the relief of rheumatism, scorbutic affections, tuberculous and nervous disorders, and skin-diseases. In the Southern Alps, which, as before stated, run along the west coast of South Island, are a number of lakes, two among them— Lakes Wakatipu and Te Anau—being especially notable for size and beauty. The scenery of these islands is grand and beautiful in an unusual degree.

The yearly rainfall is large, occurring chiefly during the winter season in North Island, but being more evenly distributed through all seasons in South Island. Observations at Wellington in 1882 showed the amount of rain for that district to be 55½ inches, falling on 166 days, and the smallest record, that for Dunedin from 1864 to 1881, was 34.6 inches. Droughts very rarely occur, and it is humid throughout the year. Dr. Hector, in his *Handbook of New Zealand*, says:

"The climate resembles that of Great Britain, but is more equable, the extremes of daily temperature only varying throughout the year by an average of 20° F.; while London is 7° colder than the North and 4° colder than the South Island of New Zealand. The mean annual temperature of the North Island is 57° and of the South Island 52°, that of London and New York being 51°. The mean annual temperature of the different seasons for the whole colony is in spring, 55°; in summer, 63°; in autumn, 57°; and in winter, 48°. The climate on the west coast of both islands is more equable than on the east, the difference between the average summer- and winter-temperature being nearly 4° greater on the southeast portion of the North Island and 7° on that of the South Island than on the northwest, on which the equatorial winds impinge. This constant wind is the most important feature in the meteorology of New Zealand, and is rendered more striking by comparing the annual fluctuation of temperature on the opposite seaboards of South Island, which have a greater range of temperature by 18° at Christchurch on the east than at Hokitika on the west."

The prevailing wind for all districts and for the entire year is westerly. The soil is lighter than that of England and more easily worked, and it is also, for the most part, very fertile. New Zealand has about 750 endemic species of flowering plants. The principal towns are **Wellington**, the seat of the Government, **Auckland**, and **Christchurch**.

CHAPTER XIX.

ISLAND CLIMATES.

The Bermudas.

THIS group of islands lies 600 miles east of the North Atlantic coast of the United States, in latitude 32° 14' to 32° 25' north. It is less than three days from New York by steamer.

Hamilton and St. Georges are the principal towns. They are provided with good hotels. There are supposed to be 365 islands in all, many of them but coral-reefs. The largest island, called Great Bermuda, is sixteen miles long by one and a half miles wide. The porous limestone rock that lies underneath the surface-soil readily absorbs water. Rainwater is used for all domestic purposes.

The climate is moist and equable. The nights are usually but 3° or 4° cooler than the day. The mean temperature for the year is 69° F., with extremes from 42° to 90°. The January mean is 61°; for July it is 79°. The coolest month is March, 61°; and the warmest August, 80°. The record for two winters, 1888–'89 and 1889–'90, shows a mean monthly maximum of 70° and a mean minimum of 51°. For the two months of March and April for two years, 1889 and 1890, the mean maximum was 72° and the mean minimum 55°.

The annual precipitation is 50 inches. October has the most rain, and the months from April to June the least.

The annual mean relative humidity for Bermuda is 80 per cent.

The islands are resorted to almost solely in winter. The summer, which holds on late, is said to be extremely debilitating, with warm fogs. The prevailing wind at that season is a damp southwest wind.

The roads on the island are good and hard.

Bermuda is an important naval and military station of Great Britain. It is the headquarters of the North Atlantic squadron, and at Ireland Island is the largest floating dock in the world.

The Bahamas.

The Bahama Islands (latitude, 25° 5' north) are under the English flag and have a governor appointed by the Crown.

The islands are of coral-formation, covered with drift. Mr. Stark, in his guide-book,[1] says that they are evidently formed on a plateau of submarine mountains of great height. Soundings on the north-east or ocean-side of Eleuthera Island show on the charts a depth of water of 12,000 feet near the shore and 16,000 feet fifteen or twenty miles out.

The island of New Providence is 960 miles south (and somewhat west) of New York—a little over three days by steamer—and about 200 miles east of the end of the Florida peninsula. From Key West the distance is nearly 300 miles. The island is about sixteen miles long, with an average width of five miles.

Nassau. The capital city faces a harbor on the northern shore, where the benefit of the constant northeast trade-wind can be felt. Its population is placed at 12,000, the negroes greatly predomi-nating in numbers. The roads around the town are very good, being hard and smooth. There are two good hotels.

From the harbor of Nassau the land rises steadily to a height of 100 feet, affording natural facilities for drainage, but there is no town-system of sewerage. The soil absorbs water rapidly, and this porous quality adds to the danger from cesspools and vaults. The usual arrangements for water for domestic use are frequently open to criticism. Rainwater is the best source of supply, and great care is necessary to keep the cisterns thoroughly clean.

The residences are usually built of limestone, with thick walls. Each house stands alone, surrounded by more or less land, with its own flower-garden and fruit-trees.

The general health of Nassau is very good. There are large marshes near the centre of the island, which, although they feel the cleansing influences of the daily tides, would undoubtedly affect injuriously the healthfulness of the town if winds from the south and west were not, fortunately, of extremely rare occurrence. The prevailing northerly and easterly winds, which are quite steady during the winter and spring, are of great value to the residents on the northern shore of the island.

Mr. Drysdale, in his entertaining collection of letters,[2] which were first published in a New York paper, says: " From November to June there is no healthier place (than Nassau). . . . I do not see how it could be otherwise. The island is a solid rock, per-

[1] History and Guide-book to the Bahama Islands, etc. James H. Stark, Boston, 1891.
[2] In Sunny Lands. William Drysdale, New York, 1885.

petnally swept by sea-breezes, and being on it is like being on the deck of a great steamer in mid-ocean without any sea-sickness. I have mentioned before the fact that tropical islands with rich soil are generally unhealthy, while the rocky islands built up by the industrious coral insect are always the reverse. No stretch of the imagination could make the soil of Nassau appear rich, and there is nothing for sickness to build itself upon."

Mr. Drysdale quotes from the *Nassau Almanac* a record of the mean monthly temperature for the year 1878, taken at 3 P.M. (the hottest portion of the day), which by seasons was as follows: winter, 73°; spring, 79°; summer, 84°; autumn, 80°. January had an average of 73°; March, 76°; July, 86°. The mercury rose above 90° for three days during the year. The maximum solar temperature was in January, 140°; March, 149°; July, 159°.

A record of the mean monthly temperature of Nassau for ten years, compiled from observations taken on week-days at 9 A.M. at the Nassau Military Observatory,[1] arranged by seasons, is as follows: winter, 71°; spring, 75°; summer, 81°; autumn, 77°. The monthly mean for January was 70°; for March, 72°; for July, 82°; annual mean, 76°.

Mr. Stark reports the average temperature for the winter months as 70° F.; for spring, 77°; and says the coldest day registered for twenty-one years was 64° (which seems a few degrees too high), and the warmest day from November to May was 82°. Mr. Drysdale, in one of his letters to the New York *Times,* referred once to a temperature of 55°, and the writer has a recollection of having seen the thermometer at 58° under the stone porch of the hotel during a cold rainstorm one February. It would therefore be wise for a delicate invalid to take the precaution of securing quarters where a fire could be had if needed, for, although frost is unknown, the damp chill of a long, hard storm is most penetrating.

As is usual with warm and humid marine climates, the annual range of temperature at Nassau is limited. The mean annual barometric pressure is 30 inches. The average relative humidity for the year is 79 per cent.; for winter, 83 per cent.; for spring, 76 per cent. (means for five years[2]). Taking the above temperature-record for ten years as a basis, this would show for the year an average of

---

[1] This temperature-record appeared originally in Governor Rawson's report for 1864. It is quoted in Ives's Isles of Summer and Buck's Handbook of the Medical Sciences.

[2] Furnished by United States Weather Bureau.

7.63 grains ; for winter of 6.84 grains, and for spring of 7.12 grains of absolute humidity or vapor present in each cubic foot of air.

The average rainfall for ten years was for winter, 7.2 inches ; spring, 13.8 inches. Annual mean, 56 inches. The rainfall at Nassau during the six months from November to May is usually about one-third of the total amount for the year.

The only danger from hurricanes is from August to November, and the heaviest and most damaging hurricanes are separated by intervals of several years.

The most attractive occupation for a visitor is boating in row-boats and sailing in small craft to the beautiful Sea Garden and to the white beaches and cocoanut-groves of neighboring semitropical islands. Sea-bathing is a luxury that can be enjoyed at any time of the year, the temperature of the sea-water being usually in the vicinity of 70° F.

Nassau has warmer winters than Bermuda, the Azores, Madeira, or Teneriffe.[1] The air at Nassau appears to be less moist and muggy than that of Bermuda.

The Royal Victoria Hotel is built of limestone, four stories in height, with wide verandas. It is the finest hotel in the West Indies, and has the advantage of being less than four days from the markets of New York. Beginning in January, 1896, a steamer will run regularly three times a week during the winter season from Palm Beach, Florida, to Nassau, making the trip in fifteen hours.

Eleuthera. Forty or fifty miles east northeast of Nassau, Eleuthera Island stretches its irregular, narrow length of seventy miles on the Atlantic, affording in a measure a breakwater for the protection of Nassau from the ocean-surges.

It is a safe and delightful cruise in the lee of islets and cays to the tropical quiet of the settlements on Eleuthera for those who can live for a time without the comforts of modern civilization. The best way of seeing the islands leisurely is to charter a small schooner with suitable accommodations, with skipper and boy, ample provisions, and a cook.

The principal harbor is on the northeast of Eleuthera, inside of Harbor Island. It is well sheltered, and can be entered by vessels drawing less than nine feet of water.

Harbor Island is about two miles long by one mile wide.

[1] See Table XIV.

Dunmore Town, the largest settlement in the Bahamas next to Nassau, is on the lee side of the island, facing the bay. Mr. Stark gives its population as 2000. "A very pleasing little place it is, encircled by beautiful cocoanut-groves, and sauntering and dreaming by its beautiful green waters in an air of solitude and peace is very enticing to one who is weary of the rush and giddy whirl of modern life, while the cool trade-winds always moderate the heat."

Besides Harbor Island there are in the district of Eleuthera eleven other settlements, the population of which varies from a dozen persons or thereabout to 500 or more.

Governor's Harbor, half-way down the island, is one of the largest and most attractive settlements, with its churches and public buildings on a "cay" (as such a small island or reef is usually called), which is connected with the "mainland" by a causeway leading to the foot of the hill. "The houses on this hillside are white and very neat appearing, each one standing alone in its own garden, and the whole place having the appearance of a beautiful tropical watering-place."

On the ocean-side of Eleuthera, a few miles south of Harbor Island, is a limestone arch, eighty-five feet above the sea, known as the "Glass Window." In 1872 an extraordinary tidal-wave, unaccompanied by wind, washed under this arch and over the island. In this vicinity is also the curious rock-formation called the "Cow and Bull" and a limestone cave extending underground several hundred feet.

Cocoanuts, pineapples, oranges, lemons, bananas, sugar-cane, figs, almonds, corn, potatoes, yams, tomatoes, and melons are all more or less cultivated on the island and shipped to Nassau and New York.

Jamaica.

The island of Jamaica is about ninety miles south of Cuba. Its total length is one hundred and forty miles, with a width varying from twenty-nine to forty-nine miles. It is the largest and most important of the British West Indian possessions. A ridge called the Blue Mountains runs through the eastern end of the island from southeast to northwest, rising at the highest point to a height of 7500 feet.

The temperature in the lowlands of the coast will average 75° F.

at night and 85° in the day.   New Castle (3800 feet) is credited
with 68° for the hottest month and 61° for the coldest. On the
highest levels it is from 40° to 50°.

The healthiest portion of the island is said to be above an elevation of 1400 feet, and the north side of the island is preferred.
The highest regions of the mountains have many clouds.   In this
cool, moist region is found vegetation belonging to colder climates.
Buck's *Reference Handbook of the Medical Sciences* states that at
an elevation of 4000 feet the temperature ranges from 44° F. in the
winter to 65° in the summer.   The difference in temperature between the north and south sides of the island is about 5° for the
year and from 8° to 10° during the first three months.

The mean annual rainfall of the island for five years, according
to the *Handbook,* was 50 inches on 116 days.

One of the healthiest portions is the parish of St. Ann, which is
in the centre of the northern side of the island.   The mean annual
temperature is 76° F.

The inland village of **Moneague** (950 feet) has a small hotel.
The roads in this vicinity are good.

Mandeville (2500 feet) is a pleasant town in the coffee and
grazing country.   It has a good hotel.

The Santa Cruz district has also a healthy climate.

The rainiest portion of Jamaica is naturally at the northeast end,
facing the trade-wind.   Hurricanes may occur between July and
October.

The population of the island in 1891 was 640,000, of which
about 15,000 were white.

Kingston, the capital of the island of Jamaica, is situated in latitude 17° 58′ north.   Population, 50,000.   The city is regularly laid
out on a gently sloping plain on the south coast near the east end of
the island, at the foot of the Blue Mountains, the loftiest ridge being
to the northeast of the plain.   The harbor of Kingston is one of
the best in the West Indies.   The town is commonplace, with
unpaved streets, frame buildings, and few trees.   It is always hot.

The soil on which Kingston is built is gravel, sloping to the harbor
and offering natural opportunities for drainage that are sadly neglected.   The water-supply is drawn from two rivers several miles
above the city, and is considered good.   The town is lighted by electricity and has a tramway.   The new Hotel Rio Cobre is fifteen

miles north of Kingston.   The island of Jamaica has an equable, hot, and moist climate.

The mean annual temperature of Kingston (ten years' records) is 78° F.   Mean relative humidity, 78 per cent.   Mean absolute humidity, 8.02 grains of vapor to the cubic foot.   Wind-movement for the year a little over 3 miles per hour from the southeast.   During the winter it is about 4 miles and during the summer 2⅔ miles per hour.   The mean of annual precipitation is 32½ inches, the greatest rainfall being in May and June and August, September, and October.   The least rain falls from November to April.   The mean of the rainfall for the island of Jamaica for the year 1893 was 86½ inches.

The following seasonal meteorological table for Kingston is adapted from a valuable table compiled by Maxwell Hall, M.A., F.R.A.S., F.R.M.S., published in the *Handbook on Jamaica*, prepared by the Honorary Commissioner for Jamaica to the World's Fair at Chicago in 1893, Lieutenant-Colonel the Honorable Charles J. Ward, C.M.G.

TABLE IV.—KINGSTON, JAMAICA.

For the ten years from June, 1880, to May, 1890.

Monthly mean.	Temperature.			Relative humidity	Absolute humidity	Cloudiness; per cent. of whole sky.	Wind; miles per day.	Total rainfall.	
	Mean	Max.	Min.					Kingston.	The Island.
				Per cent.	Grains.			Inches.	Inches.
Winter . .	74°	86°	67°	78	7.07	31	65	2.78	12.09
Spring .	76	86	70	76	7.31	41	73	9.11	15.46
Summer	80	89	73	77	8.42	53	99	11.75	18.98
Autumn . . .	78	83	72	79	8.12	54	59	8.90	19.77
Annual	78	88	71	78	8.02	44	74	32.54	66.30

Temperature, maxima and minima, based on monthly means, not on the extreme readings. The above figures represent the monthly means, except for the rainfall, which is total.

Barbados.

Barbados is the most eastern of the Caribee Islands under the English flag.   It is situated in 13° 4′ north latitude and 59° 37′ west longitude.   Authorities do not agree as to its size, but it is nearly eighteen miles long by twelve miles wide.   It is seven days from New York by direct passage on the Brazil line, and about two weeks by steamers stopping on the way at other islands. Landing is made on the leeward or southwest side of the island, on the shores of an open roadstead called Carlisle Bay, where the town of **Bridgetown** is located. The Marine Hotel—an excellent structure built by the Government—is at Hastings, two and one-half miles east from Bridgetown, connected by tramway. The soil is coral and limestone rock. It is quite bare of trees. The greatest elevation is 1200 feet on the eastern side of the island. The healthiest residence-portion of the island is in what is called Scotland. Residence in Bridgetown is to be avoided.

Barbados has a moist but salubrious climate, without rains or heavy dews during what is called the "dry" or rainless season, from December to May. It will average several degrees warmer during the winter than Nassau. Its blandness and equability of temperature are suited to a certain class of invalids, as there are no sudden or dangerous changes. The northeast trade-wind blows steadily during the day; occasionally it is unpleasantly strong. The air is pure and healthy. Temperature ranges from 76° F. at night to 83° for the day during the winter months. The mean for January is 76° and for August 80°. The rainy season is from June to November, when the island is also liable to be visited with hurricanes. August and October are usually the rainiest months. The annual rainfall is about 57 inches, of which about 2 inches fall in March, the driest month, and 11 inches in October, the dampest. The mean annual relative humidity is 72 per cent. The mean for winter is the same, 72 per cent.[1]

The sea-water temperature is about 78° F., and delightful for bathing.

The population of the island is estimated at 182,000, of which 15,000 are white, the rest African.

Barbados is one of the most thickly populated places in the world, there being an average of 1100 persons to each square mile. There is a railway which renders most of the island accessible.

The principal industry is the cultivation of sugar-cane.

The Azores.

The nine islands called the Azores are two thousand miles from Boston and over eight hundred miles from the coast of Portugal, to

which country they belong. They are of volcanic origin, and have
suffered from infrequent earthquakes of great violence. They are
rugged and picturesque, with precipitous coast-lines, rising toward
the interior to an average height of 3500 feet. On the island of
Pico is a peak 7600 feet above sea-level, and extending below to a
depth of 1600 feet.

There are no natural harbors; large vessels anchor in the open
roadstead off the principal ports. A breakwater is being constructed
off Ponta Delgada that is supposed to be capable of protecting one
hundred vessels of all sizes.

The islands are reached by steamer from New York, and by a
regular sailing packet—the barque "Sarah"—from Boston.

There are mail steamers to Lisbon twice a month and fruit
steamers to London, as well as ships putting in constantly for supplies or repairs.

Sao Miguel or St. Michael's, the largest island of the group, is
over forty miles long and from five to twelve miles wide. A pamphlet
by Dr. Emerson Warner, of Worcester (Mass.), descriptive of the
islands,[1] says the population of the principal town, **Ponta Delgada**
(latitude, 37° 45' north), is 25,000. The city is situated on the
southwest side of the island, and extends along the shore for two
miles. The streets are straight and broad. The houses are built
of basaltic lava. Stoves are seldom seen, fires being used by the
natives for cooking-purposes only.

Hotel-accommodations are limited, and are mostly under Portuguese management. There is a small hotel in Ponta Delgada kept
by an English family. At Fayal is a Portuguese hotel of some
pretensions. The expenses of living are moderate.

The roads are good. In the larger cities are fairly comfortable
carriages, usually drawn by mules, but in the country the donkey is
the chief beast of burden.

The highest point on St. Michael's is on the west and has an
elevation of 3060 feet. The plains are fertile.

Hot springs abound in St. Michael's. A palatable table-water is
bottled on the island and offered for sale. In the lovely valley or
crater of the Furnas, twenty-seven miles from Ponta Delgada, are
many hot mineral waters. A large bath-house and small hospital
adjoining have been built for the use of patients. Springs of fresh
water are plentiful on the islands.

---

[1] The Azores as a Resort. Emerson Warner, M.D.

On the island of Fayal the principal town is **Villa de Horta** (latitude, 38° 30′ north) ; population about 8000. This port trades largely with America, while Ponta Delgada is in more direct communication with England.

The Azores have an equable climate. The mean annual temperature is 62° F. The extremes are stated to be 86° and 45°. The range between winter and summer is from 10° to 15°. The night-temperature is ordinarily not more than 4° cooler than the day. The summer is enervating at 70°; one is drenched with perspiration on the slightest exertion.[1]

The mean temperature for winter is 58°, for spring 61°, for summer 68°. The three coldest months are usually January, February, and March.

In winter it sometimes feels chilly and damp, and one seldom leaves home without an umbrella. The climate is very humid. Wall-paper will not adhere, and the veneering of furniture strips off. The mean annual relative humidity is 76 per cent., and for winter it is 77 per cent.

The mean annual rainfall is 38½ inches.[2]

The winds blow with great force at times and there are frequent storms. The prevailing direction of the wind in winter is from the south, southwest, and northwest, and in summer from the northeast, east, and north. (See Table XIV.)

The Madeiras.

Madeira Island, the largest of the group of that name, is about thirty miles long by thirteen miles wide. It is traversed by a mountain-chain running its entire length from east to west. Near the middle is the highest peak, 6100 feet. The northern and southern sides of the mountains are broken by deep ravines watered by limpid streams. On the slopes are gardens and vineyards. The soil is fertile. The islands are of volcanic origin, but earthquakes are rare. Deep-sea soundings show them to be the peaks of lofty submarine mountains. The inhabitants are of Portuguese descent, with some admixture of Moorish and negro. The population of the island is 132,000. The wine of Madeira has always been famous.

---

[1] A Summer in the Azores. C. Alice Baker, Boston, 1882.

[2] Records of humidity and rainfall furnished by kindness of U. S. Weather Bureau. Records for Azores for four years, for Madeira for five years.

Funchal (latitude, 32° 38′ north; population, 20,000), the capital of Madeira, is picturesquely situated on the south coast of the island, on an open bay or roadstead, from which the mountains rise behind the town to a height of 4000 feet. Funchal is 360 miles from the coast of Africa, 535 miles from Lisbon—with which it is connected by the Brazilian cable—and 1215 miles from Plymouth. It has communication by steamer with Liverpool (four days), London, and Plymouth, Antwerp, Lisbon (two days), Brazil, Cape of Good Hope (five days), the Cape Verde Islands, Azores, and Canaries (one and a half days).

The streets of Funchal are narrow, but fairly clean. They are paved with cobblestones, are without sidewalks, and are lighted at night by oil-lamps. The shops are poor. Food-supplies are said to be good except the mutton, which is execrable. Wheeled vehicles are hardly ever used by the natives—not even wheelbarrows. The town has two public walks bordered with trees and flowers. There is plenty of excellent water.

There are good English hotels, and quintas[1] can be rented furnished. There is a good English club and library. There are no amusements except riding and being carried in hammocks through the beautiful scenery. Both English and French are generally understood. There is good fishing, but no shooting.

Madeira has long been noted for its soft, damp air and equable temperature. It was formerly held in high repute as a resort for consumptives. The mean annual temperature is 65° F., for winter 60°. The coldest months are January, February, and March, and the hottest August and September. The extreme range of temperature is from 90° to 46°, which is the lowest record. The temperature of the sea-water in winter ranges from 61° to 72°.

The mean annual rainfall is 27 inches, with extremes for different years, as high as 49 inches and as low as 16 inches. The greatest amount of rain falls from November to March. There is little rain during the summer, but the vegetation is freshened by dews. In winter snow occasionally falls on the mountain-peaks.

The mean annual relative humidity is 66 per cent.; for winter it is also 66 per cent. There are no cold winds, but occasionally a hot and dry east wind—the *teste*—is felt from the distant desert.

---

[1] *Quintas*, the name for pleasant villas in their own gardens, the rent being three hundred dollars a year and upward.

The opposite, or west wind, usually brings rain. The prevailing winds are from the north, northeast, or northwest. (See Table XIV., for seasonal details.)

The Canaries.

The Canary Islands, which are under the Spanish flag, lie sixty miles east of the coast of Africa, in the main current of the Gulf-stream and in the line of the trade-winds. There are seven inhabited islands and a few uninhabited islets.

Island of Teneriffe. It is said that nearly one thousand steamers call at Teneriffe during the year, rendering it particularly accessible from European ports. The time from England by steamer is from four and one-half to six days. Communication can also be had with Portugal and Spain. There is a good hotel on the island. The island of Teneriffe is described by Dr. G. V. Perez,[1] of Orotava, as sixty-seven miles long, sixty miles being its greatest width and sixteen miles its smallest. The peak is very near latitude 28° north, and rises 12,000 feet above the sea. It is usually partly obscured by a cap or "parasol" of clouds, about 1000 feet thick, which begins at an elevation of 3500 feet. This cloud forms during the forenoon and remains until blown away by the south wind that arises in the afternoon.

Orotava is the capital of Teneriffe. It has one good hotel in the port and a larger one higher up in the valley. The valley or amphitheatre of Orotava rises from the harbor of Port Orotava on the sea to an elevation of 2000 feet. The hills shelter it on all sides except to the north, which gives the benefit of the pure ocean-breeze. The "trades" begin to blow between 9 and 10 A.M., dying away between 1 and 2 P.M. At sunset the south land-breeze begins and continues throughout the night. The cloud "parasol" that covers the higher portion of the island tempers the strength of the sun's rays and furnishes shade for the upper part of the valley. Toward the seashore there is more sun at midday than there is higher up. During the summer the trade-winds blow more strongly than in winter, and the "parasol" then appears more regularly.

Above the cloud-layer the climate presents great extremes, ranging in July from 83° to 28° F. The air is much drier, the difference between the dry and wet bulbs even amounting to 30°. The wind

[1] Orotava as a Health-resort. George V. Perez, M.D.

which blows from the northeast on the shore of the island, at an altitude of over 10,000 feet blows from the west and southwest.

Orotava has an average temperature for January of 60° F.; for July, of 73°. Mean annual temperature 67°, with an average annual range of 14°. The temperature is said rarely to rise above 82°. By seasons it is as follows: winter, 60°; spring, 65°; summer, 72°; autumn, 70°.

From May to August there is no rainfall. The annual precipitation is 13 inches. Average number of rainy days, 51; the rain most frequently falls in the night. The average percentage of relative humidity is at 9 A.M., 65; 2 P.M., 60; 9 P.M., 72. Mean annual relative humidity, 66 per cent. The record for temperature and humidity is for two years. The temperature of the sea-water is lowest in March, 64°; highest in July, 68°.

The climate of Orotava is equable, moderately damp, and rather relaxing, while the island has this great advantage, that higher elevations and drier air are obtainable, if desired. The extraordinary luxuriance and profusion of vegetable growth are evidence not only of the equability of the climate, but also of its humidity. Oranges, figs, pineapples, bananas, dates, etc., are found on the island, but not cocoanuts or breadfruit. The air is said to be rich in ozone. The water-supply is from tunnels bored into the sides of the mountains, and is reputed to be very good. The soil is porous and dries rapidly after rain.

The editor of the *British Medical Journal,* Dr. Ernest Hart,[1] says that Orotava is not troubled by heavy dews, frosts, siroccos, miasmas, extreme heats, or even mosquitoes. Dr. Perez recommends the regions of the Cañadas, at an elevation of 7000 feet, as a site for a sanitarium.

The Hawaiian (or Sandwich) Islands.

These islands are twenty-one hundred miles southwest from San Francisco, and are usually reached by steamer in six and a half or seven days. By sailing-vessel the time may be eighteen days. The group is made up of eight inhabited islands.

Honolulu (latitude, 21° 18' north; population, 25,000) is situated on the southern shore of the island of Oahi, on a deep and spacious bay. The city is sheltered by mountains from some of the trade-

---

[1] A Winter Trip to the Fortunate Islands. Ernest Hart, in British Medical Journal, 1887.

winds and from much of the rainfall of the windward coast. Honolulu is laid out on a level plain about 25 feet above the sea, and claims a good system of municipal government; broad, clean streets lined with beautiful homes; many churches, a theatre, street railways, electric lights, telephones, and a town water-system, besides artesian wells. While the soil is of a volcanic nature and fertile when well watered, it is porous and dries quickly after rains. The town possesses a large hotel with adjoining cottages and other hotels and boarding-houses.

At **Waikiki**, a seaside-resort a few miles east of Honolulu, there is fine surf-bathing. There is a hotel at Waikiki, and the shore is lined with villas and cottages.

The island of Oahi is thirty-five miles long by twenty-one miles wide. A peak in the western range of mountains reaches a height of over 4000 feet. On the eastern shore there is a lower range of mountains running northwest and southeast, which offers its sides directly at right-angles to the northeast trade-wind. On the slopes of all these mountains are beautiful valleys, with deep ravines, cascades, and luxuriant tropical vegetation. The mean annual temperature at Honolulu is 74° F. The lowest record in the Government Survey series for ten years was 54° and the highest 89°. The average midday maximum is about 80°. The greatest daily range of temperature was 23°.

The normal record by seasons for Punahou, a station near Honolulu, is as follows:[1]

	Temperature.	Relative humidity.	Rainfall.
Winter	70°	74 per ct.	12.4 inches.
Spring	73	74 "	9.9 "
Summer	77	69 "	5.7 "
Autumn	76	71 "	9.8 "
Year	74	72 "	37.8 "

The average absolute humidity is for the year 6.53 grains, and for winter 5.91 grains of vapor to each cubic foot of air.

The wind-movement at Honolulu is from 3 to 4 on the Beaufort scale—which indicates a breeze from 18 to 23 miles per hour.

The mean barometric pressure for fourteen years was 30.038 inches. The annual mean of cloudiness for four years expressed in tenths is 4.22.

---

[1] From reports of Prof. C. T. Lyons, in charge of Weather Bureau, Honolulu.

A very good illustration of the varying quantity of rain registered in different rain-gauges not far from each other, but exposed under different conditions, is shown in the report of Prof. C. T. Lyons for 1890. In the showing of comparative rainfall the table for the year 1889 presents the records from four reporters residing on different streets in Honolulu, showing totals for the year ranging from 18 to 25 inches, a difference of 7 inches in the same town. In the same report is given a record of rainfall in Honolulu kept by Dr. R. McKibbin for thirteen years, showing an annual mean of 30.1 inches, and a record kept by Mr. W. W. Hall for sixteen years (including the same thirteen years), showing an annual mean of 39.5 inches.

Points of small annual precipitation are Mahukona, Kawaihae, and South Kona, all on the west coast of the island of Hawaii.

Hawaii, the largest island of the group—one hundred miles long by ninety miles wide—is nearly two hundred miles southeast from Honolulu. It has the great volcanic peaks Mauna Kea (13,953 feet) and Mauna Loa (13,760 feet). On the southern side of the latter, at an elevation of 4000 feet, is the famous active crater of Kilauea, fourteen miles from the sea and thirty miles south of the port of Hilo. There is a hotel at the volcano.

The annual rainfall at Hilo is said to average 144 inches a year. The tropical verdure is unusually rich on this side of the island.

West of the great peaks the town of Kona, on the west coast, and Waimea, which is further inland, are reported to have climates with comparatively low temperatures and less moisture than the windward coast.

Between the two peaks is a desert-plain lying at an elevation of 5000 or 6000 feet. It is entirely destitute of water and has not even a trail.

The Hawaiian Islands are in about the same parallels of latitude as the south coast of Cuba. The climate is equable, warm, and moist. Compared with Teneriffe the island of Hawaii is larger, and its two peaks rise nearly 2000 feet higher from the level of the sea into the region of cold and snowstorms; it is further south, and on the coast has a warmer average of temperature. The great elevation affords the possibility of an unusual range of climate, but there is no elevated interior valley on the protected western slope which is known to possess the natural advantages of the valley of

Orotava, and, although such a spot may exist, it is not as yet within easy reach of necessary supplies.

A Voyage upon Southern Seas.

The value of the climate of the Pacific Ocean has been brought into prominence of recent years by the example of that distinguished writer, the late Robert Louis Stevenson, in seeking an asylum within its influence. In his romance, *The Wrecker*, occurs a reference to this sea-life, in which a personal note can perhaps be detected. It is when the adventurers are fairly started in the schooner "Norah Creina," in search of the wreck of the "Flying Scud." The passage is as follows:

" I love to recall the glad monotony of a Pacific voyage, when the trades are not stinted and the ship day after day goes free . . . the deliberate world of the schooner, with its unfamiliar scenes, the spearing of dolphin from the bowsprit end, the holy war on sharks, the cook making bread on the main hatch: reefing down before a violent squall, with the men hanging out on the foot-ropes, the squall itself, the catch at the heart, the opened sluices of the sky, and the relief, the renewed loveliness of life, when all is over, the sun forth again, and our out-fought enemy only a blot upon the leeward sea. . . . Day after day in the sun-gilded cabin the whiskey dealer's thermometer stood at 84°. Day after day the air had the same indescribable liveliness and sweetness, soft and nimble, and cool as the cheek of health. Day after day the sun flamed; night after night the moon beaconed or the stars paraded their lustrous regiment. I was aware of a spiritual change, or perhaps rather a molecular reconstitution. My bones were sweeter to me. I had come home to my own climate, and looked back with pity on those damp and wintry zones miscalled the temperate.

" 'Two years of this, and comfortable quarters to live in, kind of shake the grit out of a man,' the captain remarked ; ' can't make out to be happy anywhere else. A townie of mine was lost down this way, in a coal-ship that took fire at sea. He struck the beach somewhere in the Navigators, and he wrote to me that when he left the place it would be feet first. He's well-off, too, and his father owns some coasting-craft down East; but Billy prefers the beach and hot rolls off the breadfruit trees.'

"A voice told me I was on the same track as Billy . . . perhaps it is the impression of a few pet days which I have unconsciously spread longer, or perhaps the feeling grew upon me later in the run to Honolulu. One thing I am sure; it was before I had ever seen an island worthy of the name that I must date my loyalty to the South Seas. The blank sea itself grew desirable under such skies; and wherever the trade-wind blows, I know no better country than a schooner's deck."[1]

[1] The Wrecker. R. L. Stevenson and Lloyd Osborne, New York, 1893.

CHAPTER XX.

METEOROLOGICAL TABLES.

TABLE V.—ANNUAL AVERAGES IN THE U. S. A.

	Elevation [1]	Temperature Month.mean January	July	Annual	Dew-point	Relative humidity	Absolute humidity	Rainfall	Cloudy days	Wind. Mean monthly movement	Normal air-pressure	Number of years of record. Temperature	Humidity	Rainfall	Wind	Pressure
	Ft.					p.c.	grs.	in.		miles.	inch.					
Adirondacks,	1500	16°	64°	40°	...	...	...	39.0	129[2]	......	......	3	...	15		
Asheville,	2250	38	73	54	44°	69	3.24	45.0	106	......	......	13	4	13		
Aiken,	550	48	...	61	...	...	...	48.0	...	......	......	8	2	25		
Thomasville,	330	52	82	68	56	65	4.86	51.5	97	1225	......	6	6			
Jupiter (Fla.),	00	65	80	73	67	82	7.21	58.0	73	6564	30.06	5	7	5	4	5
Key West,	00	71	84	77	69	75	7.48	40.0	71	6952	30.04	19	18	21	10	20
Chattanooga,	700	42	79	61	51	71	4.04	57.0	102	4116	29.28	13	10	13	10	14
Denver,	5300	27	73	50	31	50	2.04	14.4	57	4980	24.73	20	6	22	11	20
Colorado Springs,	6000	26	69	47	29	50	1.84	14.4	57	6663	24.03	16	6	16	4	2
Pueblo,	4700	29	76	52	31	49	2.14	12.0	53	5438	25.27	15	5	11	4	4
Santa Fé,	7000	28	70	49	30	48	1.79	14.6	48	4681	23.26	17	10	33	8	20
Salt Lake City,	4300	28	76	52	36	54	2.36	18.9	88	3681	25.64	18	10	24	10	20
El Paso,	3700	44	83	64	40	48	3.16	9.0	39	3941	26.21	12	10	28	10	11
San Antonio,	650	51	84	69	58	68	5.26	30.6	92	5301	29.33	13	10	21	7	15
Prescott,	5300	34	74	53	35	51	2.31	16.0	51	4898	24.74	13	5	20	10	13
Tucson,	2400	50	88	69	44	42	3.25	12.0	57	3735	27.45	14	6	14	4	2
Yuma,	140	53	92	72	49	46	3.91	3.0	21	4570	29.76	14	11	16	10	17
Los Angeles,	300	53	72	62	52	72	4.42	18.0	45	3758	29.64	14	10	21	7	15
Santa Barbara,	00	52	65	60	51	73	4.20	18.0	73	2872	......	8	8	24	7	
San Diego,	00	58	68	61	52	73	4.34	10.0	69	4105	29.92	17	10	42	10	20
St. Paul,	800	10	72	44	36	74	2.44	27.6	107	5156	29.10	19	10	22	10	20
Boston,	00	27	72	49	40	72	2.84	46.0	107	7997	29.88	20	10	22	10	20
New York,	00	31	74	52	43	73	3.19	45.0	110	6883	29.85	20	10	22	10	20
Chicago,	600	24	72	49	41	73	2.88	35.0	117	7007	29.14	19	10	22	10	20
St. Louis,	500	31	80	56	46	70	3.52	38.0	110	7011	29.45	19	5	21	5	20
San Francisco,	00	50	59	57	49	75	3.90	24.0	88	6863	29.87	18	17	42	5	20
Davos-Platz,	5200	20	55	38	31	77	2.03	33.6	63		24.86	21	21	21	...	15

[1] The elevations are given in round numbers for convenience in memorizing. They are believed to be within fifty feet of the actual measurement.

[2] Adirondacks cloudy days from record for Saranac Lake Weather Station for one year (1894).

TABLE VI.—SEASONAL AVERAGES, U.S.A.

Winter (December, January, February).

	Seasonal temperature.	Relative humidity.	Absolute humidity.	Total rainfall.	Wind. Mean monthly movement.
		per cent.	grains.	inches.	miles.
Adirondacks	18°	...	....	8.8	
Asheville	38	63	1.67	9.3	
Aiken	46	63	2.33	10.7	2489
Thomasville	50	63	2.57	10.6	
Jupiter (Fla.)	64	82	5.38	9.1	
Key West	71	82	6.56	5.5	8029
Chattanooga	44	72	2.38	17.4	4802
Denver	30	54	1.06	1.8	5245
Colorado Springs	29	50	0.94	0.7	
Pueblo	31	57	1.16	1.4	5027
Santa Fé	30	52	1.02	2.3	4909
Salt Lake City	31	56	1.14	5.5	3092
El Paso	46	52	1.84	1.3	3857
San Antonio	54	69	3.24	6.7	5438
Prescott	36	57	1.40	5.0	4502
Tucson	49	48	1.89	3.0	3567
Yuma	56	47	2.36	1.6	4416
Los Angeles	54	68	3.19	11.6	4031
Santa Barbara	52	69	3.02	11.5	2588
San Diego	54	62	2.91	5.8	3842
St. Paul	15	77	0.80	3.2	4822
Davos-Platz	23	82	1.21	6.1	
Albuquerque	41				
Silver City	37	49	1.25	1.0	
Las Cruces	43	43	1.37		
Eddy	40	...	......	1.0	
Phoenix	51	...	......	2.6	
Redlands	51	...	......	6.5	
Charleston	52	81	3.54	11.0	
Jacksonville	55	77	3.52	10.0	
Nassau (W. I.)	70	83	6.63	7.2	

TABLE VII.—SEASONAL AVERAGES, U.S.A.

Spring (March, April, May).

	Seasonal temperature.	Relative humidity.	Absolute humidity.	Total rainfall.	Wind. Mean monthly movement.
		per cent.	grains.	inches.	miles.
Adirondacks	37°	...	....	9.1	
Asheville	58	61	2.76	11.2	
Aiken	59	52	2.89	13.4	
Thomasville	67	62	4.49	12.9	
Key West	76	71	6.86	5.8	7160
Chattanooga	60	65	3.74	15.2	4790
Denver	48	49	1.88	5.8	5295
Colorado Springs	49	46	1.81	4.5	
Pueblo	51	45	1.90	3.6	6314
Santa Fé	48	36	1.37	2.3	5378
Salt Lake City	50	47	1.92	6.1	4208
El Paso	64	36	2.36	0.6	4853
San Antonio	68	66	4.94	7.8	5677
Prescott	51	44	1.86	3.1	6079
Tucson	62	37	2.27	1.2	3797
Yuma	70	43	3.43	0.3	4480
Los Angeles	60	73	4.20	3.8	3951
Santa Barbara	58	72	3.87	3.9	3847
San Diego	59	72	4.01	2.3	4496
St. Paul	43	63	2.0	7.0	5873
Davos-Platz	35	74	1.74	6.4	

TABLE VIII.—SEASONAL AVERAGES, U. S. A.

Summer (June, July, August).

	Seasonal temperature.	Relative humidity.	Absolute humidity.	Total rainfall.	Wind. Mean monthly movement.
		per cent.	grains.	inches.	miles.
Adirondacks	62°	...	......	10.8	
Bethlehem (N. H.)	65	65	4.41	14.0	
Asheville	70	71	5.67	13.7	
Aiken	77	71	7.08	13.8	
Thomasville	80	66	7.22	16.2	
Key West	83	72	8.64	13.4	5405
Chattanooga	77	74	7.38	12.7	3278
Denver	70	46	3.67	4.5	4806
Colorado Springs	67	50	3.62	7.2	
Pueblo	73	44	3.87	5.0	5402
Santa Fé	64	43	2.82	6.6	4628
Salt Lake City	72	33	2.81	3.5	4139
El Paso	83	45	5.40	4.1	3742
San Antonio	82	66	7.08	8.8	4698
Prescott	70	48	3.83	6.1	5167
Tucson	82	40	4.65	5.4	3656
Yuma	89	43	6.18	0.5	4720
Los Angeles	70	72	5.75	0.2	3616
Santa Barbara	64	74	4.86	0.1	3130
San Diego	67	76	5.52	0.3	4320
St. Paul	69	71	5.49	11.3	4697
Davos-Platz	52	74	3.29	11.9	

TABLE IX.—SEASONAL AVERAGES, U. S. A.

Autumn (September, October, November).

	Seasonal temperature.	Relative humidity.	Absolute humidity.	Total rainfall.	Wind. Mean monthly movement.
		per cent.	grains.	inches.	miles.
Adirondacks	43°	...	......	10.1	
Asheville	53	66	3.0	8.2	
Aiken	61	77	4.58	8.4	
Thomasville	66	67	4.70	11.7	
Key West	78	78	8.02	15.3	7213
Chattanooga	61	73	4.34	11.6	3892
Denver	50	50	2.04	2.3	4617
Colorado Springs	48	54	2.05	2.0	
Pueblo	52	50	2.19	1.7	4346
Santa Fé	56	56	2.81	3.5	4101
Salt Lake City	52	49	2.14	4.3	3289
El Paso	62	56	3.44	3.0	3292
San Antonio	68	69	5.16	8.5	4980
Prescott	53	49	2.21	2.7	4218
Tucson	68	43	3.22	2.4	3646
Yuma	73	46	4.04	0.7	3646
Los Angeles	64	68	4.46	2.3	3530
Santa Barbara	62	70	4.30	2.5	2422
San Diego	63	69	4.38	1.4	3760
St. Paul	45	68	2.33	6.2	5232
Davos-Platz	34	79	1.81	9.2	

TABLE X.—WINTER WEATHER, U.S.A. SUPPLEMENTARY TABLE [1]

	Elevation.	Mean temperature.																					October to April—7 months.			
		October.			November.			December.			January.			February.			March.			April.			Relative humidity Monthly mean.	Rainfall, includ'g melted snow. Total.	Cloudy days. Monthly mean.	Wind. Mean hourly velocity.
	Feet.	7 A.M.	3 P.M.	11 P.M.	7 A.M.	3 P.M.	11 P.M.	7 A.M.	3 P.M.	11 P.M.	7 A.M.	3 P.M.	11 P.M.	7 A.M.	3 P.M.	11 P.M.	7 A.M.	3 P.M.	11 P.M.	7 A.M.	3 P.M.	11 P.M.	per cent.	inches.	days.	miles.
San Antonio, Tex.	650	62°	79°	68°	52°	67°	57°	47°	61°	53°	44°	58°	50°	49°	62°	55°	56°	71°	62°	63°	78°	68°	70	14.33	10.1	4.8
Fort Davis, Tex.	4900	52	72	62	44	62	49	36	57	44	36	51	41	43	60	47	43	66	53	56	72	63	51	4.34	3.8	6.0
El Paso, Tex.	3700	53	75	61	41	61	49	32	56	46	36	53	41	41	60	50	45	67	56	61	73	63	48	4.21	2.4	5.2
Fort Thomas, Ar.	2700	47	71	61	43	60	48	34	54	48	35	50	41	41	51	46	41	61	51	45	67	57	56	5.62	3.7	3.7
Fort Grant, Ar.	4800	51	83	61	43	69	50	38	54	45	30	52	42	39	54	46	43	57	51	44	65	70	43	6.71	3.7	7.2
Yuma, Ar.	140	62	66	51	59	71	59	57	68	55	46	62	52	57	67	54	51	75	65	57	71	72	45	1.95	1.6	5.6
Prescott, Ar.	5300	40	66	51	32	56	39	27	40	33	21	43	32	25	41	34	31	49	42	37	65	55	52	8.31	2.3	6.0
Santa Fé, N. M.	7000	41	60	48	29	48	35	24	40	29	15	36	25	20	41	31	30	49	39	37	61	56	48	5.15	3.7	7.1
Denver, Col	5300	40	62	49	28	46	33	23	30	24	13	34	26	30	61	51	30	62	55	53	61	57	51	6.41	4.5	6.4
San Diego, Cal.	......	38	69	61	53	66	57	55	66	54	48	61	53	53	65	56	57	70	60	65	76	66	72	10.08	6.8	5.7
Jacksonville, Fla.	......	56	76	68	57	68	60	59	63	51	50	62	54	53	65	56	57	70	60	65	76	66	72	25.89	7.7	6.7
New York,	......	52	60	54	40	47	41	31	36	38	27	33	35	28	35	30	31	41	35	41	52	45	70	24.69	9.8	10.3
Boston,	......	49	57	50	37	41	38	27	33	29	23	30	32	21	32	27	31	38	32	42	49	42	70	20.51	11.1	10.1
Chicago,	600	49	58	52	36	42	39	27	32	29	21	24	24	21	25	25	31	38	35	43	49	45	72	19.36	10.8	9.0

[1] Condensed from tables compiled by Dr. W. M. Yandell, of El Paso, and the local observer of that place, from U. S. Signal Service records. Except in the case of Fort Thomas, which is for one year only, these tables are based on observations for several years.

TABLE XI.—NIGHT-TEMPERATURES, U. S. A.

	Winter.	Spring.	Summer.	Autumn.	Year.
Denver. . .	16°	33°	55°	38°	36°
Colorado Springs	17	32	51	34	34
Pueblo .	15	34	56	37	36
Santa Fé .	20	35	55	39	38
Salt Lake City .	22	37	58	44	40
Prescott	27	36	56	37	39
Tucson .	35	47	69	51	50
Yuma .	43	51	73	57	57
El Paso .	33	50	68	50	50
San Antonio	44	58	72	58	58
Los Angeles	44	48	57	52	50
San Diego . . .	45	51	60	55	53
Boston .	22	40	61	45	41
New York .	26	40	62	48	45
Chicago	19	36	61	45	43

NOTE.—An idea of the average temperature during the last half of the night is given by the record of the mean monthly minima, which record represents the coldest period of the twenty-four hours, usually between 2 and 4 A.M.

In the United States an excellent standard for cool summer-night temperature is furnished by the mean minima for July (for three years, 1891–1893) of Eastport and San Francisco, which both register 51°.

The above table is based on the reports of the United States Weather Bureau for three consecutive years (1891–1893) for each of the stations, except Colorado Springs, two years (1893 and 1894, except autumn one year, the record for 1894 being incomplete); Prescott two years (1886 and 1889); and Tucson two years (1892 and 1893).

TABLE XII.—SUNRISE AND SUNSET FOR THE FIRST DAY OF JANUARY.

Stations.	Elevation.	Sunrise.	Sunset.	Possible sunshine.	
SWITZERLAND.	feet.			hours.	min.
Maloja	6000	9.35 A.M.	3.45 P.M.	6	10
Wiesen .	4770	10.35 "	3.45 "	5	10
Pontresina .	5900	8.30 "	3.10 "	6	40
St. Moritz .	6080	10.00 "	3.05 "	5	5
Davos-Platz	5200	11.03 "	3.00 "	3	57
COLORADO.					
Denver . .	5300	7.30 "	4.37 "	9	7
Colorado Springs	6000	7.20 "	4.09 "	8	49

TABLE XIII.—MEXICAN CITIES. ANNUAL AVERAGES (FROM RECORDS OF THE MEXICAN GOVERNMENT).

	Lat. N.	Eleva- tion.	Temperature.			Mean relative humid- ity.	Abso- lute humid- ity.	Rain- fall.	Cloudy days.	Wind.	Popu- lation (esti- mated).	Observations taken three times a day except for the City of Mexico and Guadalajara.
			Mean.	Max.	Min.							
		feet.				pr. ct.	grains.	inch.		miles pr. hr.		
Chihuahua,	28° 35'	4700	...	...	...		......	...			20,000	No records.
Saltillo,	25 25	5350	62°	93°	27°	61	3.75	21			20,000	Records for 4 years.
Durango,	24 2	6300	62	88	20		......	21½			30,000	Records for 1 year.
Zacatecas,	22 46	8180	55	71	43	48	2.33	22			50,000	Records for 10 years.
San Luis Potosi,	22 9	6300	63	93	29	60	3.81	15			60,000	Records for 9 years.
Aguas Calientes,	21 53	6100	65	85	37		......	21			35,000	Records for 1 year.
Leon,	21 7	5900	66	96	30	66	4.63	28			100,000	Records for 14 years, except mean temperature 2 years.
Guadalajara,	20 41	5100	67	96	24	53	3.84	34	124	2.4	100,000	{ Observations taken at noon, except rainfall (taken at 6 P.M. Records for 7 years, except wind 1 year only.
Queretaro,	20 35	6060	64	91	32	59	3.87	23			50,000	Records for 3 years.
Pachuca,	20 7	8070	56	81	33	59	2.96	17			30,000	Records for 1 year.
Mexico,	19 26	7400	60	89	29	60	3.45	24	106	1.8	350,000	{ Observations taken hourly. Averages based on records for 15 years, except cloudy days 5 years.
Puebla,	19 3	7100	60	89	30	63	2.76	36			80,000	Records for 14 years.

TABLE XIV.—COMPARATIVE TABLE OF HEALTH-RESORTS.[1]

	Temperature.					Relative humidity.					Absolute humidity.		Annual rainfall.	Notes on rainfall, etc.
	Mean annual.	Winter.	Spring.	Summer.	Autumn.	Mean annual.	Winter.	Spring.	Summer.	Autumn.	Mean annual.	Winter.		
						pr. ct.	pr. ct.	pr. ct.	pr. ct.	pr. ct.	grs.	grs.	inch.	
Nice	59°	17°	56°	72°	61°	67	66	68	65	69	3.72	2.12	32	60 rainy days.
Mentone . . .	60	49	60	73	55	72	....	....	....	....	4.12	......	....	90 rainy days. Heavy rains usually in Oct. and March.
San Remo . .	60	48	57	72	61	71	67	70	75	71	4.06	2.56	28¾	
Genoa . . .	60	45	59	75	63	63	59	64	64	63	3.62	1.94	55	
Naples . . .	61	48	58	71	64	67	70	67	62	70	3.80	2.76	35	
Cairo . . .	72	58	74	85	71	56	66	47	47	63	4.77	3.01	1¼	12 to 15 rainy days.
Algiers . . .	64	55	66	77	60	73	75	72	72	72	4.93	3.64	40	87 rainy days, November to June.
Azores . . .	62	58	61	68	62	76	77	74	74	75	4.87	4.14	38½	Records of humidity and rainfall taken at Ponta Delgada.
Madeira . . .	65	60	62	70	67	66	66	63	67	67	4.48	3.80	27	88 rainy days. Records of humidity and rainfall taken at Funchal.
Canaries . .	67	60	65	72	70	66	67	64	65	69	4.78	3.85	13	51 rainy days, August to May, Island of Teneriffe.
Bahamas . .	76	71	73	91	77	79	83	76	77	80	7.63	6.84	56	17 inches from November to April, inclusive. Record for Nassau.
Hawaiian Islands	74	70	73	77	76	72	71	71	69	71	6.58	5.91	37¾	Lightest rainfall in summer. Records for Punahou near Honolulu.
Davos-Platz[2] . .	38	23	35	52	34	77	82	71	74	79	2.04	1.20	33	9.7 inches from November to March, inclusive.
Bloemfontein[3] .	76	55 Mean of min.	....	82 Mean of max. during 6 hot mos.	....	55	....	....	....	....	5.31	....	17	70 rainy days. Temperature of four hottest months averages 85° F.
Mexico[4] . . .	60	54	63	62	58	60	56	47	68	68	3.45	2.63	24	139 rainy days. 22 inches fall May to October. Hottest month May.

[1] This table is adapted partly from the table of Sir James Clark, with additions by Dr. J. Henry Bennet; and partly from records kindly furnished by the U. S. Weather Bureau, the Mexican Weather Bureau, and from other sources. [2] Elevation 5200 feet. [3] Elevation 4540 feet. [4] Elevation 7400 feet.

TABLE XV.—COMPARATIVE TABLE OF HEALTH-RESORTS.[1]

	Average temperature in winter			Rainy days.	
				Year.	Winter.
Madeira	13.5° R.	62.4° F.	Torquay	154	79
Cairo	13.0	61.25	Ventnor	152	80
Algiers	11.2	57.2	Hastings	152	
Ajaccio	9.3	·52.9	Pau	140	
Palermo	9.1	52.47	Pisa	122	62
Mentone	9.0	52.25	Madeira	94	44
Nice	7.0	47.75	Venice	84	22
Cannes	7.0	47.75	Algiers	83	72
Pau	6.4	46.4	Mentoue	80	
Pisa	6.3	46.18	Nice	72	38
Hyères	6.3	46.18	Hyères	62	17
Ventnor	6.1	45.72	Montreux	60	21
Torquay	5.6	44.6	San Remo	56	15
Venice	5.5	44.37	Cannes	52	
Merau	4.3	41.67	Meran	52	13
Montreux	3.8	40.55	Cairo	12 (?)	9 (?)

	Absolute minima of temperature.			Average differences of temperature in winter.	
Davos	−24.0° R.	−0.22° F.	Cairo	7° to 8° R.	15.75° to 18° F.
Torquay	−11.0	+7.25	Nice	4 " 5	9.0 " 11.25
Ventnor	−10 0	9.5	Venice	2 " 3	4.5 " 6.75
Pau	−10 0	9.5	Palermo	1 " 3	2.25 " 6.75
Pisa	−6.5	17.38			
Arco	−5.0	20.75			

[1] Taken from Dr. Julius Braun's Climatic Health-resorts.

INDEX OF AUTHORITIES QUOTED.

INDEX TO TABLES.

INDEX.

Katama, 198
Keene Valley, 212
Kennebunkport, 196
Kidney-diseases, 96, 168
  climatic treatment of, 170
  effects of compressed air upon, 96
  hepatic complications in, 174
  influence of climatic factors in, 169
  lardaceous, 172
  prevalence of, 168
Kilkee, 377
Kilrush, 377
Kimberly, 426
Kingston, 439
Kingstown, 377
Kissingen, 413
Kreuznach, 412

LA BOURBOULE, 393
  Ladybrand, 426
Lake Champlain, 214
  George, 214
  Placid, 212
  Tahoe, 346
  Worth, 229
Lakes, influence of, upon climate, 30
Lakewood, 217
Laryngitis, chronic, 158
  tubercular, 146
Las Cruces, 283
  Vegas, 278
  Hot Springs, 278
Laubbach, 410
Le Croisic, 390
Leesburg, 229
Lemoenfontein, 425
Leon, 363
Leysin, 417
Lisbon, 395
Lisdunvarna, 377
Liver, diseases of, 175
  climatic treatment of, 175
  functional, 175
  mineral waters in, 177
  organic, 175
Llandudno, 380
Long Branch, 199
Los Angeles, 324
  Baños Chicos, 362
  Pinellas, 229
Lucca, 403
Luhatschowitz, 405
Luxor, 421

MADEIRA ISLANDS, 443
  Maine, resorts of, 204, 205
Malaga, 395
Maloja, 418
Malta, 401
Malvern, 386
Mandeville, 439
Manitou, 259
  Park, 261

Maplewood, 206
Margate, 384
Marienbad, 405
Martha's Vineyard, 197
Maryland, resorts of, 220
Massachusetts, resorts of, 208, 209
Matlock, 386
Mauch Chunk, 218
Mediterranean Islands, 400
Mehadia, 408
Melbourne, 430
Meningitis, 163
Mentone, 398
Meran, 406
Mesilla Valley, 282
Meteorology, 21
  comparative tables of, 451
Mexico, City of, 355
  resorts of, 355–366
Migraine, 164
Mirror Lake, 212
Mists, 31
Mobility of atmosphere, 89. (*See also* Winds.)
Moffatt, 379
Moneague, 439
Monmouth Beach, 199
Mont Doré-les-Bains, 392
Monterey, 344, 359
Montreux, 414
Moosehead Lake, 205
Morelia, 365
Morocco, resorts of, 423
Mount Desert, 195, 196
  Dora, 229
  Harvard, 326
  Lowe, 326
  Pocono, 219
  Wachusett, 209
Mountain-climbing, 44
  -sickness, 43
Mountains, influence of, upon climate, 25
Murree, 427
Murren, 417

NAIRN, 378
  Nantucket, 198
Napa Soda Springs, 345, 347
  Valley, 344
Naples, 403
Narragansett Pier, 199
Nassau, 435
Nauheim, 411
Nephritis, acute, 171, 172, 173
  chronic, 172, 173
  scarlatinal, 173
Nervi, 400
Nervous disorders, climatic change in, 162
  effects of altitude on, 115
  use of compressed air in, 115
Neuenahr, 410
Neuralgia, 164

www.ingramcontent.com/pod-product-compliance
Lightning Source LLC
Chambersburg PA
CBHW031811270326
41932CB00008B/377